Natural Hazards

NATURAL HAZARDS
Local, National, Global

Edited by GILBERT F. WHITE

New York   OXFORD UNIVERSITY PRESS   London Toronto 1974

Contents

Figures

I Introduction

1. Natural hazards research: concepts, methods, and policy implications

GILBERT F. WHITE
University of Colorado

Extreme natural events illuminate one aspect of the complex process by which people interact with biological and physical systems. Every parameter of the biosphere subject to seasonal, annual, or secular fluctuation constitutes a hazard to man to the extent that his adjustments to the frequency, magnitude, or timing of its extremes are based on imperfect knowledge. Were there perfectly accurate predictions of what would occur and when it would occur in the intricate web of atmospheric, hydrologic, and biological systems, there would be no hazard. However, there would remain the question of how, given the particular aims of a social group, to respond effectively to the completely predictable order of events, e.g., the rising and setting of the sun. Ordinarily, the extreme events can only be foreseen as probabilities whose time of occurrence is unknown.

The hazard attaching to the coming of the rains for dry-land farmers or the duration of peak river flow for a floodplain manufacturer or the magnitude of the infrequent but certain earthquake for a fault-zone dweller is a significant element in decisions which many individual users of the environment must make on a daily, seasonal, or yearly basis. The more common extreme geophysical events are avalanche (snow), coastal erosion, drought, earthquake, flood, fog, frost, hail, landslide, lightning, snow, tornado, tropical cyclone, volcano, and wind.

Certain of these form an important target of many government programs to buffer social dislocations from catastrophes or to improve the productivity of resource users, as witness national programs to reduce floods, prevent drought losses, insure against hail, and carry out reconstruction following earthquakes and tornadoes.

By definition, no natural hazard exists apart from human adjustment to it. It always involves human initiative and choice. Floods would not be hazards were not man tempted to occupy floodplains: by his occupance he establishes the damage potential, and may well change the flood regimen itself. Therefore, when earthquake, flood, or tropical cyclone strikes one part of the earth it may or may not disrupt and damage the society in that area. The effects of human adjustments will depend upon the particular combination of physical and social environment that prevail at the time. It is barely conceivable and highly improbable that a group of people could manage to live free of any distress caused by rare and unexpected events in nature. It is equally improbable that they could exist wholly at the mercy of every fluctuation around average conditions. Between these poles, a wide variety of response is made to risk and uncertainty in the natural environment.

Finding out how these responses to extremes differ from place to place and from time to time helps understand the way one system affects another and how these relationships can be changed for the benefit of the people who suffer from severe events. If the means of enabling individuals to take intelligent action or governments to design and carry out effective programs of assisting individuals are to be improved, it will be essential, along with further appraisal of both physical mechanisms and social accommodations, to gain greater knowledge of the processes by which people do, in fact, cope with hazards in nature. That is the aim of the collaborative program of research on natural hazard.

The findings presented in this volume grow out of collaborative research among geographers and investigators in related disciplines in more than 23 countries over a period of 6 years. Workers in 15 countries have carried out roughly parallel investigations, using a variety of methods and arriving at conclusions that are at times more divergent than convergent. The studies presented in the following pages speak for themselves in outlining problems of methodology and in indicating the main lines of conclusions. It may be helpful, however, to introduce them by reviewing the origin and aims of the collaborative investigations, defining a few of the concepts commonly employed in the investigations, and pointing out the broader implications of this work for public policy formation at local, regional, national, and international levels.

Origins

Several earlier investigations of natural hazards by geographers centered upon the questions of why people occupy hazardous floodplains as they do and what would be the effects of public action to reduce flood losses upon local land use and national economy. These inquiries in the United States led in 1957 to the troublesome finding that the net effect of heavy federal investment in flood-control works such as dams, channel improvements, and levees in the 20 years following the passage of the Federal Flood Control Act of 1936 was to increase the total national losses from floods at the same time that more than $5 billion was expended on prevention or reduction of flood damage in areas covered by protection works (White et al., 1958).

The establishment of these facts led to more complicated questions. What was the net impact of the protection work upon national efficiency and upon the life of the communities affected? Were there alternative ways of dealing with flood losses that might be more effective

3

socially? What would be the human response to different sorts of initiatives such as information programs, land-use regulation, improved warning systems, and insurance systems? The history of the resulting line of investigation is outlined elsewhere (Miller, 1966; White, 1973), and it is mentioned here only to indicate that the prior inquiry on floods branched off in a number of paths leading in time to a more extensive investigation in different cultures with regard to different hazards.

In 1967 there was initiated at the University of Chicago (subsequently transferred to the University of Colorado), Clark University, and the University of Toronto a collaborative program of research which attempted to explore the applicability of findings from the flood studies to other geophysical hazards and to investigate the interaction of social and natural systems in a variety of environments and cultures beyond those that had thus far been covered in North America. The wider effort contemplated analysis of particular hazards in both developing and developed countries, examination of experience at the international level, and a series of detailed field investigations in a framework which would permit comparison of findings from among different hazards and a variety of cultures and physical environments. The results of the investigations of policy at regional and national and international levels are contained in a companion volume (Burton, Kates, and White, forthcoming). This volume presents primarily the results of comparative field observations and of efforts to make integrated reviews of the hazard situation in selected countries and of human responses to particular extreme events on a global scale.

The studies should be viewed as exploratory, and as intended to probe the variety of human responses that are presented in the contemporary human scene in dealing with a selection of natural extremes. The selection of particular places and hazards was to a considerable extent fortuitous; it was dependent upon availability of competent investigators and field staff to carry out the field observations.

Paradigm, concepts, and hypotheses

The basic research paradigm is relatively simple and changed very little during the course of the comparative field observations, although in the light of that experience it now would be appropriate to revise and extend it. Essentially, the effort was to:

1. Estimate the extent of human occupance in areas subject to extreme events in nature.
2. Determine the range of possible adjustments by social groups to those extreme events.
3. Examine how people perceive the extreme events and resultant hazard.
4. Examine the process of choosing damage-reducing adjustments.
5. Estimate what would be the effects of varying public policy upon that set of human responses.

To pursue these questions required frequent reference to several concepts, already noted, whose utility, in turn, was tested by the circumstances of the field investigations (Kates, 1970). The analysis was framed from the outset in terms of *interacting systems*. It recognized that social systems of resource use cannot operate independently of atmospheric, hydrologic, geomorphic, and biotic systems, and carry a growing technologic capacity to modify the biosphere.

Natural hazard was defined as an interaction of people and nature governed by the coexistent state of adjustment in the human use system and the state of nature in the natural events system. Extreme events which exceed the normal capacity of the human system to reflect, absorb, or buffer them are inherent in hazard. An *extreme event* was taken to be any event in a geophysical system displaying relatively high variance from the mean.

There is large latitude for classifying human response to extreme events, but the term commonly used to describe a human activity intended to reduce the negative impact of the event was *adjustment*.

The individual organization of stimuli relating to an extreme event or a human adjustment was defined as *perception*. This was a broader description of perception than found in much psychological literature, and it implied a major cognitive component. The interest was in finding out how people viewed the occurrence or threat of the extreme event and of the opportunities open to them in coping with the event.

As the investigations proceeded it became apparent that most of these terms needed sharpening. In particular, it was recognized that important distinctions should be drawn between human activities that involved adaptations of the use system in contrast to those which coped with risk without altering the system, and that awareness and judgment of natural events and human opportunities encompassed far more than organization of physical stimuli. However, the broader and less precise definitions have been retained in this volume with the thought that recognition of the difficulties may help shape a more precise conceptual framework.

Five major hypotheses were put forward for examination, and investigators were invited to assess, revise, and add to them. These, too, changed as field studies progressed. While observations were under way, the basic theory of decision making was reviewed from the standpoint of psychological laboratory experiments and of economic analysis (part III below). At the outset the hypotheses were formulated as follows (Natural Hazards Research Working Paper No. 16, 1970):

A. Human occupance that persists in areas of recurrent hazard is justified in the view of the occupants for the following reasons:
   1. Superior economic opportunity;
   2. Lack of satisfying alternative opportunities;
   3. Short-term time horizons;
   4. High ratios of reserves to potential loss.

B. There are three types of response to natural hazards which may be characterized as:
   1. Folk, or preindustrial, adjustments which involve a wide range of adjustments requiring more modifications in behavior in harmony with nature than control of nature, are flexible and easily abandoned, are low in capital requirements, require action only by individuals or small groups, and can vary drastically over short distances.
   2. Modern technological, or industrial, adjustments which involve a more limited range of technological actions emphasizing control of nature, are inflexible and difficult to change, are high in capital requirements, require interlocking and interdependent social organization, and tend to be uniform.
   3. Comprehensive, or postindustrial, adjustments which combine features of both earlier stages so as to involve a larger range of adjustments, greater flexibility and variety of capital and organizational requirements.

C. Variation in hazard perception and estimation can be accounted for by a combination of the following:
   1. Magnitude and frequency of the hazard;
   2. Recency and frequency of personal experience, with intermediate frequency generating greatest variation in hazard interpretation and expectation;
   3. Importance of the hazard to income or locational interest;
   4. Personality factors such as risk-taking propensity, fate control, and views of nature.
   This variation is not related to common socioeconomic indicators such as age, education, and income.

D. For individuals, the choice of adjustment is a function of:
   1. Perception of the hazard;
   2. Perception of the choice open to them;
   3. Their command of technology;
   4. The relative economic efficiency of the alternatives;
   5. The perceived linkages with other people.

E. For individuals, the process of estimating economic efficiency is related to the perceived time horizon, the ratio of reserves to anticipated loss, and the degree to which choice is required.

F. For communities, the choice of adjustment is a function of perception of hazard, choice, and economic efficiency as influenced by the stability and the power structure of government.

Methods

In the comparative field observations the primary framework for investigation was a site description and basic interview developed from experience in a variety of North American market societies and then tested in Mayan and Yucatecan peasant societies in Mexico.

Site description included the conventional attributes of land form, soil and vegetation types, land use, and population. It reviewed the sequent occupance of the area and estimated the mean and range of annual income. Attention was given to distribution in time and space of the hazard as well as to damages resulting from it over the period of historic record. An effort was made to judge the importance of the particular hazard to the economy and social organization of the region and of the nation.

The interview was intended to elicit information about the social and economic status of the household, the conditions in which it is obliged to make decisions in the face of the specified hazards, and the precise types of adjustment which are perceived as being made by others. The factors entering into the choice of particular adjustments are probed. The interview contains several measures of personality traits, including a story in which the respondent selects what he regards as the most suitable outcome, and a sentence completion test which is coded for characteristics of external-internal control, traditionalism-modernism, and feelings toward hazards.

This basic interview was modified from place to place in order to take account of differences in local environment. While attempts were made to ask the same key questions in every observation site, many investigators found it necessary to omit, revise, and add items (chap. 23 below).

The interview typically takes about an hour. Approximately 120 respondents are sought in order to permit a size of sample population large enough to carry out elementary tests of association among two or more of the characteristics recorded. In most sites all respondents were selected so as to be either agricultural or nonagricultural. Interviews were only sought with heads of households who were married and had children. In order to permit comparison between respondents according to education and family responsibilities, an attempt was made to stratify them according to whether or not children were of working age and according to years of formal schooling, if any.

The basic instrument is reproduced below, exactly as it first was introduced to participants in the investigations.

General questionnaire

1. Date   [   ] [   ] [   ] [   ]
2. Country   [   ] [   ] [   ] [   ] [   ] [   ] [   ]
3. Study site   [   ] [   ] [   ] [   ] [   ] [   ]
4. Location   [   ] [   ] [   ] [   ] [   ] [   ]
5. Severity of hazard   [   ]   High   [   ]   (Estimate by interviewer)
                                                Medium   [   ]
                                                Low   [   ]
6. Number of interviewer   [   ] [   ]
7. Number of respondent   [   ] [   ]
8. Sex of respondent   [   ]
9. Occupation (major source of income)
   F* Farmer   [   ]   Pastoralist   [   ]   Other   [   ]   Government   [   ]
   NF* Artisan [   ]   Tradesman   [   ]   Laborer   [   ]   Retired   [   ]
   Education   [   ]   Manufacturer [   ]   Unemployed   [   ]
10. Tenure of land or home (cross out one if it does not apply)
    Owner   [   ]   Tenant   [   ]   Shifting   [   ]   Communal   [   ]
    Laborer   [   ]
11. Predominant language spoken in household   [   ] [   ]
12. Religion   [   ] [   ]
13. Age of respondent in years [   ] [   ]   (Estimate, if necessary)
14. Number, age, and sex of persons in household

Age	Male		Female	
Over 21	[ ]	[ ]	[ ]	[ ]
16–21	[ ]	[ ]	[ ]	[ ]
0–15	[ ]	[ ]	[ ]	[ ]
Total	[ ]	[ ]	[ ]	[ ]

    NF If business firm, number of employees
15. Literacy   [   ]
    Cannot read   [   ]
    Read but less than 6 years formal schooling [   ]
    Formal schooling of 6 years or more   [   ]
16. What are the principal disadvantages in living in this building, or working this field? What are the principal advantages?
    a. Emphasis on: Advantages [   ] Disadvantages [   ] Neither [   ]

    b. List any hazards (social or natural) noted by respondent

17. What are the principal disadvantages in living in this area? What are the principal advantages?
    a. Emphasis on: Advantages [   ] Disadvantages [   ] Neither [   ]

    b. List any hazards (social or natural) noted by respondent

18. Do the people of this place have any trouble with (drought, hurricanes, etc.)?
    Yes [   ] Doubtful [   ] No [   ] Don't know [   ]

    (Insert sentence completion test here)

19. If by some misfortune this area is affected by (a drought, a hurricane) in what way do you think it would affect your household?
    a. List any effects volunteered by respondent

    Structure   [ ]   Anxiety   [ ]
    Nothing   [ ]   Contents   [ ]
    Other property   [ ]   Activity   [ ]
    Other (specify)

    b. Are the damages considered to be:
    Total (80%–100%)   [ ]   Substantial (21%–79%)   [ ]
    Slight (1%–20%)   [ ]   Nonexistent (0%)   [ ]
    c. What are the major damages?

    F   House and buildings   [ ] NF   Workplace   [ ]
        Crops   [ ]   House   [ ]
        Animals   [ ]   Community   [ ]
        People   [ ]   People   [ ]
20. How many times have (droughts come, there been hurricanes, etc.) in this area in the years since you were born? [ ] [ ]
    List years

21. When was the worst year for (drought, hurricanes, etc.)? [ ] [ ] [ ] [ ]
22. When was the last year when there was trouble with (drought, hurricanes, etc.)? [ ] [ ] [ ]
    [ ]
23. a. Do you think the (drought, hurricane, etc.) will come again in your lifetime?
       Yes   [ ]   Don't know   [ ]   No   [ ]
    b. If yes, soon? [ ]   In a few years? [ ]   In many years?   [ ]
24. Do people in other places have trouble with (drought, hurricanes)?
    Yes [ ] Doubtful [ ] No [ ] Don't know [ ]
25. Are there other places with fewer (droughts, hurricanes) where you could earn as good a living?
    Yes [ ] Doubtful [ ] No [ ] Don't know [ ]
26. Here is a story on which we would like your comments.
    Once after (a drought, hurricane) four men spoke about (the rains coming late, a hurricane coming again).
    The *first* said that the (late rains, hurricane) would *come again soon* because when (late rains, hurricanes) happen, more are soon to come.
    The *second* thought that (late rains, hurricanes) would *come again but did not know when* because (late rains, hurricanes) can happen in any year.
    The *third* said that *he knew when* the (late rains, hurricanes) *would come* for there is a regular time and that time must pass before it comes again.
    The *fourth* thought that the (late rains, hurricanes) *would not come again*.
    Which man had the best idea about the coming of (late rains, hurricanes)?
    First [ ] Second [ ] Third [ ] Fourth [ ] Don't know [ ]
27. How many years have you worked in this place? [ ] [ ]
28. a. Do you think you will live in this building (or work in this place) many more years?
       Yes [ ] Doubtful [ ] No [ ] Don't know [ ]
    b. If the answer is "Doubtful" or "No," where do you think you would move?
       Same hazard zone   [ ]   Different region   [ ]
       Different zone   [ ]   Different country   [ ]

29. During the years you have worked here how many years would you say have been for your (harvest, business, etc.)?
   Good   [ ] [ ]   By "good," we mean in terms of your income.
   Bad   [ ] [ ]
   Regular   [ ] [ ]
30. When you want to talk over an important community problem, to whom do you speak? (Give an example of such a problem in that community)
   Family   [ ]   Friend   [ ]   Special group or person [ ] (Specify   )
   Government [ ]   No one   [ ]   (Example)
31. If you have problems after (a drought, a hurricane, etc.) occurs, who can you go to for help in recovering from the losses?
   Family   [ ]   Friend   [ ]   Special group or person   [ ]
   Government [ ]   No one   [ ]   How do they help?
32. How successful have such people been in helping you recover your losses?
   Completely   [ ]   Don't know [ ]   Somewhat unsuccessful [ ]
   Somewhat successful   [ ]   Unsuccessful   [ ]
33. Do you know anyone who has been helped by the government because of (drought, hurricane, etc.)?
   a. Yes [ ] No [ ] Don't know [ ]
   b. If yes, explain
   a. No [ ] Yes [ ] b. If yes, explain how
34. Are there any signs or ways of knowing when a (drought, hurricane, etc.) will come again? a. Yes [ ] No [ ] Don't know [ ]
   b. If yes, explain
35. When a (drought, hurricane, etc.) comes, what do you do?

(List Adjustments)**	Mentioned by Respondent	When Asked		Different Next Time		Warning Given	
		Yes	No	Yes	No	Yes	No
	(1)	(2)		(3)		(4)	
A_1	[ ]	[ ]	[ ]	[ ]	[ ]	[ ]	[ ]
A_2	[ ]	[ ]	[ ]	[ ]	[ ]	[ ]	[ ]
A_3	[ ]	[ ]	[ ]	[ ]	[ ]	[ ]	[ ]
A_n (as many as required)							

First, let respondent volunteer and record in column 1. Then ask about others and record in column 2.
36. Next time, would you do anything any different than you did last time? (Record in column 3 above.)
37. If a warning were to be given that a (drought, hurricane, etc.) is coming this year, would you do anything any different? (Record in column 4 above.)
38. What do your neighbors do? Record in column 1, than ask and record in column 2.)

	(1) Mentioned by Respondent	(2) When Asked Yes No		(3) Good Bad		Don't Know	(4) Why? 1 2 3 4 5 6 7 8
A_1	[ ]	[ ]	[ ]	[ ]	[ ]	[ ]	[ ] [ ] [ ] [ ] [ ] [ ] [ ] [ ]
A_2	[ ]	[ ]	[ ]	[ ]	[ ]	[ ]	[ ] [ ] [ ] [ ] [ ] [ ] [ ] [ ]
A_3	[ ]	[ ]	[ ]	[ ]	[ ]	[ ]	[ ] [ ] [ ] [ ] [ ] [ ] [ ] [ ]
A_n	[ ]	[ ]	[ ]	[ ]	[ ]	[ ]	[ ] [ ] [ ] [ ] [ ] [ ] [ ] [ ]

39. a. About these things that you and your neighbors do, do you think they are good or bad? (Record in column 3.)
    b. Why? (See code below for response. Record in column 4.)
       1. Has not heard of adjustment.
       2. Thinks the local environment is favorable or unfavorable.
       3. Doesn't know how to do it.
       4. Thinks it would pay or would not pay.
       5. Cannot afford it.
       6. Is encouraged or discouraged by the effect it would have on other people.
       7. Doesn't think it will work.
       8. Other

40. a. Do you try to carry any (F crops—NF money) over from one year to the next year? Yes [ ] No [ ] Don't know [ ]
    b. If yes, about how much do you carry over in a good year?_____
    c. In a bad year?_____ (Express as percentage of income)

41. a. Is there anything that the government or people, your friends and neighbors, can do to *prevent* damage from a (drought, hurricane, etc.)?
       Yes [ ] Don't know [ ] No [ ]
    b. If yes, specify

42. On the basis of your annual income, in which of the following classes do you consider yourself?
    Currency Unit
    _000 − _000 [ ] (High)   (Classes to be set by local conditions)
    _00 − _000 [ ] (Medium)
    _   − _00 [ ] (Low)
    No response [ ]

Interviewer's Comments (to be completed after interview)
    Estimated income in relation to mean for group
    High   [   ]
    Medium   [   ]
    Low   [   ]

Interview situation:
    Alone   [   ]
    Group   [   ]

Respondent's   attitude:

    Hostile   [   ]   Neutral   [   ]   Helpful   [   ]
    Unreliable [   ]   Reliable [   ]

---

*F = applicable to farmers; NF = applicable to nonfarmers.
**For example, "Move out of building," "Buy provisions"

As an integral part of the interview a sentence completion test was administered where appropriate. It was intended to invite relevant responses on hazard questions, to encourage unstructured responses in an economical manner, and to use a universal characteristic of language structure—the sentence. The sentence stems were designed to elicit information on responses to a specific hazard experience (stems 1, 4, and 9), the emotions felt in experiencing a specific event (stems 3, 6, 7, and 10), and the sense of internal versus external control by the respondent (stems 2, 5, 8, and 11).

The sentence stems are reproduced below.

Sentence completion test

   Directions: I'm going to read you the beginnings of some sentences. I would like you to complete each sentence with whatever comes to mind. There are no right or wrong answers. Just say whatever comes first into your head. For example, how would you finish this sentence: "The thing I like best to eat is . . ." Good, that's the idea. Now, here is the first one.

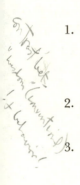

1. If a (drought, hurricane) is predicted, I

    IF NO EXPERIENCE WITH HAZARD: If a (drought, hurricane) were predicted, I would

2. The main thing that helps people to get ahead in the world is

3. When a (drought, hurricane) is coming, the first emotion I feel is

    IF NO EXPERIENCE WITH HAZARD: If a (drought, hurricane) were coming, the first emotion I would feel would be

4. During a (drought, hurricane) I

    IF NO EXPERIENCE WITH HAZARD: During a (drought, hurricane) I would

5. As far as my own life is concerned, God

6. The emotions I feel while I am going through a (drought, hurricane) are

    IF NO EXPERIENCE WITH HAZARD: The emotions I would feel while I was going through a (drought, hurricane) would be

7. In a (drought, hurricane) the people I feel some responsibility for are

    IF NO EXPERIENCE WITH HAZARD: In a (drought, hurricane) the people I would feel some responsibility for are

8. I believe that luck

9. When a (drought, hurricane) is over, I

    IF NO EXPERIENCE WITH HAZARD: When a (drought, hurricane) was over, I would

10. When a community experiences a (drought, hurricane), the feelings among its people

    IF NO EXPERIENCE WITH HAZARD: If a community experienced a (drought, hurricane), the feelings among its people would be

11. As far as the future is concerned, I

   The basic interview was accompanied by suggestions about procedure, including pretesting, the ethics of interviewing, training of interviewers, and sources of error and bias.

International cooperation

The action could not have taken place had it not been for the sponsorship of the Commission on Man and Environment of the International Geographical Union. This commission, established at the International Geographical Congress in New Delhi in 1968, became the instrument through which geographers in member countries were made aware of the investigation and invited to participate. The commission gave general guidance to the international aspects of the inquiry and encouraged a series of national reviews, four of which are included in this volume.

   The number of investigations which were stimulated by the total effort, including some which worked quite independently of either the Commission or those participating in collaborative research but which drew in part upon the concepts and methodology outlined here, are summarized in Tables 1–1 and 1–2.

   Only a selection of these studies is reported in this volume. Others have been or will be published separately.

   From the comparative field observations have come a number of new insights into the ways in which people deal with hazards in the environment. At the same time, the field observations combined with national reviews and global summaries have provided background for positive recommendations concerning public policy.

Policy implications

Man is responding to the risks and uncertainties of natural events such as floods, hurricanes, and earthquakes in a way which leads to increasing property losses. This is striking in many countries where economic growth is rapid, and especially so in areas where

Table 1—1. Studies of natural hazard perception and choice: comparative field observations

Hazard	Area	Year	Investigator(s)	No. of inter-views	No. of sub-sites	Publication	Land use[a]
Avalanche (earthquake)	Huayles Valley, Yungay, Peru	1970	Carlos Penaherrera de Aguila Univ. of San Marcos and National Planning Institute	86			R
Coastal erosion	Bolinas, Calif.	1971	Rowan A. Rowntree Dept. of Geography Syracuse Univ.	120			U/R
Drought	South Australia	1971	R. Leslie Heathcote School of Social Sciences Flinders Univ. of South Australia	181	4	Heathcote (1970)	R
	Northeast Brazil	1970	Reuben H. Brooks Dept. of Geography George Peabody Coll.	397	4	Brooks (1972, 1973)	R/U
	Eastern Kenya	1971	Benjamin Wisner Faculty of Medicine Univ. of Dar es Salaam Philip M. Mbithi Inst. of Developmental Studies Univ. of Nairobi	610	17	Mbithi and Wisner (1972)	R
	Ticul, Yucatan, Mexico	1971	Carlos G. Parra Earth Sciences Dept. New Mexico State Univ.	118	1	Parra (1971)	R
	Northwestern Nigeria	1971	Wolf Roder and Herbert Dupree Dept. of Geography Univ. of Cincinnati	150	3		R
	Northeastern Tanzania	1971	Robert W. Kates Graduate School of Geography Clark Univ. Joop D. Heijnen Geographical Inst. Univ. of Utrecht	254	13	Berry and Kates (1972)	R
	Sukumaland, Tanzania	1969; 1971	Thomas D. Hankins Graduate School of Geography Clark Univ.	220	10	Hankins (1970a, 1970b, 1973)	R
Earthquake	San Francisco, Calif.	1971	Edgar Jackson Dept. of Geography Univ. of Toronto Tapan Mukerjee Dept. of Economics Univ. of the Pacific	120			U
	Cornwall, Ontario	1972	Edgar Jackson	118			U

Continued

Hazard	Area	Year	Investigator(s)	No. of inter- views	No. of sub- sites	Publication	Land use[a]
Flood	Sri Lanka	1971; 1972	Daya Hewapathirane Dept. of Geography Univ. of Colorado	479	7	Hewapathirane (in progress); Hewapathirane and White (1973)	R/U
	Ganges, India	1971	R. Ramachandran and S. C. Thakur Dept. of Human Geography Delhi School of Economics Univ. of Delhi	66	3		R
	Malawi	1972	Rodney J. Cheatle Chancellor Coll. Univ. of Malawi	120			R/U
	Rock River, Ill.	1970; 1971	Norman Moline Dept. of Geography Augustana Coll.	257			U/R
	Shrewsbury, England	1971	Donald M. Harding Dept. of Geography Univ. College of Swansea Dennis Parker Middlesex Poly. at Enfield	132			U
Snow	Northern Michigan	1971	Fillmore C. F. Earney Dept. of Geography Northern Michigan Univ.	120			U
Hurricane or Tropical cyclone	Galveston, Tex.	1971	Duane Baumann Dept. of Geography Univ. of Southern Ill. John H. Sims Committee on Human Development Univ. of Chicago	240		Baker (1972)	U
	Pass Christian, Miss.	1971	Duane Baumann John H. Sims	120			U
	Puerto Rico	1970	Duane Baumann John H. Sims	147	2		R
	Tallahassee, Fla.	1971	Duane Baumann John H. Sims	120			U
	Virgin Islands (U.S.)	1970	Martyn Bowden Graduate School of Geography Clark Univ.	93			R/U
	Coastal Bangladesh	1971; 1972	M. Aminul Islam Dept. of Geography Univ. of Dacca	125	4		R

Continued

Hazard	Area	Year	Investigator(s)	No. of interviews	No. of sub-sites	Publication	Land use[a]
Volcano	Costa Rica	1972	Gilles Lemieux Dept. of Geography Univ. of Calgary	170			R/U
	Kilauea, Haw.	1971	Brian J. Murton Dept. of Geography Univ. of Hawaii	101			R
			Shinzo Shimabukuro Okinawa, Japan				
Wind	Boulder, Colo.	1971	Donald Miller Boulder, Colo.	120		Miller (1972)	

[a]R = Rural; U = Urban

modern technology is spreading vigorously. Certain of the hazards are created by man through his alteration of land and water or by his invasion of risky areas; others are exacerbated by his efforts to reduce the risk.

At least two lessons from this experience are vital to a sound public approach to environmental management. The first lesson is that if costly threats to life and property from the extremes of natural phenomena are to be minimized, there must be careful sharing of the skills, experience, and research capacity of the family of nations.

The second lesson is that modern societies cannot expect to cope effectively with hazards in the environment by relying solely upon technological solutions. A crucial aspect of any long-term accommodation to the human environment is the skillful, sensitive use of a wide range of adjustments, including engineering devices, land management, and social regulation. To depend upon only one sort of public action is to court social disaster, environmental deterioration, and enlarged public obligations.

At the local level, studies of the type reported show that government programs can go far astray if they ignore the perceptions of the people concerned on the hazard and ways of adjusting to it. This applies to flood-control programs for the United Kingdom (chap. 6 below) or New Zealand towns (chap. 8) or Indian farming areas (chap. 5), to erosion control on the California coast (chap. 9), to drought in Sukumaland (chap. 12) or the Oaxaca Valley (chap. 15), to volcanoes in Hawaii (chap. 19) and to earthquakes in San Francisco (chap. 20). Some of the studies have had direct application, as in the case of planning for drought in Tanzania (chap. 13) or winds in Boulder (chap. 10), or for floods along the Rock River (chap. 7). Others demonstrate the practicability of investigations which would add both integrated approach and human dimension to conven-

tionally narrow technological investigations. This is the case with urban snow problems (chap. 21), agricultural frost (chaps. 17 and 18), and tropical cyclones (chap. 30).

At the regional level, it is possible to recognize major limitations and incentives which would affect new government initiatives to hopefully reduce distress. The analysis of drought experiences in Kenya (chap. 11), northeastern Tanzania (chap. 13), northwestern Nigeria (chap. 14), and southern Australia (chap. 16), and of avalanches in Norway (chap. 22) provoke significant suggestions as to the direction of effective remedial programs.

At the national level, geographical studies already have served to shape new government policies in the United States (U.S. House of Representatives, 1966) and Canada (Sewell, 1969). Reviews of the general character of those reported for those two countries (chap. 27) and for Japan (chap. 28) and the USSR (chap. 29) help to place particular policies in more balanced perspective. The New Zealand insurance experience (chap. 26), for example, carries a solemn warning for every other nation contemplating the insurance adjustment. And the Bangladesh tragedy of 1970 (chap. 2) raises basic questions about the net effect of the national program for coastal protection.

At the international level, there is a similar tendency for governments to deal with the social effects of natural hazards by turning to one or another specialized solution: a disaster relief program after avalanches, an insurance program for earthquakes, a dam-building program for floods, and similar measures. For almost all of these types of activities it is possible to obtain specialized technical assistance through United Nations agencies. For example, the Food and Agriculture Organization provides experts on crop insurance and water planning, UNESCO gives highly competent assistance on

Table 1-2. Studies of natural hazard perception and choice: field studies

Hazard	Area	Year	Investigator(s)	No. of interviews	Publication	
Drought	Oaxaca, Mexico	1970	Anne V. Kirkby Dept. of Psychology Univ. of Bristol	45	Kirkby (1972)	U
Earthquake	Kochi City, Japan	1971	T. Nakano Tokyo Metropolitan Univ.	116		
Flood (typhoon)	Kochi City, Japan	1971	T. Nakano M. Kusaka Ritsumeikan Univ.	116		
Frost	Florida (six counties)	1969	Robert M. Ward Dept. of Geography Eastern Michigan Univ.	163	Ward (1971)	
	Wasatch Front, Utah	1972	Richard Jackson Dept. of Geography Brigham Young Univ.	100		
Landslide	Kochi City, Japan	1971	T. Nakano Y. Ichinose Hosei Univ.	116		
	Ottawa-Hull Region, eastern Ontario	1968	John G. M. Parkes Dept. of Geography Faculty of Graduate Studies Univ. of Western Ontario	80	Parkes (1971)	
Snow	Toronto, Ontario	1968-69	Paula Archer Dept. of Geography Univ. of Toronto	100	Archer (1970)	
	Canton, Ohio; Duluth, Minn.; Evansville, Ind.; Greensboro, N.C.; Nashville, Tenn.; Rockford, Ill., Utica, N.Y.; Worcester, Mass.; Regina, Saskatchewan	1969	Duane Baumann Clifford Russell Resources for the Future, Inc. Washington, D.C.	401[a]	Baumann and Russell (1971)	
Tornado	Alabama, Connecticut, Illinois, Kansas, Massachusetts, Oklahoma	1968-70	Duane Baumann John H. Sims	420	Sims and Baumann (1971, 1972)	
General	London, Ontario	1970	Karen D. Moon Dept. of Geography Univ. of Toronto	328	Moon (1971)	
	London, Ontario	1970	Paul Wilkinson Dept. of Geography Univ. of Toronto		Wilkinson (1972)	
	Southwestern Ontario	1970	Kenneth Hewitt and Ian Burton Dept. of Geography Univ. of Toronto		Hewitt and Burton (1971)	

[a]Individuals only.

seismic risk and earthquake-resistant structures, the League of Red Cross Societies has information on disaster relief, and the World Meteorological Organization has scientific advice on weather forecasts and weather modification.

With the creation of its office of the Disaster Relief Co-ordinator in 1971, the United Nations took a long step forward in providing for an integrated approach which begins with relief and goes beyond the immediate emergency needs to assist in developing programs to prevent disasters. In this activity the perspective offered by the global summaries of human response to floods (chap. 31 below), tropical cyclones (chap. 30), and earthquakes (chap. 32) is an invaluable aid. These three hazards rank highest, in order given, among natural hazards in the toll they take in loss of life and damage to human habitation. That form of analysis was, in fact, incorporated in early measures initiated by the Disaster Relief Office in 1973.

Thus, the studies reported in this volume record a series of loosely coordinated efforts to deepen the understanding of social-physical interactions, to begin to construct a more general theory of such behavior in extreme situations, and to apply the findings to public action. They are exploratory rather than definitive, but they promise new understanding and practical influence.

Acknowledgments

The international collaboration in natural hazards research was carried out under the auspices of the International Geographical Union through its Commission on Man and Environment. During 1968–72 its members were David Amiran, Hebrew University of Jerusalem, Israel; Ian Burton, University of Toronto, Canada; I. P. Gerasimov, Academy of Sciences of the USSR, Moscow; T. Nakano, Tokyo Metropolitan University, Tokyo; Ernst Neef, Geographisches Institut der Techn. Univ., Dresden, German Democratic Republic; R. Ramachandran, University of Delhi, India; Gilbert F. White, Chairman, University of Colorado, Boulder.

The National Science Foundation supported the basic research effort that centered at Clark University, the University of Colorado, and the University of Toronto.

UNESCO provided partial support for a seminar sponsored by the Commission in Godollo, Hungary, in August 1971 and for a meeting in Calgary, Alberta, in July 1972. These served to bring together many of the participating scientists.

The editing of the papers in this volume was greatly assisted by Nancy Simkowski. Rebecca Boyle and Hazel Visvader carried out the checking of manuscript and proof. Jacque Myers handled the typing of the manuscript.

References

Archer, Paula. (1970) "The urban snow hazard: a case study of the perception of, adjustments to, and wage and salary losses suffered from snowfall in the city of Toronto during the winter of 1967–1968." Toronto: University of Toronto, unpublished M.A. thesis.

Baker, Earl J. (1972) "Cognitive factors relating to response and adjustment to hurricane hazard." Tallahassee: Florida State University, unpublished M.S. thesis.

Baumann, Duane, and Russell, Clifford. (1971) "Urban snow hazard: economic and social implications." Urbana: University of Illinois, Water Resources Center, unpublished.

Berry, L., and Kates, R. W. (1972) "Views on environmental problems in east Africa." *African Review* 12:299–314.

Brooks, Reuben H. (1972) Drought perception as a force in migration from northeast Brazil." Boulder: University of Colorado, unpublished Ph.D. dissertation.

——. (1973) "Drought and public policy in northeastern Brazil: alternatives to starvation." *Professional Geographer* 25:338–349.

Burton, Ian, Kates, Robert W., and White, Gilbert F. (1968) "The human ecology of extreme geophysical events." Toronto: University of Toronto, Department of Geography, Natural Hazards Research Working Paper No. 1.[1]

——, Kates, Robert W, and White, Gilbert F. (forthcoming). The Environment as Hazard. New York: Oxford University Press.

Hankins, Thomas D. (1970a) "Crop hazards and adjustments in Sukumaland." Worcester, Mass.: Clark University, Drought Conference.

——. (1970b) "Drought and famine in eastern Tanzania." Worcester, Mass.: Clark University, Drought Conference.

——. (1973) "Drought, cotton planting times and yields in Sukumaland, Tanzania." Worcester, Mass.: Clark University, Ph.D. dissertation.

Heathcote, R. Leslie. (1970) "Agricultural drought in Australia: a preliminary design." Worcester, Mass.: Clark University, Drought Conference.

Hewapathirane, Daya. (in progress) "Flood hazard in Sri Lanka: human adjustments and alternatives." Boulder: University of Colorado, Ph.D. dissertation.

——, and White, G. F. (1973) "Obstacles to consideration of resources management alternatives: south Asian experience." In Evan Vlachos (ed.), *Transfer of water resources knowledge.* Fort Collins, Colo.: 1st International Conference on Transfer of Water Resources Knowledge, Water Resources Publications.

Hewitt, Kenneth, and Burton, Ian. (1971) *The Hazardousness of a Place: A Regional Ecology of Damaging Events.* Toronto: University of Toronto Press.

Kates, Robert W. (1970) "Natural hazard in human ecological perspective: hypotheses and models." Boulder: University of Colorado, Institute of Behavioral Science, Natural Hazards Working Paper No. 14.

Kirkby, Anne V. (1972) "Use of land and water resources in past and present valley of Oaxaca, Mexico." Ann Arbor: University of Michigan, Memoirs of the Museum of Anthropology No. 5.

Mbithi, Philip M., and Wisner, Benjamin. (1972) "Drought and famine in Kenya: magnitude and attempted solutions." Nairobi: University of Nairobi, Institute for Developmental Studies Discussion Paper No. 144 (unpublished).

Miller, David H. (1966) "Cultural hydrology: a review." *Economic Geography* 42:85–89.

Miller, Donald. (1972) "Human perception of and adjustment to the high wind hazard in Boulder, Colorado." Boulder: University of Colorado, unpublished M.A. thesis.

Mitchell, J. Kenneth. (forthcoming) "Natural hazards." Washington, D.C.: Association of American Geographers, Report on Environmental Research and Education.

Moon, Karen D. (1971) "The perception of the hazardousness of a place: a comparative study of five natural hazards in London, Ontario." Toronto: University of Toronto, unpublished M.A. research paper.

Natural hazards research Working paper No. 16(1970) "Suggestions for comparative field observations on natural haz-

---

ards".[2] Toronto: University of Toronto, Department of
Geography.

Parkes, John G. M. (1971) "Awareness of, and adjustment to, a
natural hazard: sensitive clays in the Ottawa-Hull region."
London, Ontario: University of Western Ontario, unpublished M.A. thesis.

Parra, Carlos G. (1971) "Drought perception and adjustments in
Yucatan, Mexico." Boulder: University of Colorado, unpublished M.A. thesis.

Sewell, W. R. (Derrick) (1969) "Geographical research in water
management in Canada: inventory and prospect." In
Sewell, Judy, and Ouellet, Lionel, *Water Management
Research: Social Science Priorities*. Ottawa: Department
of Energy, Mines, and Resources.

Sims, John H., and Baumann, Duane. (1971) "Socio-psychologi-
cal dimensions in coping with tornado hazard." Boston:
Association of American Geographers, 67th annual meeting (unpublished).

_____, and Baumann, Duane. (1972) "The tornado threat: coping
styles of North and South." *Science* 176:1386–92.

U.S. House of Representatives. (1966) Task Force on Federal

Flood Control Policy, "A unified national program for
managing flood losses." 89th Cong., 2d Sess., House Document No. 465.

Ward, Robert H. (1971) "Cold and wind hazard perception by
orange and tomato growers in central and south Florida."
Ann Arbor: University of Michigan, unpublished Ph.D.
dissertation.

_____. et al. (1958) *Changes in Urban Occupance of Flood
Plains in the United States*. Chicago: University of Chicago, Department of Geography, Research paper No. 57.

White, Gilbert F., et al. (1958) *Changes in Urban Occupance of
Flood Plains in the United States*. Chicago: University of
Chicago, Department of Geography, Research paper No.
57.

_____, (1970) "Recent developments in flood plain research."
Geographical Review 60:440–43.

_____. (1973) "Natural hazards research". In Richard J. Chor-
ley, ed., *Directions in Geography*. London: Methuen.

Wilkinson, Paul. (1972) "The adoption of damage-reducing adjustments in relation to experience and expectation of
natural hazards in London, Ontario." In Adams, Peter,
and Helleiner, Frederick M., eds., *International Geography, 1972, I*. Toronto: University of Toronto Press,
published for the 22d International Geography Congress,
Montreal.

2. Now published at the Institute of Behavioral Science, University of
Colorado.

II   Individual and community response

2. Tropical cyclones: coastal Bangladesh

M. AMINUL ISLAM
University of Dacca

Since time immemorial the coastal areas of Bangladesh have witnessed the worst kind of disasters owing to cyclonic storms and associated coastal flooding. In the last decade the southern coastal areas have been ravaged by severe cyclones seven times. The cyclonic storm of November 12–13, 1970, which struck the offshore islands located to the south of Bangladesh, could be regarded as unprecedented in world history in terms of damage caused. The frequency of occurrence and magnitude of the cyclones have a profound bearing on the economic development and utilization of about 18,000 square kilometers (7,000 square miles) of the coastal regions with a population of nearly 6 million.

This study was undertaken to offer a more accurate and detailed knowledge of coastal flood hazards as perceived by respondents and the choices open to them in making adjustments, if any, under the prevailing circumstances at Galachipa and in other similar places.

Galachipa was chosen for investigation because this area has frequently been ravaged by cyclones and coastal inundations, and because Galachipa is an area which has experienced a steady increase of population. Population rose by 32.4 percent in 1951–61; density in people per square kilometer increased from 1,095 in 1951 to 1,430 in 1961 (424 and 561 people per square mile, respectively; 1971 figures for population and density are not available.) Galachipa *thana*, once very remote from the point of view of communication, particularly from the heart of Bangladesh, has assumed importance as it has become one of the major trade centers in extreme southern deltaic Bangladesh (Fig. 2–1), thanks to improvement in the system of inland water transport.

The questionnaire survey

A sample of 66 residents, including businessmen and teachers in Galachipa, was interviewed with regard to experiences and expectancies of cyclonic storms and flooding. Residents included in the sample had past experience of moderate to severe inundations and were potential victims of cyclone and coastal inundation.

Sample age, income level, education, and professional status were recorded, and the general category of socioeconomic class was calculated. In addition, household residents were interviewed regarding their own and family experiences and behavior during the cyclone of November 12–13, 1970.

Physical adjustment and the present flood protection scheme

As a part of the active delta with relatively seaward position, Galachipa shows no pronounced relief contrasts; soils are generally some kind of clay. The rivers, with a tidal range of about 1.5 meters (5 feet) but not too saline, flow about 2.1–2.7 meters (7–9 feet) below the land surface in winter. The secondary drainage system is sluggish owing to the lack of pronounced slopes. Except along the main channels, flooding is due largely to local rainfall being unable to flow away. But cyclone flooding assumes enormous proportions when inundation is caused by the rapid inflow of water from the nearby large rivers, particularly the Agunmukha, which is practically a part of the Bay of Bengal. During the cyclone of November 12–13, 1970, the depth of inundation as revealed by the interviews varied from 3 to 9 meters (10 to 30 feet). The local Bangladesh Water and Power Development Authority (BWAPDA) officials at Galachipa bazaar recorded a rise of 4.5 meters (15 feet).

Remarkably, the Galachipa site, although devastated a number of times in the past, does not have an integrated system of private or public adjustments. People are aware of the hazard, but the adoption of adjustments is not a function of hazard frequency.

During the November 1970 cyclone roughly 38 percent saved their lives by climbing trees; about 5 percent took shelter in the "community center"; nearly 8 percent preferred to remain on top of the embankment; others, particularly those working in the open fields, had no chance to save their lives. It is hard to explain the death of 22 persons in a family located 0.4 kilometer (¼ mile) away from a two-story community center. That people are reluctant to leave their homesteads and property despite grave danger may help explain it. Risk-taking propensity is an attribute, then, of the people of the study site.

An awareness of the hazard in its local perspectives may have induced some managers to adopt incidental, rudimentary adjustments. The layout of houses may act as a buffer against flood loss, and emergency adjustment may be simply leaving the house to save oneself.

House pattern

Most of the rural people in the coastal areas of Bangladesh live on the patriarchal family pattern. Peasant farmers usually build their homesteads on their own fields. According to the laws of inheritance family plots are subdivided among the heirs. As a consequence, houses occupied by families parentally related are grouped in clusters providing shelter for 20 to 30 persons, but higher figures are not infrequent.

The house clusters are usually surrounded by dense tropical vegetation and trees. Sometimes these trees contribute to the increase of the family income, and

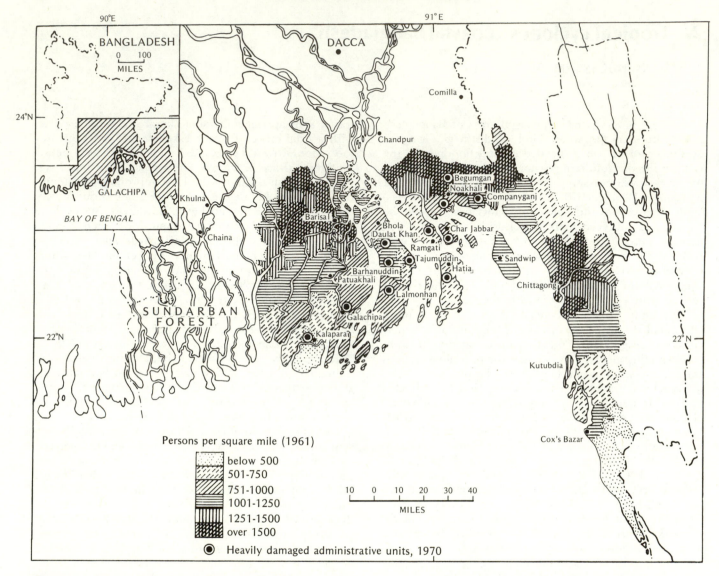

Fig. 2–1. Coastal areas of Bangladesh. Population density and major cyclone-affected areas, November 1970

they also afford good protection to the houses during cyclones—rarely, however, houses are damaged by falling trees. Most important, many lives are saved from inundation and tidal surge when occupants of the houses climb the trees. Again, many people float for a long time by holding onto pieces of wood or tree stumps.

Each cluster of houses has at least one water pond that serves for bathing and washing. Villagers normally travel long distances to obtain potable tube-well water. Some of them, however, prefer not to walk and use pond water even for drinking. In a society of orthodox beliefs and ideas, the women are not expected to be seen in public, which necessitates a pattern limiting them to the household area. Thus, the pond becomes very significant.

Quality of homesteads

Losses to property and human lives are attributable not only to unusual intensity of the cyclone but to poor quality of the houses. The cost of building a house for the average-income family does not exceed Tk. 100 (i.e. 100 takas; U.S. $13.50) in normal conditions. The house generally consists of four thin tree or bamboo trunks, slightly anchored in the floor, on which a number of straw mats or old tin sheets are fixed and on which is supported a primitive light bamboo grate and the thatch roof.

A better type of house consists of mud walls and corrugated iron sheets or tile roofing. *Tarija* fencing (i.e., woven split bamboo panels on a wooden frame) is also extensively used, either with iron sheets or with a shingle roof. Houses of well-to-do people consist either of sheet walls and roofs, or of brick walls and sheet roofs or reinforced concrete roofs.

Since the local area has slight relief, houses are built on plinths and raised platforms and in some cases on timber piles, particularly around the bazaar area. Some land is protected against all but the worst tidal flooding by a .6–.9 meter (2–3-foot) *bund* which allows fresh rainwater to accumulate in the fields for necessary agriculture purposes.

In view of recurring flooding, attention recently has been given to large-scale construction of flood protection embankments along the major rivers. In 1966 the BWAPDA undertook an integrated plan for control of floods in the nearby rivers. The project envisages construction of embankments all around the study site by 1974. A corresponding afforestation program to protect the slopes of the embankment is expected to be completed by the same time. Though the embankment program was underway, portions around Galachipa thana were almost completed, with a few gaps here and there. At the time of the investigation (March 1971) the embankments had been severely devastated by onrushing surge and were breached at many places. The steep side of the embankment was particularly affected and in many places broke down completely, demonstrating that the surge came from the western direction, i.e., from the Agunmukha River and not from the east as was expected. The afforestation program, the principal aim of which was to protect the sloping edge of the embankment on the riverfront, seemed to be a dismal failure. Owing to the onrush of water from the wrong direction the dense vegetation at places was completely washed out. Investigation revealed that casualties were maximum on those breached portions and existing gaps.

Socioeconomic class index

Criteria for assigning sample residents to social classes were based on the fact that Galachipa is a community whose prosperity has always depended largely upon the products of the land, whether for subsistence or for small-scale commerce or industry. The pattern of population distribution is closely correlated with land use and type of farming. Possession of land is the basis of economic wealth in the area.

Having this in mind, the sample residents of the households were initially categorized as (1) owner farmers; (2) tenant farmers; (3) owner farmers having other sources of income; (4) tenant farmers with subsidiary income; (5) landless laborers with other income; and (6) landless laborers. Subsidiary occupations include teaching, small business, government employment, retail trading, and fishing.

Since paddy is the main crop and the average yield per acre is a little more than 10 maunds (1,000 pounds per acre, or 1,110 kilograms per hectare), a family size of five possessing at least 10 acres (4.05 hectares) of land with income derived from subsidiary occupation was taken as the minimum requirement for the upper socioeconomic class. Owner or owner-cum-tenant farmers, and professionals such as teachers, government employees, and other persons in trades and business, with more than 10 acres of land under cultivation, are upper class. Tenant farmers with at least 10 acres of cultivated land without subsidiary occupation are middle class, as are respondents with at least 5 acres (2.1 hectares) of land and income derived from subsidiary occupation. Respondents with less than 5 acres under cultivation with or without subsidiary occupation, unless they are major merchants or businessmen, are lower class.

The total sample is distributed as follows: upper, 27.2 percent; middle, 42.4 percent; lower, 30.4 percent. The occupational groups include farmers, 65.2 percent, government employees, small merchants, primary and secondary school teachers, 16.6 percent; and others, 18.2 percent.

Formal schooling is not always a major factor in economic well-being in Bangladesh: for example, a farmer who cannot read or write may be better off economically than a schoolteacher with a college degree. However, in this study's sample there was a significant correlation between socioeconomic class and educational level (Table 2–1).

Analysis of the data

Perception of storm hazard

The existence of a substantial coastal flood hazard is known to all the respondents. Each experienced personally the flooding during the cyclone of November 12–13, 1970. Nearly 90 percent of the sample expect a future flooding. Of those, nearly 70 percent expect it "soon." Only 10 percent of those who believe that a cyclone will strike the area soon ascribed the occurrence of the future cyclone to "God's will." Perception of storm hazards does not vary appreciably by educational level or by occupation. Also, there is no significant statistical difference in attitude toward future cyclones and associated flooding by socioeconomic class (Table 2–2). Pessimism prevails among the respondents; the upper class is especially pessimistic, with 100 percent expecting reoccurrence of cyclones and flooding.

Table 2–1. Socioeconomic class and educational levels (row percentages; absolute numbers in parentheses; $N = 66$[a])

	Formal schooling	Reads but < 6 years	Cannot read	Total
Upper	83.3 (15)	0 (0)	16.7 (3)	100.0 (18)
Middle	53.5 (15)	10.7 (3)	35.8 (10)	100.0 (28)
Lower	5.0 (1)	25.0 (5)	70.0 (14)	100.0 (20)

[a] $X^2 = 24.6998$; $df = 4.00$; $p < 0.005$.

Table 2–2. Chi-square statistics for relationships of socioeconomic factors and attitudes toward floods

1st Variable	2nd Variable	Significance
Socioeconomic class	Literacy	0.005
Socioeconomic class	Future floods	NS
Socioeconomic class	Time of expected floods	NS
Socioeconomic class	Future habitation	NS
Socioeconomic class	Living condition	NS
Socioeconomic class	Earning potential elsewhere	0.05
Literacy	Future floods	NS
Literacy	Time of expected floods	NS
Age	Future floods	NS
Age	Time of expected floods	NS
Age	Future habitation	NS
Age	Living condition	NS
Age	Earning potential elsewhere	NS
Occupation	Future flood	NS
Occupation	Time of expected floods	NS
Occupation	Future habitation	NS

Knowledge of existing protection against cyclones and flooding in the area cannot be defined operationally because there has been so little protection. The incomplete protective works (embankments) failed to prevent damage from the fury of the onrushing surge in November 1970. The majority of the respondents believed that the incomplete embankment project had actually caused an increase in loss of life and damage to property. A careful review of the embankment project of the BWAPDA in the coastal areas is needed to determine whether or not this is true.

Choice of settlement

Experiencing the November 1970 cyclone appears to have had little influence on respondents' attitudes on choosing future habitation. Nearly 85 percent of the respondents said they were willing to continue to live in the same place, although two-thirds of the total sample indicated that their living conditions, on the whole, were disadvantageous.

There does not seem to be any correspondence between awareness of the past and choice of future habitation. Although nearly 90 percent of the respondents expect cyclones in the near future, willingness to live where they now live is universal, with no significant variation according to socioeconomic class, education, occupation, or age. This passivity should be interpreted in the light of government assurance of protection through the embankment-afforestation program, improved public facilities like the dissemination of warning systems, better communication facilities, and, above all, the construction of community shelters and public buildings including deep tube wells.

Half of the residents did not know how they could earn a living in places with fewer cyclones. It is well known that the proportion of residents knowing other places to earn a living is higher in the upper socioeco-

nomic class and lower in the lower class. At a statistically significant level, the upper socioeconomic class regards earning a living elsewhere, as more feasible (Table 2–3). But there is no significant relationship between age and attitude toward earning a living elsewhere.

A third of the respondents thought they could earn as good a living elsewhere, but felt they had incentives to remain. It is difficult to conclude that the inhabitants have no choice but to stay in the hazard area, although they may subjectively believe so.

Incentives to remain

Economic pressures and social preferences add to the desire of the residents to remain in their homes after a cyclone hazard rather than move to a less vulnerable area. Nearly 65 percent of the residents who had experienced several cyclone and associated flooding events declared in the interviews: "It has always been my home"; nearly 30 percent declared: "The land is good and I can easily feed my family in a good year." The family tie and above all the community feeling is so strong that respondents are willing to take risks and depend entirely upon the vagaries of nature. "Almighty God knows everything" is the prevailing mood. It is no exaggeration to state that unless they are forced to vacate the area, many will remain in the hazard zones for subsistence.

Relief and rehabilitation

One major factor encouraging people to stay where they are is availability of aid and relief, of which a substantial though always not adequate amount has been dispensed. Loans in various categories are provided to those who remain, and attempts are made to generate employment locally, particularly through rehabilitation and public works.

There should be a comprehensive study of the impact of aid and relief on the community. The respondents did have misgivings about the nature and extent of emergency relief, but the provision of relief without any reciprocal obligation may reduce the inhabitants' concern about hazards. Provision of such relief may well increase hazard loss potential.

Some communities look a little better immediately

Table 2–3. Attitude toward earning a living elsewhere, by socioeconomic class (row percentages; absolute numbers in parentheses; $N = 66$[a])

	Yes	Don't know	No	Total
Upper	55.6 (10)	5.5 (1)	38.0 (7)	100.0 (18)
Middle	28.6 (8)	25.0 (7)	46.4 (13)	100.0 (28)
Lower	10.0 (2)	25.0 (5)	65.0 (13)	100.0 (20)

[a]$X^2 = 10.4648$; $df = 4.00$; $p < 0.05$.

after rehabilitation. Thanks to aid from international agencies and help from many friendly countries, beneficial changes are brought about. The communication system has been improved to facilitate relief and rehabilitation. New clothes, household materials, and building materials such as corrugated iron sheeting give an improved look to the community, but this is only true where relief and rehabilitation have been undertaken with proper objectives and honest intentions.

Economic activity at the study site recovered quickly with the completion of an improved road. The metaled (macadam) road facilitated rapid transport of men and material from distant places and indirectly caused an increase in the number of people visiting Galachipa, a major rural marketplace. The prospect of getting aid easily caused an inflow of people from other districts, as did the opportunity for employment generated by the relief and rehabilitation program. A large number of unemployed, landless laborers and in some cases beggars claiming to be local residents from surrounding districts tried to obtain relief and thus crowded the area. The lower income groups rallied around those who by virtue of their influence managed to obtain undue relief and rehabilitation aid.

The Coastal Embankment Project, launched initially to protect the land against saline water intrusion, was not complete when the cyclone occurred. There were gaps here and there. The storm surge overtopped the embankments in many places, causing washouts and severe erosion. Many breaches occurred next to sluices. Fatalities were highest where settlements were located near the not yet completed embankment line. The embankment projects after the cyclone gave priority to (1) closing gaps in the embankments below high tide to prevent further scouring; (2) repair of sluices to prevent further damage; and (3) repair of embankments of polders which will regain precyclone protection completely or nearly completely before the monsoon season. The reconstruction program would generate interim employment for people in the devastated areas, since most of the project work was manual labor.

In spite of expectations engendered by the embankment reconstruction project, the field interviews reveal that the program did not get underway effectively because of lack of laborers from the area. The respondents considered the labor wage inadequate, although they were assured of the wage paid before the cyclone. Persons affected by the cyclone were busy taking care of their own household properties and building new structures with whatever small aid they received. They were not interested in earning money by working as laborers in the embankment reconstruction program. The practice of giving loans, including test relief and outright gift of work animals, has further reduced the concern for making extra money off the farm. Without aid, it may be emphasized, the economy of the area would have been completely shattered because time and again the people have relied on aid in rebuilding their economy.

Implications of land ownership categories

As previously mentioned the natural land-building process in the deltaic plain creates more and more opportunities for economic activity. The newly formed land areas, locally known as "char," are considered to be favorable for crop production. The salinity of the newly formed char lands does not impede cultivation, as the salt is washed away by the heavy monsoon rainfall in May/June every year. Though there are specific government regulations for the distribution of these lands, in practice big landowners who have enough resources or other influence can acquire a sizable portion of it at the expense of the local landless laborers and small tenant farmers.

Landowners belong to several classes, of whom big landowners, and, particularly, absentee landowners, play a major role in the management of land resources. Their number and the amount of land they own are not known. The field survey did not give a basis for estimating the size of holdings of big landowners. However, the study site, with its small population, is characterized by large landholdings compared to the surrounding densely populated regions. Roughly one-third of the farmers included in the sample have less than 1 hectare (2½ acres) of land; many, not included in the sample, have holdings too small to support their families without subsidiary income. Some of the big landowners and absentee landlords control more than 400 hectares (1,000 acres) of land.

To cultivate large holdings these landholders must hire agricultural laborers; thus the system of land ownership generates employment locally. The local landless or small tenant farmers, being assured of employment locally, appear not to be interested in leaving their community for a place with fewer cyclones.

They do not care, often, to hear information on the availability of employment opportunities elsewhere. Apart from a traditional inborn fatalism, limited freedom of movement and local availability of employment are factors favoring staying put after a hazardous storm.

In-migration and the high hazard potentiality

As there are limited economic opportunities in the surrounding districts or even within the district, we find a flow of in-migrants, mostly landless laborers, at sowing and harvesting time. The potential for death from cyclonic storms is intensified by the inflow of migrants during the sowing and harvesting seasons of *aman* paddy, the major subsistence crop of the area, which coincide with the cyclonic seasons. Though migrant harvesters were not included in the sample, group interviews revealed that they all knew of the cyclonic hazard in the area. In addition, in view of luxurious grass in the char areas many big farmers hire laborers to graze herd animals. Inland and offshore fishing enterprises are run by big farmholders or absentee landlords.

The migrant laborers are paid in cash or in kind and are housed in temporary huts located in the fields far away from the village or small marketplace. Although the hired laborers are so much in demand in the local economy, the basic amenities provided them are anything but satisfactory. It is no exaggeration to say that in a cyclone these migrant harvesters, having to work and stay in the open fields, suffer mostly from lack of communication and timely warning. In 1970, many unidentified dead bodies were found floating along the edges of the embankments or high grounds during the high tide period. Field interviews confirmed high casualty figures among the migrant harvesters.

In-migration during the sowing and harvesting season not only contributes to an increase of hazard potential but also makes hazard estimation difficult. Due to lack of proper data on in-migrants there is virtually no way to find out the percentage of deaths among them, and in most cases they remain unaccounted for. Notwithstanding the danger of cyclone hazard in April-May and October-November, the sowing and harvesting season of *aman,* the area attracts people from many walks of life and many places, many of them businessmen, traders, and middlemen.

Conclusion

Appraisal of the cyclone hazard in Galachipa requires an understanding of the economic and cultural background, particularly the area's land ownership system. The residents themselves are not very concerned about the cyclone hazard. The three main socioeconomic classes perceive the hazard in much the same way, though the upper class seem to favor earning a living elsewhere more.

Contrary to general assumption, there is no direct link between awareness of the past experience and the decision-making process. Decisions to settle and maintain settlement in the area are made by managers who perceive future losses consequent upon their decisions. The intricate relationship between man and nature in Galachipa manifests an adjustment process where decision making has not been a function of the natural events systems.

Available evidence suggests that in an established community where the history of cyclone hazard is as old as settlement, the recent public promise of aid, including relief and rehabilitation, has added a new dimension to the perception of hazard. Although the respondents have misgivings about the nature and extent of emergency relief, such relief without any obligation on the part of the managers has further reduced their concern about the hazard. The provision of relief has caused an increase in flood-loss potential.

There is no clear correspondence between perception of cyclone hazard and adoption of specific adjustments by the managers of the study site. They are aware of the hazard, but the adoption of adjustments is not a func-

tion of the hazard frequency. The latest cyclone created a fervor for brick houses. This complements the factor of limitation with regard to freedom of movement. The major economic advantage of staying in the area is the rising of char, giving subsistence to so many for years to come, as this is the area of sedimentation. Aside from local economic aspects, social preferences and strong family and community bonds have their influence. It is evident that the resident's rational choice will be to use the present land even if he is aware of the hazard.

Public perception of cyclone hazard has effected a series of adjustments which constitute a piecemeal approach to the crisis whenever it appears. When a cyclone of severe intensity is expected at longer intervals, the public mind tends to underestimate the severity of damage. Between two disasters there is a carefree period of inactivity. The approach in dealing with the cyclone hazard has been corrective rather than preventive. No attempt has been made to organize a community program against coastal hazards taking into consideration local conditions.

With the public promises of protection and relief, there is every likelihood of further encroachment in the hazardous coastal areas. Risk-taking propensities are a major attribute of the population. Galachipa, a major market center, is in a better position today because of its link with the major river ports of the country. Land resource management, including the present land ownership system, needs careful review. There is a special need for careful guidance in human occupance of hazardous areas such as the study site.

Basic research on resource use vis-à-vis human occupance in the coastal areas has not been undertaken. We know very little about the resource potentialities and the system of land management. For example, ideas on cooperative farming including zoning and land-use regulations have been suggested for the coastal areas, but how far these measures can be applied in a land-hungry country like Bangladesh is anybody's guess. It is imperative before any action is taken that the government specify the broad objectives of the adjustment process for areas subjected to natural hazards in Bangladesh.

The formulation of a national policy to cope with natural hazards, keeping in view the resource appraisals, is long overdue. A policy of social welfare has already been declared by the government. It is hoped that the problems of the hazardous areas will be given due consideration and that solutions will be found and implemented at the earliest opportunity.

On the basis of this study of perception and choice of the coastal dwellers, the following optimal or desirable set of adjustments can be suggested for the Galachipa area or places similar to it:

1. In view of the peculiar socioeconomic characteristics of the area, it would be difficult to relocate the present settlements under normal conditions. But empirical evidence suggests that following a disaster,

communities, though inherently conservative, are prepared to accept some changes. Thus, a planned land use, including relocation of settlements keeping in view the extreme hazardous zones, could be given a priori consideration.

2. Instead of the present dispersed pattern, the future settlements could be clustered around selected nuclei. The nucleus could be a multistoried community center capable of giving shelter to the local residents during cyclonic events. Receiving and disseminating storm warning signals should be among the important functions of the community centers. In normal times the center should provide educational and recreational facilities to the members of the community.

3. On the structural side, brick-built houses seem to be the favored adjustment of most coastal residents. However, construction of brick houses or other low-cost housing (e.g., CINVARAM blocks) usually requires government subsidy. Houses on piers and community housing should also be encouraged.

4. Another important aspect of the plan should be a change in the present system of land holding. Reforms of the prevalent land tenure system should be an integral part of the plan. The government is committed to a socialist economy; it should persuade the residents of the hazardous areas to adopt a cooperative system of farming in the greater interest of the community.

5. One promising adjustment, subject to detailed research, is attempting rescheduling of the crop season or the introduction of new varieties of crops in the coastal belt.

6. An improved road network connecting the coastal areas with the interior should be constructed. This would help in developing a local system of relays to disseminate meteorological forecasts. Above all, there must be provision for quick evacuation of highly vulnerable areas when an extremely severe disaster is predicted.

7. Protective works, including embankments, are needed to check saline intrusion on croplands. Large scale construction of embankments to buffer against storm surges needs comprehensive review considering the morphology of the area.

In sum, what is required is a comprehensive plan which would be implemented step by step on the basis of priorities set with a view to financial constraints.

3. Human response to the hurricane

DUANE D. BAUMANN
Southern Illinois University

JOHN H. SIMS
George Williams College

This paper attempts three tasks: first, to outline in general terms the possible range of human response to the hurricane threat in contemporary America; second, to review briefly what have been and continue to be the policies and actions of the federal government concerning hurricane damage in the United States; and finally, to report and discuss a modest but provocative cross-cultural study of public response to the hurricane hazard in the United States and Puerto Rico. These three aims are interrelated. On the one hand, current federal programs can be shown to focus upon but a narrow sphere of the possible repertory of preventive and adaptive behavior to hurricanes; on the other hand, research into public attitudes suggests new directions for what could possibly be more effective efforts. The analysis is presented alongside that of coastal adjustments by Baker and Patton (chap. 4 below).

Range of adjustments

One thing is certain—human occupancy of coastline susceptible to hurricanes will incur costs. Damage from hurricanes in the United States has steadily increased since the mid-1930s. As Dacy and Kunreuther (1969) point out:

> Since 1934 hurricane damage has increased at a rate of 4.5 percent per year, and since 1950 the rate of increase has been 10 percent per year. Average annual damage for the period 1925–1949 was $83 million; since 1950 the average has increased to $217 million. And in the last five year period (1960–1964) . . . hurricane damage exceeded flood damage for the first time . . .

Essentially, adjustment to the hurricane hazard can be of two kinds: one either does nothing and bears the loss, or one takes some kind of action in an attempt to modify the potential loss. Aside from bearing the loss, responses to the hurricane hazard attempt to reduce loss either by lessening the destructive power of the hurricane itself, or by strengthening the resistance of the physical-structural environment, or by directing man's behaviors in strategic ways which adapt to rather than combat the hurricane's force. This paradigm of adjustment possibilities is illustrated in Table 3–1.

Bearing the loss

Simply bearing the loss from hurricanes is probably the most common type of individual response (Burton, Kates, and Snead, 1969). In a study of the eastern seaboard of the United States, over 50 percent of the 371 respondents had borne the cost of some type of damage during their occupance of the shore. Specifically, 15 percent reported light damage, primarily to grounds; 15 percent incurred damage as a result of inundation to buildings and contents; 22 percent indicated major losses such as structural damage to buildings.

Reducing the loss by moderating the hurricane

Since the first large-scale experiments on seeding hurricanes in 1961, we are still uncertain whether any observed changes were the result of seeding or would have resulted regardless of the experiment. Clearly, hurricane modification is not presently a viable adjustment, although the potential appears promising.

Reducing the loss by strengthening the environment

Floodproofing adjustments range from the stockpiling of materials to structural changes. In some places, land filling and elevation have raised buildings above the tidal floodwaters; in addition, measures have been designed not only to keep water out but to allow water to pass through the building in an effort to reduce the hydrostatic pressure. However, even where residents are aware of flooding as a potential accompaniment of hurricane, individual adoption of floodproofing techniques is uncommon (Burton, Kates, and Snead, 1969).

Protective works can be categorized in two basic types: those that attempt to impede the passage of the waves by man-made structures—groins, breakwaters, bulkheads, seawalls, revetments—and those which are designed to use the natural defenses of the beach and dunes—land stabilizations and beach nourishment.

The existence of such protective structures is widespread and characteristic of much of the national coastline subject to hurricanes. Of the 1,095 miles of outer coastline studied by Burton, Kates, and Snead (1969), roughly 23 percent, or 384 kilometers (240 miles) had some type of protective structure. In some states, such as Connecticut, nearly 50 percent of the coastline had some type of protective work.

As in riverine flood hazard, the federal approach to the hurricane hazard has clearly been one which has favored protective works where they have been economically feasible or, as previously noted, the provision of relief (in effect, loss bearing) where the criterion of economic feasibility could not be met.

An example of this emphasis on protective works is in the recommendation for Port Arthur and vicinity, Texas:

> Pursuant to authority of Public Law 71 . . . a survey was made to determine the need and economic feasibility of providing hurricane flood protection to Port Arthur and vicinity, Texas. It was found that a serious problem of hurricane tidal flooding exists in the populated and heavily industrialized Port Arthur area. . . . Portions of the area are protected to some degree by existing levee and floodwall systems, but would be subject to inundation from wave overtopping and breaching of the levees during major hurricanes. Other large developed and underdeveloped areas are without protection . . . The improvements proposed would afford large benefits by preventing most of the damages that would occur from the flooding of existing properties *and future growth and development either by hurricane tides or rainfall runoff.* [U.S. Corps of Engineers, 1962; italics added]

In the above illustration, the survey was undertaken to determine the economic feasibility of providing hurricane flood protective structures. Alternative strategies to flood-loss reduction, such as floodproofing or preventive zoning of the undeveloped land, were not even considered. Indeed, protection was proposed *for* undeveloped areas with the expectation of future growth. Hence, as in the experience of riverine flood damage reduction, the potential damage from hurricane flood may actually be increased through the encouragement of highly vulnerable development behind protective works.

Reducing the loss by adapting man's behavior

Considerable effort has been directed toward the development of a national hurricane warning system. Sugg (1967) estimates the annual cost of detecting and monitoring hurricanes at between $2 and $4 million. The technology of early detection and monitoring of hurri-

Table 3–1. Types of adjustment to the hurricane

Bear the loss	Reduce the loss (examples)	
Moderate the hurricane	*Strengthen the environment*	*Adapt man's behavior*
Cloud seeding	Floodproofing, protective works: groins, jetties, breakwaters, beach nourishment, land stabilization, seawalls, revetments	Warnings, evacuations, land-use zoning, land easements

canes has already been developed to a point where more accurate prediction is not expected in the immediate future (U.S. Department of Commerce, 1970). However, the need for more precise knowledge is clear. For example, emergency actions regarding tidal flooding are seriously hampered because of the need for additional but currently unavailable information on local situations. The coastal resident needs to know not only normal tide levels for a specific time period, but the elevation and the expected run-up.

Little attention has been given to assessment of the formulation and adoption of individual private plans for emergency action. However, one study found minimal types of strategies of emergency actions involving personal safety and the securing of property to be fairly widespread in the coastal occupance of the eastern seaboard; 60 percent of the respondents had employed such adjustments at some time in the past (Burton, Kates, and Snead, 1969).

The use of easements and zoning to reduce potential tidal flood damage from hurricanes is not widely practiced (Burton, Kates, and Snead, 1969). However, there are some examples. The development of high-risk areas for recreation has sometimes been adopted as a policy at both local and national levels of government. The National Park Service, for example, has established national seashores which inadvertently stifle development by private owners that would increase potential hurricane damage. However, such a strategy may not necessarily be the economically optimum use of the shoreline. That is, it remains an open question whether private development of the coast might not maximize net benefits even when taking the costs of hurricane damage into consideration.

Another example of the difficulties inherent in zoning programs is Burton, Kates, and Snead's (1969) proposal for zoning ordinances back from the shoreline. One recommendation is that no building should be lower than 4.5 meters (15 feet) above mean sea level. The assumption, of course, is that for buildings below this level the potential "costs" outweigh the potential "benefits." But evidence, though critical for zoning decisions of this kind, does not exist.

Vagaries of human response

This brief review of the major categories of anticipatory response to the hurricane hazard in the United States has correctly shown the major efforts of government, both past and present, to be in the area of strengthening the environment. To date, only minimal attempts have been made at moderating the force of a hurricane, and, with the important exception of warning systems, few efforts have been directed toward affecting man's behavior. This is surely not fortuitous; it reflects this country's habitual trust in technological know-how, and at the same time, it points out the current state of ignorance concerning man's response to natural hazards. Per-

haps experience with warning systems best illustrates this latter point.

The effectiveness of warnings is usually considered to be improving. The loss of life from hurricane hazard has gradually been declining from the tragedy of 1900 at Galveston. In one instance, nearly 350,000 persons were evacuated along the Gulf Coast in preparation for Hurricane Carla—only 45 deaths were recorded. Yet, over 250 deaths occurred in response to Hurricane Camille in 1969, the highest figure since 1957 when Hurricane Audrey accounted for 390 fatalities. The point is that although average annual deaths from hurricanes have declined, experience with a few hurricanes suggests that the warning process is not sufficiently understood to be more consistently effective. Why are residents sometimes reluctant to evacuate, and why do they persist in living in hazardous zones? What factors influence an individual's decision to adopt some adjustments and exclude others, to heed or ignore a warning?

Such questions have rarely been asked and only recently are efforts being made to answer them. If we counterpose Alexander Pope's line "Fools rush in where angels fear to tread" with an old saying, "Nothing ventured, nothing gained," social scientists are currently somewhere in the middle struggling to get hold of those sociopsychological dimensions that operate in determining man's adjustments to natural hazards.

Some knowledge is being accumulated. For instance, it is clear that neither awareness of the existence of the hurricane hazard, nor indeed past experience with it, are sufficient to produce effective precautionary actions. In two studies, one of Galveston, Texas, the other of the eastern seaboard, it was shown that virtually everyone knew of the hurricane threat, and further, that the great majority of residents had experienced a storm; but there was enormous variation among them as to the types of adjustments consequently adopted (Arey and Baumann, 1971; Burton, Kates, and Snead, 1969).

Why? The available evidence only inches toward an answer—for reasons as yet poorly understood, persons vary greatly as to their perceived vulnerability to future hurricanes. Thus, almost a third of the Galveston residents perceive themselves as being missed by future storms—hence, they see no reason to make preparations that would reduce loss. Assessing the future from another perspective, 80 percent of the Galveston sample acknowledged that a hurricane would come again, but believed that a certain amount of time must pass before it could recur. Thus, as many as 16 percent of them concluded that a hurricane would not come again in their lifetime, another "reason" for lack of preparation.

The presence of such beliefs about personal vulnerability and such expectations of hazard occurrence raises serious questions about the efficacy of warning systems. The emphasis to date has been on improving the prediction and dissemination of the warning. There is no questioning the beneficial consequences; however, such efforts may be approaching diminishing returns with respect to reduction of both loss of life and property damage. What is needed now is some understanding into what determines variation in the human *interpretation* of warnings.

Where will such research take us? Certainly into psychology, sociology, and cultural anthropology. The remainder of this chapter will present a small venture into these uncharted waters as an illustration of the direction in which some natural hazards research is beginning to move.

A cross-cultural study

The Natural Hazards Questionnaire was administered to 360 respondents in three continental U.S. sites—Tallahassee, Florida; Pass Christian, Mississippi; and Galveston, Texas—and to 147 respondents in two sites, one coastal, one interior, in Puerto Rico. These data include completions to ten sentence stems designed for two purposes: (1) to explore hypothetical hazard behavior (here, regarding the hurricane) at three times—before, during, and following the hazard event; (2) to measure the psychological dimension referred to as "internal versus external locus of control," that is, the degree to which a person feels he controls his own life (Rotter, 1966; Lefcourt, 1966; Joe, 1971).

This sentence completion test was included in the hazard battery of instruments because prior research had shown that two intra-U.S. cultural groups—Southerners and Midwesterners—differed both in their responses to the tornado hazard and in their scores on locus of control in such a way as to suggest a possible causal connection. That is, Midwesterners, convinced of their own autonomy, tended to adopt active-rational types of coping behaviors in dealing with the threat of a tornado; whereas Southerners, persuaded of the power of outside forces (God, fortune, luck) on their lives, tended to adopt passive-fatalistic coping behaviors (Sims and Baumann, 1972).

The present study on response to hurricanes compares a generally middle-class, Protestant, moderately educated (12 or more years of schooling) sample drawn from the highly technological, economically rich, western society of the United States, with a generally lower-class, Catholic, less educated (8 or less years of schooling) sample drawn from the economically poor, Latin society of Puerto Rico. The adjectives in the preceding sentence should be read simply as shorthand cues to the enormous cultural differences between the two samples. The continental U.S. sites are described in Chapter 4. Three general questions were asked of the data: First, to what extent will hurricane behaviors and locus of control vary cross-culturally? Second, will that variance make sense in terms of known cultural differences, such as, the difference in religion? Third, will the two variables being measured—hurricane behavior and locus of control—be related to one another according to the same logic in both cultural contexts?

Table 3–2. Measures of hurricane behavior

Sentence stems and completions	Puerto Rico (N = 141) (%)	U.S. (N = 360) (%)
Before the hurricane		
If a hurricane is predicted, I . . .		
make preparations (unspecified)	55	31
keep on the alert	10	19
feel fear/anxiety	7	4
seek refuge	14	29
[other]	14	17
When a hurricane is coming, I feel . . .		
fear	48	31
anxiety	9	18
concern for the consequences	26	15
desire to take precautions	2	14
[other]	15	22
During the hurricane		
During the hurricane, I . . .		
make preparations (unspecified)	38	17
pray	10	18
communicate with others	13	9
feel fear/anxiety	9	16
protect myself	13	18
protect others	12	8
proceed normally	5	7
[other]	0	7
Going through a hurricane makes me feel . . .		
fear	38	31
anxiety	11	23
negative emotions (unspecified)	6	10
concern for the consequences	36	15
[stays calm]	9	4
[other]	0	17
In a hurricane the people I feel some responsibility for are . . .		
family of procreation (husband or wife and children)	40	37
children	9	17
parents	6	7
family and nonfamily	24	25
nonfamily	14	12
[other]	7	2
After the hurricane		
When a hurricane is over, I . . .		
feel positive emotions	25	26
check results	23	18
thank God	10	10
begin restoration	12	29
aid victims	14	4
feel negative emotions (fear, anxiety)	6	6
[other]	10	7
When a community experiences a hurricane, the feelings among its people . . .		
are of mutual cooperation	31	48
are of fear and anxiety	6	6
are of sadness	42	11
are shared	11	2
are positive	4	17
[other]	6	16

The data presented in Table 3–2 show that persons in Puerto Rico and the United States differ importantly in the ways they anticipate responding to a hurricane both *before* and *after* its occurrence but are very similar in the behavior reported *during* the storm.

Two of the stems direct attention to experience prior to the event of a hurricane: (1) "If a hurricane is predicted, I . . .," and (2) "When a hurricane is coming, I feel . . ." The majority of respondents in both groups are active in their response; thus only small percentages respond with negative emotions or by doing nothing. On the other hand, there is a cultural difference in the *nature* of the activity. In Puerto Rico, 55 percent make a general statement of preparation, as opposed to 31 percent of the respondents in the U.S. Contrariwise, for two *specific* kinds of activity, the American gives significantly higher percentages of response—staying alert (and attentive to the communication media) and seeking refuge.

Analysis of the second stem, "When a hurricane is coming, I feel . . .," reveals a major difference between the two groups in the nature of the negative emotions felt. Fear is experienced among 48 percent of the Puerto Rican respondents, compared to only 31 percent in the United States respondents. And, for anxiety, the figures are Puerto Rico, 9 percent; U.S. 18 percent. It is tempting to interpret fear as paralyzing and anxiety as functional (that is, as provoking action), given the responses to the first stem *and* the other responses to this stem. Thus, 25 percent in Puerto Rico express "concern over the consequences," as opposed to 15 percent in the U.S. And, on the other hand, 14 percent of the respondents in the U.S. give a response coded as "takes precautionary action" as opposed to only 2 percent in Puerto Rico.

Analysis of the completions to the three stems which focus on behavior during a hurricane reveals few salient differences between the two cultural groups, and indeed, what differences do appear are repetitive. Again, more Puerto Ricans make a *general* statement of preparatory intent, but do not specify just what they will do; again, more Puerto Ricans respond with fear and more Americans with anxiety; and finally, more Puerto Ricans are concerned with the consequences. There are no differences between the two samples regarding who it is they would feel concerned about.

Two stems—"When a hurricane is over, I . . ." and "When a community experiences a hurricane, the feelings among its people . . ."—are concerned with responses after the hurricane event. Two striking differences are observed in the completions to the first of these. First, 29 percent of the Americans as opposed to 12 percent of the Puerto Ricans are concerned with restoration. Secondly, the Puerto Ricans, 14 percent, are more concerned with aiding victims than are the Americans, 4 percent.

Completions to the second stem also reveal two salient differences: in Puerto Rico, 42 percent portray the community as suffering from grief, as opposed to 11 percent in the U.S. (Moreover, in the U.S. there are 17 percent, versus 4 percent in Puerto Rico, who express their optimism with positive affect.) Secondly, the emphasis upon mutual cooperation for restoration and aid is given greater emphasis in the U.S., 48 percent, than in Puerto Rico, 31 percent.

Table 3–3 presents the completions to three sentence stems measuring internal versus external locus of control. These are: (1) "Getting ahead in the world results from . . ."; (2) "As far as my own life is concerned, God . . ."; and (3) "I believe that luck . . ."

There were no substantial differences in the responses of both groups to the stem "Getting ahead in the world results from . . ." In the U.S. there was a somewhat greater emphasis on education while Puerto Rico respondents stressed economic development of the area. Perhaps education may be viewed as "internal," that is, as a mobilization of inner resources, whereas economic development may be seen as "external," as assistance from greater powers than the self.

A major difference does emerge between the two groups in the responses to the stem, "As far as my own life is concerned, God . . ." In Puerto Rico, 54 percent of the respondents see God as "powerful and important" with respect to their lives, as opposed to 17 percent in the U.S. Contrariwise, 55 percent of those in the U.S. versus 21 percent in Puerto Rico see God as a benevolent protector.

The important difference in the responses to the stem, "I believe that luck . . ." is the assertion by more Americans that luck is random.

Differences on completions to these three stems are consistent: in each case more Puerto Ricans give responses which recognize the importance of external forces—luck, God, economic development—in control-

Table 3–3. Measure of locus of control

Sentence stems and completions	Puerto Rico (N = 141) (%)	U.S. (N = 360) (%)
Getting ahead in the world results from . . .		
work	27	22
drive, ambition	17	12
education	6	19
moral, religious behavior	9	15
economic development of area	9	0
[other]	32	32
As far as my own life is concerned, God . . .		
is benevolent and protective	21	55
is powerful and important	54	17
[other]	25	28
I believe that luck . . .		
is good	26	8
is from God	10	10
is random	23	34
[has experienced in own life]	17	25
[other]	24	23

ling their lives. In contradistinction, more Americans stress autonomy. Thus, for example, although equally religious, the two groups have quite different conceptions of the role of God in man's world. Puerto Ricans see him as directly involved in what happens to them, whereas Americans see him as benign but removed; their motto would be "God helps those who help themselves."

How might these differences in locus of control be related to differences in anticipated hurricane behavior, and, in turn, how might both be related to cultural differences? It is possible here only to make a few initial suggestions. It appears reasonable to argue that those who feel more in control of their lives would be more specifically active in coping with the threat of an approaching hurricane and similarly that they would be more instrumental in dealing with the storm's aftermath. Following the same logic, one would argue that those who see themselves as less autonomous, or who acknowledge the power over their lives of outside forces, would be more fatalistic, more accepting, and yet more frightened of the hurricane threat and more undone by its consequences.

Relating these different patterns of psychology and hazard response to cultural differences between the two groups is a task for the cultural psychologist-historian. But surely the traditions of Spanish Catholicism and American Protestantism have consequences here as would differential commitments to technological control of the environment.

The final point to be made is a simple one: namely, that willingly or unwillingly, it is into such complex and resistant areas as psychology and cultural anthropology that scientists must venture if they are to understand and improve man's ability to cope with the unleashed powers that constitute natural hazards.

References

Arey, David, and Baumann, Duane. (1971) *Alternative Adjustments to Natural Hazards.* Washington, D.C.: U.S. National Water Commission.
Burton, Ian, Kates, R., and Snead, R. (1969) *The Human Ecology of Coastal Flood Hazard in Megalopolis.* Chicago: University of Chicago, Department of Geography, Research Paper No. 115.
Dacy, Douglas, and Kunreuther, Howard. (1969) *The Economics of Natural Hazards.* New York: Free Press.
Joe, V. C. (1971) "Review of the internal-external control construct as a personality variable." *Psychological Rep.* 28: 619–40.
Lefcourt, H. M. (1966) "Internal versus external control of reinforcement." *Psychological Bulletin* 65:206–20.
Rotter, J. (1966) "Generalized expectancies for internal versus external control of reinforcement." *Psychological Monograph: Gen. App.* 80:1–28.
Sims, John, and Baumann, Duane. (1972) "The tornado threat: coping styles of the North and South." *Science* 176: 1386–92.
Sugg, Arnold. (1967) "Economic aspects of hurricanes." *Monthly Weather Review,* 95:143–46.
U.S. Corps of Engineers. (1962) *Port Arthur, Texas and Vicinity.* Washington, D.C.: Government Printing Office, House Document No. 505.
U.S. Department of Commerce. (1970) *National Hurricane Operations Plan.* Washington, D.C.: ESSA.

4. Attitudes toward hurricane hazards on the Gulf Coast

EARL J. BAKER   DONALD J. PATTON
University of Colorado   Florida State University

In world perspective, coastal storms constitute one of mankind's most serious natural hazards. When the maritime fringes of the continents are buffeted by exceptional storms, high costs in loss of human life and property damage often ensue. As Baumann and Sims point out (chap. 3 above), comparatively little research on perception of, and adjustment to, one type of coastal storm—the tropical hurricane—was undertaken before the early 1970s. Most attention given the hurricane-prone Gulf Coast of the United States was concerned with human behavior in disaster situations (Rayner, 1953; Killian, 1954; Bates et al., 1963; Moore, 1963, 1964). More recent are studies conducted in posthurricane environments which in part dealt with attitudes toward hurricanes and adjustments (Wilkinson and Ross, 1970; Schaffer and Cook, 1972).

In 1971 the understanding of the problem was extended by research in certain attitudes toward hurricanes among sample households at three separate locations along the Gulf Coast of the United States. The focus was on attitudes toward hurricanes as a natural hazard and toward the possibility of adjustments to the hazard.

Rationale

What utility is there in attitude research? La Piere (1934) was probably the first to point out the apparent inconsistency between verbalized attitude and behavior. Several other studies have reached the same conclusion, and recent discussions of the problem by Deutcher (1966) and Wicker (1969) have appeared. Still other studies, however, have indicated a close agreement between attitude and behavior (Fishbein, 1972). Why are the findings contradictory?

Rokeach (1969) maintains that most studies take into account only "attitude-toward-object," while we actually possess various attitudes which interact to influence behavior. Particularly important but often ignored, according to Rokeach, is "attitude-toward-situation." Tending to support the Rokeach argument is Campbell, who suggests the existence of "situational thresholds" of behavior (1963).

Kiesler, Collins, and Miller (1969) explain away the "inconsistency studies" by attributing those results to poor experimental design. The authors convincingly argue that too many studies either have failed to use comparable measures of attitude and behavior (e.g., varying item difficulty) or else have tested for a behavior which should not have been expected from the attitudinal indicators.

Our contention is that if a subject has a given attitude, he indeed possesses *a* predisposition to respond as a result of that attitude. Thus, although it may be only one of several components toward behavior, possession of that attitude makes a given behavior more likely than would a differing attitude. Therefore, if a certain behavior is "desirable" (e.g., evacuating an area which is expected to be flooded by storm surge), then it is worthwhile from the social perspective for a subject to possess an attitude which would tend to elicit that behavior.

The three sites

The decision first was made to study attitudes toward hurricane hazard at a site on the Mississippi coast, which had been severely affected by Hurricane Camille in 1969. Analysis of Camille storm data revealed that the peak storm intensity occurred in the vicinity of Pass Christian, Mississippi, and it was accordingly chosen as the locale for field investigation (Fig. 4—1). Before data were collected at Pass Christian, however, the scope of the study was broadened to complement a collateral study being planned elsewhere. Two other sites were included: one, a low-risk site—Tallahassee, Florida; the second, Galveston, Texas. An early assumption that Galveston would represent an intermediate-risk site in contrast to Pass Christian as a high-risk site and Tallahassee as a low-risk site was not borne out either in terms of data on hurricane occurrences and probabilities or in terms of attitudes of the population sampled. However,

Fig. 4—1. Map of study area

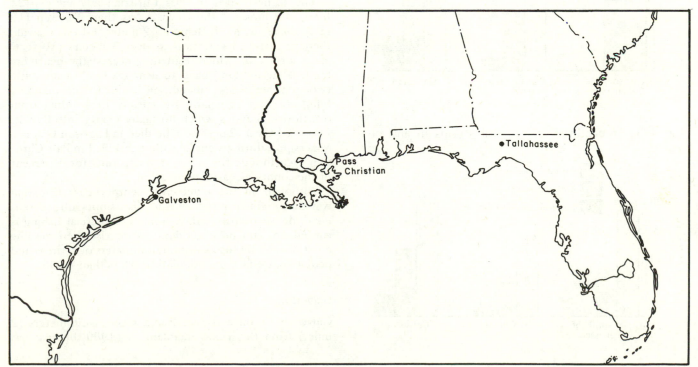

each site has a hurricane history quite distinct from the other two.

Figures 4–2 and 4–3 compare the three sites both in terms of their respective recorded hurricane histories, and their probabilities of experiencing tropical storms in any one year. Data are based on the Simpson and Lawrence (1971) technique for recounting storms that affect given coastal segments and determining the likelihood of occurrence of storms at these segments. The nomenclature of storms differs from that used by the World Meteorological Organization, which designates any storm of tropical origin, regardless of wind velocity, as a "tropical cyclone."

Tallahassee

While the Simpson and Lawrence method of studying hurricane frequencies is revealing for coastal areas, it is deceptive for Tallahassee, which is approximately 40 kilometers (25 miles) inland from the coastal zone for

Fig. 4–2. Frequencies of tropical storms, 1886–1971

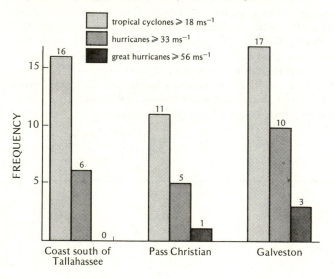

Fig. 4–3. Probabilities of tropical storms in any one year

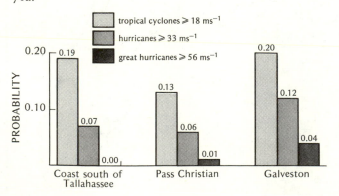

which the data apply. Wind velocities diminish before reaching Tallahassee, and the threat of inundation posed by a storm surge does not exist. Newspaper accounts indicate hurricane force winds, if only in gusts, occurred at Tallahassee in 1851, 1877, and 1886. Since Weather Bureau observations began in Tallahassee, records show 33-meters-per-second (75-mph) gusts with 29-meters-per-second (65-mph) sustained winds in 1941 and 28-meters-per-second (62-mph) gusts with 20-meters-per-second (44-mph) sustained winds in 1966. Most damage was done to roofs or occurred as a result of wind-blown objects. Flooding was localized in poorly drained areas as a result of heavy rains accompanying the disturbances. On the other hand, Hurricane Agnes, which moved inland over the coast immediately southwest of Tallahassee in June 1972, destroyed approximately half of the seafront residences along an exposed section of the coast. The city itself was little affected by Agnes, however, in part owing to the path of the storm center, which passed inland well to the west of Tallahassee.

Tallahassee is a city of approximately 80,000, with state government and higher education being the dominant activities, although central place functions are also important.

Pass Christian

At the time the data for this study were collected (Summer 1971) Pass Christian had been struck by Hurricane Camille 2 years earlier. Camille was meteorologically the most violent hurricane event to strike the United States since the initiation of weather records in the nineteenth century; at the time of the survey, scars of Camille's impact were still apparent.

The highest level in Pass Christian is 7 meters (24 feet), and most of the town is less than 3 meters (10 feet) above sea level. Behind a 2-meter (6-foot) seawall, there is a rise in elevation to over 3 meters (10 feet) along a ridge on which residences nearest the beach are built. A rapid slope back to near sea level occurs only two or three blocks inland, and residences in this low-lying area are especially susceptible to flooding from overflowing bayous which normally empty into Bay St. Louis. Of the 112 persons who died in Harrison County, Mississippi, during Camille, 60 were killed in Pass Christian even though the town only accounts for 4 percent of the county's coastal population.

Although in the nineteenth century Pass Christian was the foremost resort on the Mississippi Gulf, there is no evidence of this today. Some commercial fishing is carried on, and most residents work elsewhere on the coast. There are also substantial numbers of retirees and people on welfare. Its population is less than 4,000.

Galveston

Galveston is on a barrier island some 3 kilometers (2 miles) from the Texas mainland. In 1900 the city was virtually destroyed by a hurricane with winds of 49 meters per second (110 mph). At that time there was no seawall, and the highest land elevation on the island was only 2.4 meters (8 feet). Much of the island was covered with 1.2 meters (4 feet) of water, at least 6,000 persons were killed, and 8,000 were left homeless.

Two massive engineering projects were undertaken soon after the disaster. First, a sea wall 5 meters (16 feet) wide at the base, 1 to 5 meters (3 to 16 feet) wide at the top, and 5 meters (17 feet) high was constructed to guard the city from storm surge. Second, the elevation of the city was raised to the height of the seawall adjacent to the seafront and allowed to slope gradually back to the opposite shore. The efforts proved successful in 1915 when a hurricane struck the island and the mainland coast in the vicinity, causing 8 deaths in Galveston but killing over 100 on the mainland. Today most of the city is over 2 meters (6 feet) in elevation.

Galveston is a city of 62,000, combining resort functions and port activities. Commercial fishing is also prominent.

Procedure

At each of the three sites 120 interviews were obtained from heads of households. In Galveston and Tallahassee random nested samples of city blocks within a coarser grid were chosen. In Pass Christian random sampling of city blocks was employed. All three samples were stratified with regard to three classes of literacy and divided into two classes according to whether or not there were children of working age. The Tallahassee sample was further stratified by excluding areas known to be dominated by university student residences. No deliberate areal stratification was made in Pass Christian or Galveston, but it should be noted that in Pass Christian roughly two-thirds of the sample was in areas more subject to flooding. This was a result of the above stratification by literacy, which tended to vary with income. Low-income residential areas usually corresponded to those parts of the town most likely to flood. Surveying was accomplished in July and August 1971, before the hurricane season affected the Gulf Coast.

Hypotheses

Previous research had shown that indicators of cognitive factors concerning natural hazards are associated with other variables. Determination of the associations can aid in directing efforts to reduce undesirable consequences in the event of hurricane occurrence. Fifteen hypotheses of relationships between variables relating to attitudes toward, and adjustments to, hurricane hazards were tested. In the analysis the three dependent variables were (1) the evaluation on the part of the respondents of the hurricane hazard in their locality, (2) their prediction concerning the future occurrence of hurricanes, and (3) their attitude toward the possibility of preventing hurricane-related damages. It was hypothesized that each of the three dependent variables is associated with five independent variables: site, literacy, income, age, and tenure.

To ascertain how respondents evaluated the hurricane hazard, the question "Do the people of this place have any trouble with hurricanes?" was asked. Answers were classed as "Yes," "Doubtful," "No," or "Don't know."

To investigate expectation of hurricane recurrence, the question "Do you think a hurricane will come again during your lifetime?" was asked. If the response was "Yes," the individual was asked, "When—soon, a few years from now, or many years from now?" In some cases the answer "Yes" was given, followed by "But I don't know when." Answers of "No" and "Don't know" were also recorded.

In order to elicit an indication of attitudes toward the preventability of destruction caused by hurricanes, the question was posed: "Is there anything that the government or people, your friends and neighbors, can do to *prevent* damage from a hurricane?" Responses were "No," "Don't know," and "Yes." Specific damage prevention measures were probed for if the subject answered "Yes."

Among the five independent variables which were treated, it was recognized at the outset that site would very likely be of special importance. It was hypothesized that sharp differences could be expected between the three sites in terms of the way the hurricane hazard was perceived because of recent hurricane history, which respondents could be expected to be aware of even if they had had no direct hurricane experience. Moreover, Pass Christian and Galveston contrasted with each other in the nature of their hurricane defense works and in elevations of residential sites. Thus the selection of three sites allowed a considerable variation in the natural hazard itself to be incorporated in the analysis.

The other four independent variables served as indicators of general access to information relating to hazard adjustments. It was assumed the respondents' level of formal education would be an accurate indicator of literacy. For the income variable, three classes of income were used: low was $6,000 or less; medium ranged from $6,001 to $12,000; and high was over $12,000. Respondents were also grouped into three categories of age range: (1) less than 40; (2) at least 40 but less than 65; and (3) at least 65. Finally, tenure referred to whether the respondent owned his place of interviewing or not. In almost all cases nonowners were renters.

Analysis of data

Responses to the basic questionnaire were summarized and subjected to preliminary statistical analysis at the University of Colorado Institute of Behavioral Science. The 15 hypotheses were subsequently tested using chi-square tests of independence. The null hypothesis in all cases was that the probability of a response in a given column of the contingency table was the same for each cell in the column (i.e., H_O: $p_{11} = p_{21} = p_{31} = \ldots$, $p_{12} = p_{22} = p_{32} = \ldots$, etc.). A level of significance of 0.05 was generally required to reject H_O. A measure of association, the Goodman-Kruskal τ_b, was also calculated for each contingency table. (Note: τ_b is analogous to a correlation coefficient but is not interpreted in the same manner. See Goodman and Kruskal, 1954, for discussion.)

Sample size varied from hypothesis to hypothesis primarily because response categories were combined differently when necessary to avoid unacceptably small expected cell frequencies. In some cases, the final sample size was altered by deletion of data in the process of data compilation. This was particularly true for the data of the Galveston site. Primary causes for deletion were recording abnormalities in the field and incorrect coding.

Results

A summary of the results from an analysis of the respondents' answers both in terms of the significance level of chi-square test results and in terms of τ_b is presented in Table 4–1.

Hazard evaluation

Respondents' evaluation of the hurricane hazard was shown to vary strongly with site. In both Pass Christian and Galveston there was near-unanimous agreement in evaluation of the hazard, responses indicating absolutely no relationship to any of the other independent variables. Associations between hazard evaluation and income or age appeared significant only in the low-hazard area, Tallahassee. At that site, the subjects' response to the hazard evaluation question did tend to vary with income, and the null hypothesis was rejected at $p < 0.025$. There was a tendency for negative responses to increase with income; all the "Yes" responses came from low-income respondents. A significant difference ($p < 0.05$) was also found in terms of the age of the respondent, with the oldest age group answering negatively a disproportionately high number of times.

Expectation of hurricane recurrence

Respondents' answers to this question varied only according to site, and they accord with actual probabilities of recurrence. Taken jointly or by pairs the sites showed significant differences in their respondents' answers. The Tallahassee questioning elicited unexpectedly high numbers of "Don't know" and "No" answers. Galveston respondents had the greatest tendency to answer "Yes" or "Don't know." Pass Christian exhibited greater uniformity of responses across all categories. No significant differences in responses in terms of any of the four other independent variables appeared.

Table 4–1. Summary of relationships between variables

Dependent variable by independent variable	Significance level of x^2 test result	τ_b
Hazard evaluation by		
Site	0.005	0.678
Literacy (Tallahassee only)	NS[a]	0.017
Income (Tallahassee only)	0.025	0.051
Age (Tallahassee only)	0.05	0.077
Tenure (Tallahassee only)	NS	0.017
Expectation of hazard recurrence by		
Site	0.005	0.116
Literacy (all)	NS	0.017
(Tallahassee)	NS	0.024
(Pass Christian)	NS	0.022
(Galveston)	NS	0.020
Income (all)	NS	0.009
(Tallahassee)	NS	0.001
(Pass Christian)	NS	0.029
(Galveston)	NS	0.011
Age (all)	NS	0.080
(Tallahassee)	NS	0.002
(Pass Christian)	NS	0.017
(Galveston)	NS	0.197
Tenure (all)	NS	0.004
(Tallahassee)	NS	0.009
(Pass Christian)	NS	0.008
(Galveston)	NS	0.174
Attitude toward damage prevention by		
Site	0.005	0.092
Literacy (all)	0.005	0.096
(Tallahassee)	0.005	0.097
(Pass Christian)	0.005	0.102
(Galveston)	0.005	0.149
Income (all)	0.005	0.049
(Tallahassee)	0.005	0.110
(Pass Christian)	0.025	0.095
(Galveston)	NS	0.031
Age (all)	NS	0.014
(Tallahassee)	NS	0.015
(Pass Christian)	0.05	0.044
(Galveston)	NS	0.033
Tenure (all)	NS	0.001
(Tallahassee)	NS	0.006
(Pass Christian)	NS	0.010
(Galveston)	NS	0.002

[a]NS indicates no significance at the 0.05 level.

Attitude toward prevention of hurricane damage

This dependent variable was found to be strongly associated with site, but it also appeared to be associated with literacy and income. In terms of site, Tallahassee accounted for the majority of the difference, varying from each of the other sites at $p < 0.005$. Very little difference in response appeared between Pass Christian and Galveston. Tallahassee was distinguished by its very high frequency of "Don't know" answers.

Every testing of literacy (i.e., each site as well as the three taken jointly) indicated strong association between literacy and attitude. Differences were significant at $p < 0.005$. Increasing education was accompanied by increasing propensity to answer "Yes." Income also tended to show strong relationships with particular categories of response. The differences were significant for the three sites taken together ($p < 0.005$), for Tallahassee alone ($p < 0.005$), and for Pass Christian ($p < 0.025$). Little variation of response with income appeared in the case of Galveston, however, although there was a tendency for middle-income subjects to answer "Yes."

The strong association between literacy and attitude toward the possibility of hurricane hazard adjustments to prevent damage may provide a key to potential adjustment modifications. Some form of public information program designed to provide residents with appropriate guidance on hurricane hazard adjustments would appear promising.

Further statistical analysis of data was undertaken in order to define more precisely the variables which should be considered in organizing such an informational program. The technique used was the construction of a multidimensional contingency table (MCT) for four variables: (1) damage prevention attitude, (2) site, (3) literacy, and (4) income.

The table was then analyzed using a procedure discussed by Feinberg (1970). In essence the method is a multivariate stepwise regression tool applicable to categorized data. Damage prevention attitude was treated as a dependent variable while site, literacy, and income were considered independent variables. The goal was a log-linear model for the generation of expected cell frequencies for the MCT. A sequence of nested hierarchical models containing main effects and multiple-interaction effects was computed by the Iterative Proportional Fitting Procedure. The likelihood-ratio chi-square statistic was used to test the hypothesis that each model fitted the MCT of observed cell frequencies. Partitioning the chi-square of the simplest model into its components led to identification of the effect of each successive interaction component in reducing residuals (Goodman, 1971; Feinberg, 1970). The results appear in Table 4–2.

The two-factor interactions of literacy-attitude and site-attitude were found to be sufficient to account for residuals relating to the dependent variable. Thus our model for the MCT is:

$$\ln e_{ijkl} = [1] + [SLI] + [LD] + [SD]$$

where e_{ijkl} is the expected value of cell entry (i, j, k, l).

Conclusion

The MCT analysis indicated that whereas other relationships involving attitude toward the preventability of hurricane damage exist, two variables, literacy and site, are associated with it so strongly that they alone may be used in explaining most of the variation in this attitude. It would thus seem that efforts to remedy false beliefs and conceptions which result in negative attitudes toward damage prevention measures might consider the individual's education level and his site above other variables. Better-educated respondents are more likely to have a positive attitude toward damage prevention adjustments. To the extent this is true, more intensive efforts should be directed toward less-educated segments of the public in order to counteract their tendency to lack confidence in damage prevention adjustments. Such an effort could have two components:

1. Low-education groups should be given priority as targets of information.
2. Pamphlets or messages disseminated might well contain somewhat differently worded information depending on the general education level of the target audience.

A positive attitude toward the possibility of avoiding hurricane-related damage varied directly with frequency of past hurricane occurrence. The greatest difference among the three sites was in the "Don't know" category. The less collective experience the population of a site had with hurricanes, the more indecisive its residents appeared to be concerning the efficacy of damage prevention actions. Thus, in addition to designing messages differently for varying educational levels, the messages should also vary depending on the site's history of hurricane experience.

In conclusion, public awareness programs in hurricane hazard sites may tend to rely too often on standard pamphlets and brochures published by the National Oceanic and Atmospheric Administration. These efforts should be tailored to the site and to the education level of various target groups.

Table 4–2. Nested hierarchy of models[a]

	Models	X^2	df	X_d^2	df_d
(1)	[1] + [SLI] + [D]	141.26	52		
(2)	(1) + [LD]	82.57	48	58.69	4
(3)	(2) + [SD]	31.92	44	50.65	4
(4)	(3) + [ID]	27.23	40	4.69	4
(5)	(4) + [SID]	18.07	32	9.16	8
(6)	(5) + [SLD]	13.10	24	4.97	8
(7)	(6) + [ILD]	5.94	16	7.06	8

[a]Interactions were also entered in other sequences to assure these results held regardless of ordering.
S = site.
L = literacy.
I = income.
D = damage-prevention attitude.
1 in model (1) = "grand mean."
[ ] indicates effects due to given parameter(s).
X_d^2 indicates the difference between X^2 for model (i) and X^2 for model (i + 1).
df_d indicates the difference between df for model (i) and df for model (i + 1).

Acknowledgments

We would like to express our gratitude to the following persons who aided in some aspect of our research: Douglas A. Zahn, Florida State University; and Duane D. Baumann and James Lorelli, Southern Illinois University.

References

Bates, F. L., et al. (1963) *The Social and Psychological Consequences of Natural Disaster: A Longitudinal Study of Hurricane Audrey.* Washington, D.C.: National Academy of Sciences—National Research Council, Disaster Study No. 18.

Campbell, D. T. (1963) "Social attitudes and other acquired behavioral dispositions." In S. Koch, ed., *Psychology: A Study of a Science,* vol. 6. New York: McGraw-Hill.

Deutcher, I. (1966) "Words and deeds: social science and social policy." *Social Problems* 13:235–54.

Feinberg, S. E. (1970) "The analysis of multidimensional contingency tables." *Ecology* 51:419–33.

Fishbein, M. (1972) "The prediction of behavior from attitudinal variables." In K. K. Sereno and C. C. Mortensen, eds., *Advances in Communications Research.* New York: Harper & Row.

Goodman, L. A. (1971) "The analysis of multidimensional contingency tables: stepwise procedures and direct estimation methods for building models for multiple classifications." *Technometrics* 13:33–62.

———, and Kruskal, W. H. (1954) "Measures of association for cross classifications." *Journal of the American Statistical Association* 49: 732–64.

Kiesler, C. A., Collins, B. E., and Miller, N. (1969) *Attitude Change: A Critical Analysis of Theoretical Approaches.* New York: Wiley.

Killian, L. M. (1954) *Evacuation of Panama City Before Hurricane Florence.* Washington, D.C.: National Academy of Sciences—National Research Council, Committee on Disaster Studies, unpublished report.

La Piere, R. T. (1934) "Attitudes vs. actions." *Social Forces* 13:230–37.

Moore, H. E. (1963) *Before the Wind: A Study of the Response to Hurricane Carla.* Washington, D.C.: National Academy of Sciences—National Research Council, Disaster Study No. 19.

———. (1964) *And the Wind Blew.* Austin, Tex.: Hogg Foundation for Mental Health.

Natural Hazards Research (1970) *Suggestions for Comparative Field Observations on Natural Hazards.* Toronto: Dept. of Geog., Univ. of Toronto.

Rayner, J. F. (1953) *Hurricane Barbara: A Study of the Evacuation of Ocean City, Maryland.* Washington, D.C.: National Academy of Sciences—National Research Council, Committee on Disaster Studies, unpublished report.

Rokeach, M. (1969) "The nature of attitudes." In D. Sills, ed., *International Encyclopedia of the Social Sciences,* 1: 449–57.

Schaffer, R. C., and Cook, E. (1972) *Human Response to Hurricane Celia.* College Station: Texas A & M University Environmental Quality Program.

Simpson, R. H., and Lawrence, M. B. (1971) *Atlantic Hurricane Frequencies Along the United States Coastline.* U.S. Department of Commerce NOAA Technical Memorandom. Washington, D.C.: U.S. NWS SR-58.

Wicker, A. (1969) "Attitudes vs. actions: the relationship of verbal and overt behavioral responses to attitude objects." *Journal of Social Issues* 25:41–78.

Wilkinson, K. P. and Ross, P. (1970) *Citizens' Responses to Warnings of Hurricane Camille.* State College: Mississippi State University Social Science Research Center, Report No. 35.

5. India and the Ganga floodplains

R. RAMACHANDRAN AND S. C. THAKUR
Delhi University

Policy with regard to flood hazard in India is related to three important needs: (1) to produce more food grains to meet the existing deficit and the increasing demand in future years, (2) to find ways in which the floodplain resources may be further utilized to offset the increasing pressure of population on land, and (3) to explore ways in which flood damages may be reduced effectively in view of the nation's limited capital resources. The third consideration directly focuses on aspects of human adjustment to flood hazard and the range of alternative theoretical and perceived forms of adjustments.

Flood damage in India 1953–71

In India, flood hazard is a spatially extensive phenomenon and all the states are affected in varying degrees. The least affected are the states of Madhya Pradesh and Mysore, while the most affected are Uttar Pradesh, Punjab, West Bengal, and Bihar, in diminishing order of total value of damages. According to regional delineation, the fertile agricultural tracts of Ganga and Brahmaputra valleys and the deltaic areas in Andhra Pradesh and Orissa are the most severely affected.

Floods in India affect, on an average, Rs. (rupees) 870 million (U.S. $116 million) annually (Fig. 5–1). The major damage is to crops, accounting for 73 percent of the total; the rest is to houses, livestock, and public utilities (Fig. 5–2, 5–3, and 5–4). The average annual loss of food production is estimated at over 1 million

Fig. 5–1. Flood damages in India, 1953–69 (constant 1965 prices)

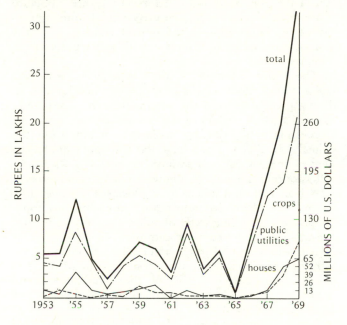

Fig. 5–2. Cattle lost by floods, 1953–69

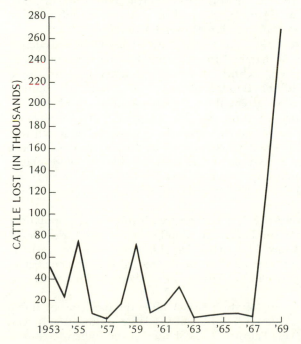

Fig. 5–3. Crop area affected by floods, 1953–69

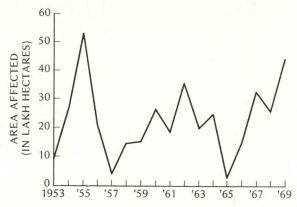

tons. In an average year, over 16 million people are directly affected by floods and about 700 human lives are lost (Fig. 5–5). The totals reached new heights after 1966. Preliminary estimates for 1970 and 1971 are 1,076 and 1,023 lives lost.

Flood control in India

After independence in 1947, concrete steps were taken to control floods and minimize the devastating havoc. The government made gradual progress with its program of flood control, allocating money in four five-year plans beginning in 1954.

In relation to the annual average flood loss of Rs. 870 million (U.S. $116 million), the fourth five-year plan envisages expenditure on flood control of the order Rs. 60–100 crore (U.S. $80–$133 million). Attention to flood problems was extremely low in the first five-year plan, with only Rs. 10.8 crore (U.S. $14.4 million), an

Fig. 5–4. Houses damaged by floods, 1953–69

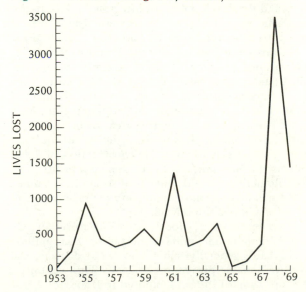

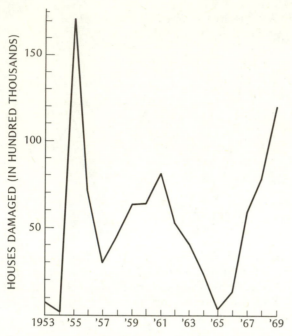

Fig. 5–5. Human lives lost by floods, 1953–69

insignificant amount, allotted. When the plan was drawn up, schemes regarding flood control were envisaged as part of the irrigation program and there was no separate program for flood control.

The floods of 1954–55 were severe and catastrophic. They highlighted the need to deal with this problem in a coordinated and planned manner, independent of irrigation and power. This led to the formation of the Central Flood Control Board in the same year, flood problems being an interstate problem because many rivers traverse more than one state. The Central Water and Power Commission (CWPC) was strengthened by the addition of a Flood Wing to assist in preparation of flood-control schemes and the development of integrated plans. It also considers flood-control projects forwarded by the state flood control boards. Eleven states have such flood control boards along with four river commissions.

The national program of flood control was initiated in 1954 and problems related to flood control then began to receive intensive study. Because of financial difficulties and the nature of the problem, some progress has been made in protecting areas from vagaries of floods. The overall progress made at the end of the third five-year plan was as follows: (1) construction of 6,900 kilometers (4,300 miles) of new embankments; (2) 9,200 kilometers (5,700 miles) of drainage channel; (3) 178 town protection schemes; and (4) raising of over 4,582 villages above flood level.

As a consequence of flood-control measures already practiced, about 5.9 million hectares (14.6 million acres) of land, generally subject to flood damage, had received reasonable protection by 1967. This was only a small proportion of the land exposed to flooding.

The Ganga floodplains

Some aspects of adjustment to floods in such areas are revealed by study of a narrow 16 kilometer (10-mile) hazard belt along the upper Ganga floodplains. Floods are the annual feature along this belt, the peak flood season occurring in the months of July and August. Those events are regarded as relatively less severe, not so much because they cause less damage, but because there are other more extensive areas where floods occur at about the same time and which receive greater publicity through press and radio. The problems of human occupance of the floodplain and human adjustment to the hazard have much in common with other parts of the country.

The Ganga floodplains are the adjacent low-lying areas of the River Ganges subject to inundation due to heavy precipitation in the catchment area. They lie in the Great Plains, which was a trough at one stage. This depression was filled by sediments mainly from the Himalayas. Today it is a large flat alluvium plain having one of the densest populations on the earth. The older (Pleistocene) alluvium is known as *bangar* and it occupies higher ground adjacent to *khadar,* the newer alluvium. The plains are remarkably homogeneous topographically: for hundreds of miles the only relief discernible is the floodplain escarpment, locally known as *khola.* Minor natural levees and badland topography are other landform characteristics found in the plains. The Great Plains cover an area of about 652,000 square kilometers (247,760 square miles), of which one-third lies in Uttar Pradesh.

The maximum depth of the alluvial soil in the Great Plains is reported to be about 4,500 meters (15,000 feet) near its southern edge. The depth varies from place to place but the soil is very fertile consisting of clay, loam and silt. It is an ideal soil for agriculture and supports a population of over 100 million.

Historically, the Ganga floodplains were considered mainly as marginal for permanent human occupance. It was only a century ago that sedentary land use took place in the floodplain proper. Formerly, the adjustment to recurrent flood hazard was primarily shifting in harmony with nature. For example, the floodplains were used as grazing grounds by pastoral communities such as the Gujars, who practiced a form of transhumance between the floodplains and the older plains not subject to flood. A low level of resource use was appropriate when population densities were relatively low.

During the last century, with increasing population pressure on land, the floodplains, despite the flood hazard, provided attractive sites for permanent agricultural settlements. For some time, agriculture coexisted with pastoralism. Now crop cultivation dominates the plains.

Their exploitation began after the partition of India and Pakistan in 1947 when a large influx of refugees from West Pakistan took shelter in the Indian territory.

The relatively less densely populated floodplain areas provided new sites for refugee rehabilitation. One such area was discovered in the new alluvium area near Hastinapur in Meerut District (U.P.). Despite their lack of experience in meeting flood problems, the refugees established a strong foothold, chiefly by fully utilizing the credit facilities offered by government relief agencies, by investing in modern farming techniques, and by emphasizing commercial crops, especially sugarcane and wheat.

The contrasting attitudes and values of the refugees and the traditional farmers in the floodplain provide an interesting background to study of perception and adjustment to floods in this region. Social conflicts arising from the relative prosperity of the refugees and the poverty of the traditional farmers are important. However, the traditional farmers have a perennial interest in the floodplain. The refugees, who are more enterprising, are only too eager to seek opportunities elsewhere in the older alluvium, and have a short time horizon in terms of the exploitation of the floodplain resources.

The study area

The study was undertaken in Bastura Narang village and the analysis made on data obtained from 66 respondents. Information collected there was supplemented by reconnaissance studies of several neighboring villages.

The area on the Ganga floodplains near Hastinapur, 120 kilometers (75 miles) northeast of Delhi, offers an excellent site for the study of human perception and adjustment to floods (Fig. 5–6). Within the area there are a number of settlements, large and small, and a few abandoned village sites which had been destroyed by past floods. Villages near the floodplain escarpment (kola) have greater accessibility, while villages in the new alluvium are least accessible. Hastinapur, a small town of about 10,000 persons, stands on the escarpment.

The area near Hastinapur was chosen for major studies on the basis of the following criteria:

1. It is easily accessible by road;
2. The area is subject to severe flooding;
3. The area consists of villages having a long history of occupance.

The strategy of the study is based on careful field observations and intensive interviewing of the selected respondents of the village of Bastura Narang. The main study centers on human behavior in relation to the perception and adjustment to flood hazards.

Bastura Narang is a small nucleated village located about 8 kilometers (5 miles) south of Hastinapur in the new alluvium proper. It is connected by a paved road for a distance of 6 kilometers (3.8 miles) and an unpaved road covering the remaining distance on the right bank of the Ganges about 2 kilometers (1.2 miles) away from the floodplain escarpment. Situated on relatively elevated ground between Burhi Ganga (Old Ganga) and the existing river channel, it is a traditional village having a long history of occupance. The villagers are mainly agricultural farmers engaged in the production of sugarcane, wheat, rice, maize, and fodder. People are hardworking and yet poor because of the meager production of crops due to no canal irrigation and poor soil.

The village had a population of approximately 900 persons in 1961. By 1971 the population could be safely estimated to be around 1,200. It has a small sugar factory producing brown sugar, a telephone connection, but no electricity. In general the dwelling standard of the villagers is poor. Due to economic factors they are unable to prosper and improve their living. The majority live in dilapidated houses made up of thick mud walls and thatched roofs. There are, however, some brick houses with flat roofs which obviously act as an adjustment to flood. Another adjustment vividly noticed is the raised floor of the houses. The village has about half a dozen chaupaals (raised platforms), each having a tall evergreen neem tree. These trees, during an emergency, are of course used for saving life and property. The streets are narrow and deep like drains allowing for the easy flow of floodwaters.

Bastura Narang is mainly dominated by Rajputs, who obviously have a major hand in its administration. Poor Harijans (untouchables) are usually suppressed, especially when loans are distributed after floods.

Geomorphic setting

Physiographically there is little diversity in the structure and surface of the study area. It has monotonous flat land with abandoned meanders and oxbow lakes. The Ganga flows meandering north to south through the study area over the floodplain that stretches about 8 kilometers (5 miles) on either side. To the west, the floodplain is bounded by a highly eroded escarpment or bluff which separates the newer floodplain from the older alluvial plain.

Within the floodplain proper, there are vestiges of abandoned channels, some of which get dry during the dry season and become good grazing ground for animals. The wet channels, called jhils, cause serious waterlogging, but their marginal marshy lands are used for grazing. There is freshwater fishing in these permanent lakes, which serves as the permanent occupation of the lower-caste people. The main river channel shifts often. At one time the Ganga roared down along the escarpment but now it has shifted about 8 kilometers (5 miles), leaving the original course called Burhi Ganga. Fluctuation in river course is another devastating hazard for the floodplain dwellers. During floods, Bastura Narang is cut off from the floodplain escarpment by the Burhi Ganga; nevertheless, emergency evacuation is possible using the road that leads to Hastinapur, across the Burhi Ganga.

The people of Bastura Narang perceive the conversion of the existing mud road into a paved road as an

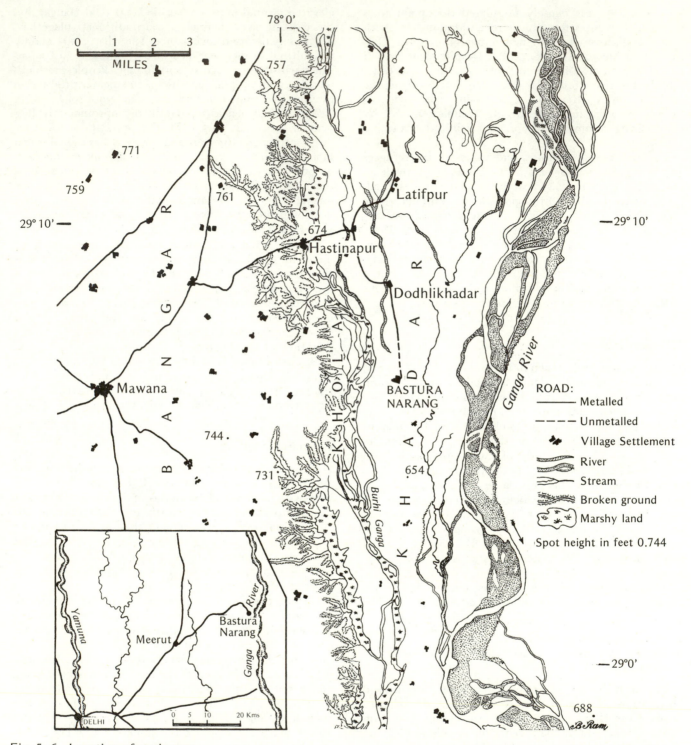

Fig. 5–6. Location of study area

important adjunct to emergency evacuation. A further extension of this road to the south would also serve as an embankment protecting the village site from low and medium floods.

Soil and land-use pattern

Generally, floodplains are distinguished by fertile alluvial soil. Unfortunately this is not the case in the Bastura Narang study area. It receives insufficient rainfall,

69–100 centimeters (27–40 inches) annually, and is subject to severe flooding during the rainy season. Through the coming of a few tube wells and two tractors, a very gradual transformation in agricultural practices is under way.

The drainage system is extremely poor and the water table is high. As a result, capillary action during the dry winter brings water to the surface where its evaporation leaves soluble salts and alkalis in the upper soil layer. The soils, therefore, range from sandy on the inner bank of the river to alkaline in the middle and heavy clay close to the floodplain escarpment. But by and large the whole of the new alluvium area is seriously affected by alkali and since the area is devoid of canal irrigation, nothing substantial can be done to remove it and make the soil more productive and useful.

The abandoned channels in the floodplain are subject to prolonged flooding and hence do not support vegetation, except the tall elephant grass. The edges of the floodplain escarpment are mostly devoted to forests, managed by the forest department. The forests protect soil from erosion and are also partly used for grazing cattle during high floods.

Agricultural land uses dominate the floodplains. In practically all the villages, sugarcane and wheat are the two major crops. The major crops in Bastura Narang are sugarcane, 37 percent; wheat, 34 percent; rice, 10 percent; and millets, 6 percent.

Agricultural activities on the new alluvium are seasonal and thus automatically adjust to flood hazard. Sugarcane, being an annual cash crop, cannot escape the vagaries of floods. Crops, according to seasons, are classified into three chief categories: *rabi, kharif,* and *flez.* The easily perishable crops are harvested before the onset of the monsoon. However, during the monsoon months (June to September) sugarcane and fodder crops, such as maize and millets, are allowed to remain in the field. Fodder crops are easily damaged by floods, while sugarcane can withstand floods for short duration and low magnitude.

Methodology

The research was based on field observation and intensive interviews. The former included the reconnaissance survey of villages in the study area including landforms and human occupation. In Bastura Narang, the observations also included flood adjustment in each household, house type, land-use pattern, and the general reaction of people toward flood hazard.

To collect some general information about the area, the first reconnaissance study was carried out over 15 villages. A second reconnaissance study took place when the area was under floods to see how flood victims made adjustments. The final reconnaissance was conducted after the floods to study flood damages.

Sampling

Three villages were selected during the first reconnaissance survey, each falling in high-, medium-, and low-hazard zones. The sampling was stratified on (1) severity of flood hazard, (2) accessibility, (3) population size, and (4) education status of the people. Bastura Narang belongs to the medium-hazard zone and was stratified according to the above criteria. Only heads of households who were married and had children were interviewed. Fourteen to 20 respondents were interviewed from each of four categories: illiterate with children below working age, illiterate with some children above working age and two similar literate categories; the latter categories had fewer respondents. In all, 66 respondents were obtained. The respondents included 56 farmers, 9 agricultural laborers, and an artisan. Of the total respondents, 60 are Hindus, the rest Muslims.

Field interviews

The interviews involved a general questionnaire including a sentence completion test. After trial with ten households, necessary modifications were made to suit the local conditions. In addition, a few questions on the size of landholdings and land use within holdings were asked to provide an independent base for the assessment of income and area devoted to different crops.

The interview of each respondent ranged from 1 to 2 hours including the inspection of flood adjustment in each house. The questions were translated into Hindi and, where necessary, explanations were given to elicit correct answers. Respondents found it difficult to answer the sentence completion test, the reason being that they are mostly illiterate or perhaps inexperienced.

Perception and adjustment to floods in Bastura Narang

Perception of flood hazard

The floodplain dwellers in the study area are aware of the flood hazard. All 66 respondents of Bastura Narang report flooding as an annual feature, though their opinions regarding its magnitude and frequency differ. Floods inundating cropland are, of course, a regular event; but opinion is divided concerning devastating floods. With their past knowledge and experience, all but two expect the occurrence of future flooding, but they divide on when it will come. About 24 percent are uncertain, while 45 percent expect it "soon," and 15 percent in a few years.

The respondents are also aware of other flood-affected areas. Sixty-four of the 66 respondents recognize that most of the villages in the new alluvium are flood-prone.

The people of Bastura Narang suffer from flood damages regularly. Damages are primarily to crops and, to a lesser degree, to animals and houses, including life and household property. Most respondents (65 percent) regard the flood damages as being substantial; less than one-third thought that it was 80 percent or more.

The floodplain managers are aware of flood damages but are very dissatisfied as far as its recovery is concerned. Most of the damages are borne by themselves. Fifty percent of the respondents said that they would not seek help from anyone; others said they would seek help from the government, and only a few sought help from family and friends.

During floods the government agencies of the flood-control board keep alert to help flood victims. Flood warning, it is perceived, if given efficiently in time, may greatly reduce flood damage. Therefore, the government is very particular about it and makes every effort to convey to people the impending danger. For our study area, the flood message originates from the gauging station at Hardwar. As soon as the flood level reaches the danger mark, a message to its effect is conveyed telegraphically to the collector at Meerut. He immediately passes the message to the flood officer who in turn conveys the same to the *tehsildar*. The tehsildar then issues special flood-warning orders and requests those likely to be affected by the flood to evacuate immediately.

At the time of floods, the government renders help in cash and kind. Among the 66 respondents, 45 knew of other persons who had been previously helped by the government in recovering from flood losses. The help was primarily in the form of loans (*taccavi*) with a fixed rate of interest and emergency supplies during the flood and soon after. Food, clothing, and blankets are also supplied at the time of emergency.

Due to various reasons the floodplain managers are unhappy with the flood relief rendered to them. The majority of them feel that it is insufficient and very poorly organized.

Despite regular flooding and insufficient flood relief from the government, opinion was divided on the question of the relative advantages and disadvantages of living in the floodplain area and more specifically in Bastura Narang. While a larger number of respondents emphasized the disadvantages of living in the floodplain, fewer persons thought so about living in the village itself. When asked about their expectation of living in the same area and site, a vast majority thought that they would continue to live in the same place. Only a small minority of the respondents expected to migrate to a different zone or region.

Adjustment to floods

Recent floods reached up to and encompassed the village, but did not affect the houses which had raised floors. The last experience of such a flood was in 1970. The worst floods, in the experience of the older villagers, occurred in 1924 and 1942.

Adjustments to floods, practiced in Bastura Narang,

are of the folk or preindustrial type. They are, therefore, corrective rather than preventive in character. Loss bearing is the traditional alternative and is borne by every manager of the floodplain property. Other flood adjustments such as emergency evacuation, structural change, and land-use change are theoretically possible in the study village. Some of these alternatives are feasible and are used during crises.

Among the 12 adjustments listed in Table 5-1, adjustments 1 to 6 are almost commonly adopted, while adjustments 7 to 11 are adopted by a few people. There was no adopter of the last adjustment, the use of a private boat. Adjustment 10 is mainly for the identification of field boundaries after floods.

The villagers have diversified perception about flood adjustments. They are less enterprising or have very strong faith, so they leave everything to the will of the God. Community adjustments such as emergency evacuation using hired boats, construction of embankments and roads, and regulation of river flow are perceived as being entirely the responsibility of the government. It is notable that adjustments 5 and 6, which are indicative of socioeconomic status, are not perceived as adjust-

Table 5-1. Adjustments to floods in Bastura Narang

Adjustments	Mentioned by the respondent	Yes, when asked	No, when asked
1. Evacuation by foot	47	3	16
2. Reach for rooftop/treetop	39	16	11
3. Storing food grains above ground level (in *tands*)	15	46	5
4. Go to (a) temple, (b) *dharamsala* (dormitory), (c) relative's place, (d) others, outside floodplains	14	11	41
5. Raised floors	4	42	20
6. Build brick houses with flat roofs for refuge	5	36	25
7. Wait for government evacuation	3	10	53
8. Evacuation before start of floods (of household items, cattle, children, old persons, etc.)	1	2	63
9. Go to own house outside flood plains	1	2	63
10. Planting trees near house and in fields	1	15	50
11. Prior arrangement with private boatman	0	12	54
12. Use private/neighbor's boat	0	0	66

ments unless asked. Field reconnaissance of other villages in the upper Ganga floodplains reveals a recurring pattern of perception and adjustments to the flood hazard. Differences in the mix of adjustments adopted are largely explained by local variations in the severity of the hazard and the duration of floodplain occupance.

Opinion among floodplain dwellers indicate universal dependence on government for more effective emergency relief, flood-warning, and preventive measures. Nevertheless, any enlargement of the role of government in flood problems is likely to result in the abandonment of at least some of the existing mix of individual adjustments.

Conclusion

Flood hazard in India is an annual feature and it occurs in a fairly large area affecting nearly all the states. Usually floodplains are inundated in the monsoon season due to heavy precipitation in the catchment areas.

Of all the regions in India, the Ganga floodplains experience the most devastating floods, causing enormous damage and loss of life and property. It is one of the largest flat alluvium plains in the world with very dense population. It is an agricultural tract having monotonous topography from the Bay of Bengal to the foothills, near Hardwar.

In the study area on the upper Ganga floodplains about 120 kilometers (75 miles) northeast of Delhi near Hastinapur, research included reconnaissance survey of over 15 villages and an intensive study of Bastura Narang village on the basis of a detailed questionnaire. A small nucleated settlement having roughly 1,200 persons falls in the medium flood zone, between the floodplain escarpment and the existing river channel. People are generally poor, uneducated, and humble. Their percep-

tion of flood occurrence is clear. But unfortunately their perception regarding flood adjustment is highly diversified and corrective rather than preventive. Loss bearing seems to be the only choice universally adopted in the village.

Acknowledgments

The authors are grateful to Professor V. L. S. Prakasa Rao, for his persistent encouragement. Grateful thanks are due to the Central Water and Power Commission, New Delhi, and the District Flood Control Agency at Meerut for their kind cooperation in providing data and necessary information. Thanks are also due to Miss Bina Srivastava, Miss Anjana Saim, and Mr. B. Ram for scrutinizing the first draft, processing data, and making maps, respectively.

References

Central Water and Power Commission. (1957) *Flood Atlas, Part 1 (Descriptive) 1956 to 1960: Uttar Pradesh.* New Delhi: Ministry of Irrigation and Power.
_____. (1966) *Flood Atlas, Part 1 (Descriptive) 1954 and 1955, Uttar Pradesh.* New Delhi: Ministry of Irrigation and Power.
_____. (1970) *India: Irrigation and Power Projects (Five Year Plans) (Revised).* New Delhi: Ministry of Irrigation and Power.
Kates, Robert W. (1963) "Perceptual regions and regional perception in flood plain management." *Regional Science Association Papers* 2:217–27.
Kumra, P. N. (1968) "Flood protection methods in India." *Geographical Review of India* 30 (4):45–54.
Mansharamani, M. M. (1970) "Flood forecasting and flood warning in India with special reference to Yamuna Basin." In Central Water and Power Commission, *Silver Jubilee Souvenir*, pp. 83–91.
Shori, B. R., and Ramesh Rao, K. (1970) "A review of flood control and river training in India." In Central Water and Power Commission, *Silver Jubilee Souvenir*, pp. 53–58.
Spate, O. H. K., and Learmonth, A. T. A. (1960) *India and Pakistan.* London: Methuen.

6. Flood hazard at Shrewsbury, United Kingdom

DONALD M. HARDING
University College of Swansea

DENNIS J. PARKER
Middlesex Polytechnic

Floods have become a significant natural hazard in the United Kingdom in the last three decades. In the 1940s, a series of major floods caused much public concern, which was raised further by a very serious local flood at Lynmouth in the southwest of England in 1952. In this flood, some 28 people died. The following year, sea flooding on the east coast of England caused about £50 million (U.S. $120 million) of damage and 250 deaths. Floods on the River Severn in 1960 (estimated damage £11 million—U.S. $26 million), 1964, and 1965 (estimated damage £5 million—U.S. $12 million), and two flood events in 1968 have had a great effect on public concern. The first flood in 1968, in the west country, caused serious damage estimated at £12 million (U.S. $29 million) (Porter, 1970) over a wide area. Some months later, floods in the Thames catchment and its tributaries caused damages estimated at £6 million (U.S. $14 million). In this case, the losses were a direct reflection of the continuing increase in occupance of the floodplains of the Thames catchment, and in many parts of the country potential losses have risen considerably in the recent period.

Public concern after the 1968 floods was reflected in government action with the establishment by the Natu-

ral Environment Research Council of a "Flood Studies" team at the Institute of Hydrology at Wallingford, Berkshire. However, the research at the Institute has been mainly directed at hydrological design studies and this has been the character of most research on floods in the United Kingdom. Such research has been related to the natural event system, while the human responses to the flood hazard have been largely ignored. There is a strong need for developing research in this field at the present time and the study reported here, concerned with one hazard zone, is a contribution toward the development of flood hazard research in United Kingdom catchments.

Fig. 6–1. Regional location map of Shrewsbury

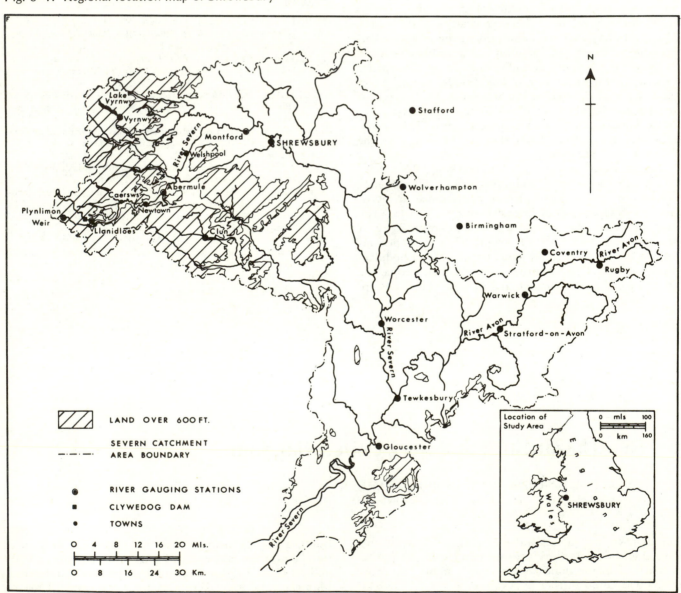

The study site—Shrewsbury, Shropshire

Shrewsbury is located on the River Severn in Shropshire, England, 246 kilometers (153 miles) northwest of London and 69 kilometers (43 miles) from Birmingham. It is a historic settlement, located near the English border with Wales, set in an agricultural region, and with a population of 56,000. In recent years, the town has developed important industrial and commercial functions.

The River Severn, with a catchment area of approximately 11,400 square kilometers (4,400 square miles) is one of the major river systems in the United Kingdom, originating in the central uplands of Wales, and with tributaries draining a large part of the densely populated and industrialized Midlands (Fig. 6–1). Shrewsbury lies in the upper reaches of the river, the catchment area above the town being about 2,100 square kilometers (800 square miles), of which a substantial proportion consists of upland areas above 180 meters (600 feet). A number of towns upstream and downstream of Shrewsbury also experience flood problems.

The settlement was chosen for study because the authors had considerable knowledge of the flood hydrology of the Upper Severn catchment (Howe, Slaymaker, and Harding, 1967), and in addition, the settlement was of suitable size. There are approximately 500 buildings on the floodplain, of which about 190 can be classed as residential.

Increased occupance of hazard zone

The original settlement of Shrewsbury was located on a relatively high well-drained area in the core of a meander of the River Severn (Fig. 6–2), which consists of glacial sands and gravels. The floodplain varies in width from some 90 meters (300 feet) at its narrowest, to greater than 760 meters (2,500 feet) in width in the area to the northwest of the town. Directly south of the central area of the town, the floodplain is narrow but opens out in the upstream and downstream areas. It is here that the floodplain has been encroached upon by residential and commercial development.

Originally, the settlement of Shrewsbury had a defensive function. The castle formed the focus for settlement and the Town Walls formed a restriction on the expansion of the town. Important market functions were developed in the early phase of growth, and later the textile industry was established in the town. With the coming of the railways in the early part of the nineteenth century and the associated development of the road network, the communications function of Shrewsbury became important and there was an associated expansion in the commercial and industrial aspects of the town's economy. The need to develop housing to meet the rising population related to the growing economic importance of Shrewsbury led to development beyond the Town Walls, and housing asso-

ciated with commercial and industrial zones at Coleham and Frankwell can be identified on maps dated as 1860. Extension of the housing areas on the floodplain occurred in the latter parts of the century and in the early part of the twentieth century. The development of housing on the floodplain, mixed with commercial and industrial uses, has led to the creation of high potential damage.

The passing of the 1947 Town and Country Planning Act and the 1948 River Boards Act meant that an element of control was introduced in floodplain management. Under the acts, the Local Planning Authority was required to refer any applications for building in a hazard zone to the River Board for consideration. Normally, if there was a hazard, the Board would propose that planning consent be refused or that permission be granted subject to certain conditions, such as elevation above known maximum flood level. The result has been a decline in the rate of expansion of the area of floodplain developed and relatively few flood-prone properties date from the post-1947 period.

Historical occurrence of floods

Fortunately a long record of river flood levels is available for the Welsh Bridge at Shrewsbury (Fig. 6–3). The pattern of historical occurrence of floods is, in many ways, a remarkable one (Howe, Slaymaker, and Harding, 1967). There have been two periods when flood levels have been high and between these periods there was a period of low floods. In the mid-nineteenth century there were four floods over 5.5 meters (18 feet). Between the 1880s and 1940, the only flood to exceed 5.2 meters (17 feet) was that of 1889. The period since 1940 has seen a series of major floods and the highest of these, in 1946, was followed by two further major floods in 1947 and 1948. The other recent period of high floods was in the 1960s, when four floods exceeded 5.2 meters (17 feet) and caused considerable damage. The results of a flood frequency analysis indicate that the flood of 1946 has a recurrence interval of at least 120 years while that of 1960 has a recurrence interval of around 20 years.

The hazard zone

Two outlines depicting the areas subject to flooding are available and are of considerable value. The first relates to the 1946 flood, which was the highest in the recent period, and the other shows the area inundated by the 1960 flood which attained a peak height some 0.30 meter (1 foot) lower at the Welsh Bridge. For the purposes of this study, the hazard zone has been defined as the area inundated by the 1946 flood (Fig. 6–2).

The hazard zone comprises three main areas. First, to the north of the Welsh Bridge, there is a particularly wide area of floodplain. This includes an area of agricultural occupance and a major zone of residential occupance at Coton Hill. Mixed commercial uses are found in the Mardol/Smithfield area nearer the Welsh Bridge itself. Frankwell, on the opposite bank of the river, comprises an important area of potential damage and uses in this zone are mixed residential and industrial.

The second part consists of the reach of river through the Quarry part of the town to St. Julians Friars. In this reach, the floodplain is much narrower, with a steeply rising slope on the southern side of the river. This area is mainly one of recreational use and so potential losses in this zone are much lower than in the area described above.

Finally, at St. Julians Friars and also on the opposite bank of the river at Coleham, the floodplain opens out again to form the third part, and potential losses are very high. The Coleham and Abbey Foregate areas are some of the more seriously affected parts of the town at times of major flood. On the Coleham bank of the river the occupance type is largely residential, with some areas of public buildings; on the northern bank of the river around English Bridge uses are more mixed, being retail and commercial as well as residential.

In the questionnaire survey, most interviews were undertaken in the two main parts of the hazard zone at Shrewsbury where potential damage is high.

Fig. 6–2. Extent of the flooded area in Shrewsbury. Sources: Severn River Authority and Local Authority

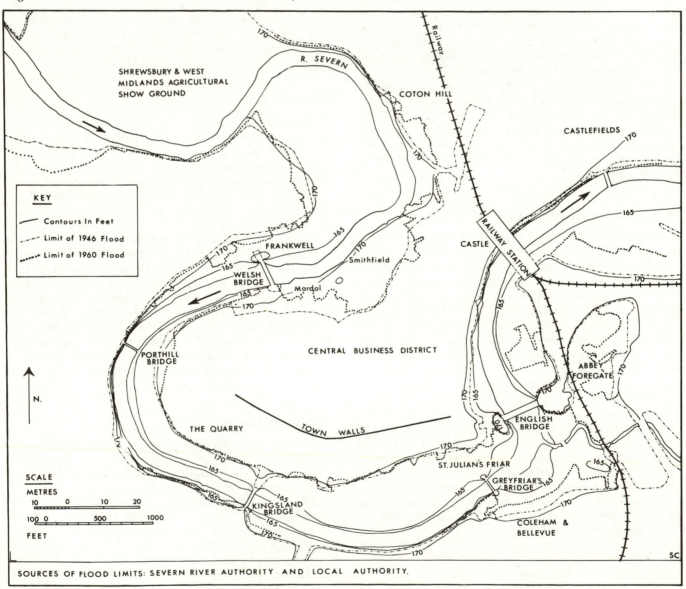

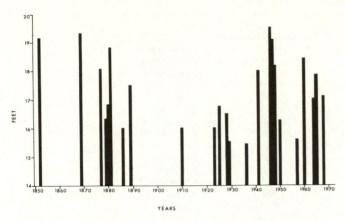

FLOOD HEIGHTS AT WELSH BRIDGE, SHREWSBURY, 1850 TO 1970.

Fig. 6–3. Flood heights at Welsh Bridge, Shrewsbury, 1850–1970

The questionnaire survey

A land-use survey in which the use of every building in the hazard zone was mapped revealed that there were approximately 190 houses on the floodplain, and a 100 percent residential sample was attempted. The survey was undertaken in late July and early August 1971 and the interviews were generally satisfactory, with a fairly low "no response" rate of 9.5 percent, although the detailed nature of the interview and its duration (approximately 45–60 minutes) meant that some refusals were inevitable. Some opposition to the survey was encountered in areas where local authority plans to rehouse families had been proposed, and there were some refusals in these areas. The refusal rate was 21.05 percent and 132 interviews were satisfactorily completed. The distribution by hazard zone was 32 percent located in the high-hazard zone, 40 percent in the medium, and 28 percent in the low.

Results of the survey

Respondent characteristics

Shrewsbury is located in a fairly rich agricultural part of the United Kingdom and has an industrial and commercial economy superimposed on a long-standing market function. The town therefore represents a society in which educational attainment levels are high, as are income levels. The floodplain areas of Shrewsbury, however, represent less desirable locations for residential purposes. They tend to be ones where income levels are low compared to those generally found in the town, and where there are also a substantial number of old people.

Very few of the floodplain occupants fall within the "high-income" category and 44 percent have incomes below £1,000 (U.S. $2,400) per annum. A fair number of these are retired people, although there are many

widows and spinsters living in the area. The most common occupational group were those associated with trade and service industries, while the only other significant occupational group apart from the retired people were laborers. The majority of people (62 percent) were tenants. The language spoken was English, but there were three Welsh-speaking respondents. By far the most common religion was Protestant (85 percent), although there were significant minorities of Catholics (8 percent) and atheists (5 percent). Almost all respondents had at least 6 years of formal education, only 3 of the older people having less than this period at school.

One sampling constraint not possible to build into the Shrewsbury survey, however, concerned the need to interview parents. Because of the age structure of floodplain occupants and also because many houses were small and occupied by a single person, it was not possible to sample from a population of families with children. In fact, only 28 percent of respondents had two or more children in the age group 0–21.

Shrewsbury therefore represents an advanced community, but with the occupants of the floodplain being generally the poorer members. This is probably characteristic of many flood-hazard zones in towns in the United Kingdom.

Results of an analysis of question responses

The level of awareness of the flood hazard among floodplain inhabitants was generally low—a significant finding considering the relatively high degree of dependence in the Shrewsbury floodplain on individual adjustments to reduce flood damages. In response to the question; "Is there a flood problem in this area?" more than half of the 132 floodplain respondents answered "No" (37 percent), "Don't know" (14 percent), or that they were doubtful.

Although it was expected that the majority of those respondents giving a negative response would be from the medium- and low-hazard zones, this was not the case. The results indicate that respondents living in the high-hazard zone were just as unaware of the hazard as those in the medium- and low-hazard zones. In responding to the question "Do you think there will be another flood in the future?" only 26.7 percent of the respondents replied positively with over 50 percent saying that they did not expect a flood in the future. Again, those living in the high-hazard zone showed no greater expectation of future floods than those in the other zones. The low proportion of respondents expecting a future flood may be partly explained by the misguided belief that the Clywedog Dam upstream of Shrewsbury would prevent flooding in Shrewsbury in the future, and partly by the lapse of time since the last flood affecting the town.

Analysis of responses to the question "What are the principal advantages and disadvantages of living (1) in this building, (2) in this area?" showed that everyday

Table 6–1. Types of adjustment practiced at Shrewsbury

Private adjustment	Type of action (I = Independent; C = Collective)	Percentage of respondents who "mentioned" that adjustment was adopted	Encouraged or discouraged by government
Nothing	I	11.4	Discouraged
Pray	I	1.5	Neither
Evacuate upstairs	I/C	56.8	Encouraged
Evacuate premises	C/I	12.8	Encouraged
Move furniture	I/C	43.2	Encouraged
Permanent floodproofing	C/I	3.0	Encouraged
Contingent floodproofing	I	3.0	Encouraged
Emergency floodproofing	I	15.1	Encouraged
Prepare to move	I	9.8	Encouraged
Buy provisions	I	12.8	Neither
Listen to weather	I	4.5	Neither
Ring fire, police	I	3.6	Neither

"social hazards" such as vandalism, traffic noise, and lack of entertainment—all with a high frequency of occurrence—were uppermost in respondents' minds. Just over half of the respondents mentioned social hazards, while in contrast only about 16 percent mentioned the flood hazard. The fact that so few people mentioned the flood hazard is probably a function of the relative frequency of occurrence of the various hazard events. Relatively infrequent hazards, such as floods, are typically soon forgotten by most people and are consequently subordinated in the minds of respondents by the more frequent and common everyday problems. Traffic noise and vibration, for instance, was a particularly annoying everyday problem in several parts of the Shrewsbury floodplain.

There appeared to be no significant difference between responses to the two variants (1) and (2) of this question—there being no evidence to support the hypothesis that respondents tend to exclude themselves and their home from the general problems that were characteristic of the area. In both cases, well over half of the respondents (56.1 percent and 61.3 percent, respectively), emphasized the advantages of both their own home and the area in which they lived.

Chi-square tests of association undertaken on some variables in the analysis showed that there were associations, significant at the 5 percent level, between the total number of adjustments made by each floodplain respondent and the following variables; (1) respondent's expectation of extent of damage to home and household if area is affected by floods; (2) number of perceived hazard events; and (3) length of time respondent expected to live in the place. In a further analysis, there was a strong association between hazard perception and respondents' experience of flood-hazard events.

These tests appear to indicate that the level of adjustment, measured in terms of the number of adjustments made, is related to the flood experience of respondents

and to their expectancy of future damage. In addition, the number of adjustments made is linked to the time period the respondent expects his property to be exposed to the hazard. Income and tenure variables show no significant association with the number of adjustments.

Adjustments to the flood hazard

Adjustments to the flood hazard may be conveniently divided into two groups (Tables 6–1 and 6–2): those adopted by individuals and those adopted by a collective body such as the Local Authority or River Authority. Some adjustments such as "moving furniture" have both individual and collective variants. For example, an individual may move his furniture to a higher part of the house, or the local authority may organize local furniture removers to evacuate furniture from houses to a central corporation store.

Of the adjustments open to individuals very few respondents mentioned that they "prayed," undertook "permanent or contingent floodproofing," or "phoned the fire brigade, police, etc.," when asked, "When a flood comes what do (or would if they had not experi-

Table 6–2. Collective public adjustments

Collective public adjustment	Remarks
Dams	Clywedog
Floodwalls, levees, etc.	Minor earth embankment only
Structural change/elevation	Planning Office, e.g., Riverside shopping
Flood insurance	Household, commercial, light industrial available
Flood warnings	Flood-warning scheme
Emergency evacuation	Flood emergency plan
Land-use regulation and change	Planning Office, e.g., parklands
Public relief	Mayor's Flood Fund

enced a flood) you do?" By far the most commonly mentioned adjustments were "evacuation upstairs" and "moving furniture," indicating that a majority of flood-plain housedwellers relied mainly on receiving a flood warning prior to being flooded.

As far as collective action is concerned, the most important adjustment in reducing flood damage is the flood-warning system linked with an emergency evacuation and relief plan. A schematic diagram of the stages of this procedure is shown in Fig. 6–4. The importance of the warning system in dealing with serious flooding in Shrewsbury is indicated by the high number of respondents mentioning "moving furniture" and "evacuating premises."

Flood warning and emergency plan

Flood warnings are issued in Shrewsbury on receipt of a warning from the Severn River Authority, which usually is able to give about 24–36 hours warning of a flood. The Severn River Authority bases its warnings on consideration of exceedance of critical river stages (predetermined) and current and predicted meteorological conditions in the catchment area above Shrewsbury. Both the Local Authority and the police are notified by the River Authority, but the police are responsible for directly alerting the town clerk, Shrewsbury Division surveyor, county welfare officer, borough surveyor, and listed flood wardens. The borough surveyor then declares a flood alert; this is given either as a preliminary measure before a flood danger warning is given on more information—or is given if the river is expected to reach bankfull, but not overflow. Yellow alert flags are erected by the Highways Department on English Bridge and Welsh Bridge. When a flood danger warning is issued, red danger flags are flown.

As soon as a flood warning is received, a Control Center is set up in the borough surveyor's office in the

Fig. 6–4. Schematic diagram of Shrewsbury flood warning and emergency plan

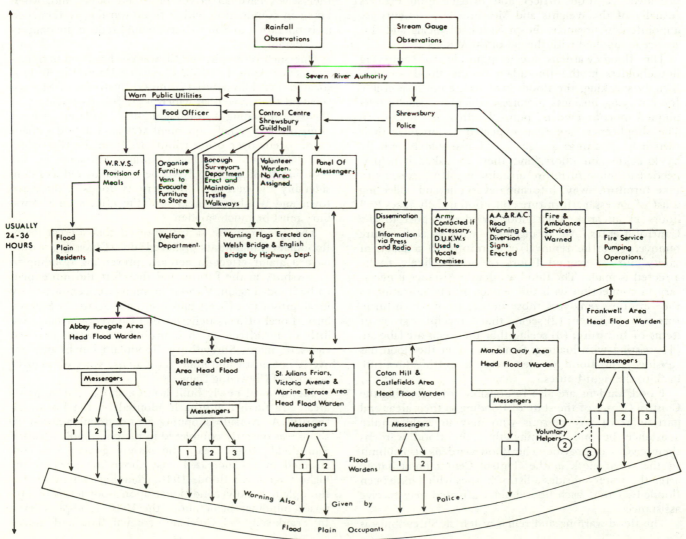

Guildhall: all matters relating to flood-relief work are dealt with through this center. A panel of messengers is established to maintain contact with head flood wardens and flood wardens in each of the six areas in the event of breakdown in telephone communications. Flood wardens arrange for their own messengers to be on duty, as well as voluntary helpers.

At the same time the police contact the Automobile Association and Royal Automobile Club and arrange for road warnings and diversion signs to be erected. Parking restrictions are normally lifted and traffic rerouted. Notices are displayed in flood-prone car parks where permanent notices giving alternative parking places during a flood are displayed. The Parks Department arranges for gates on parks adjacent to the river to be locked. In addition, the chief constable contacts the director of social services and the town clerk, who in turn contacts the Women's Royal Voluntary Service (WRVS) leaders. This action brings the Shrewsbury Flood Welfare Scheme into operation. Under this scheme two rest centers are set up and accommodations and meals are provided. A food officer and district food officers (usually of the Weights and Measures Department) are appointed to organize the provision of meals and this service is supplied with the aid of the WRVS.

The flood wardens are responsible for notifying householders in the floodplain of the flood warning. Strictly speaking, the flood wardens are not responsible for notifying business premises, which are usually telephoned or contacted by police loudhailers (bullhorns). The shopkeepers are then expected to organize their own help. Warnings are given to householders verbally by knocking on doors, and they are asked if help is needed to move furniture upstairs or onto boxes, or to take furniture away. Informing everyone and collecting a list of necessary furniture collection usually takes 2–3 hours in an area. Then flood wardens report to the Control Center. An arrangement with local furniture removers for the removal of furniture to a place of safety out of ground floors of houses likely to be affected is made. The flood wardens' assistance is necessary in connection with this service in: (1) indicating to the furniture removers those houses from which furniture is to be taken; (2) seeing that a receipt is given for items of furniture; (3) seeing that the items are labeled; (4) assisting with loading. Similarly, after the flood has subsided, the flood warden helps in getting furniture back to the right houses.

Flood wardens are also responsible for informing the Control Center of the state of flooding in their areas and particulars of any persons who may urgently require assistance beyond anything that the flood wardens themselves can provide. The flood wardens also submit to the town clerk in the Control Center immediately after the water subsides a list of houses which have been flooded so that each householder affected may receive assistance.

The flood-warning and relief system at Shrewsbury is therefore a fairly complex and well-developed one and forms a most important element in loss reduction on the Shrewsbury floodplain and an adjustment which almost all floodplain occupants utilize at times of flood.

Flood control and flood alleviation schemes

Engineering solutions to the Shrewsbury flood problem have often been suggested, especially after the 1946, 1947, and 1948 floods. Local protection schemes were proposed but never implemented except in the case of a very limited length of earth embankment recently constructed on the downstream side of the Shrewsbury Football Ground. Flood-control dams and reservoirs upstream of Shrewsbury were suggested as a partial or total solution to the problem. Flooding would only be prevented if it were possible to provide a reservoir large enough to contain floodwaters, and to be kept empty in readiness for that purpose. Such a scheme does not seem practicable, but it has been appreciated that development of smaller reservoir sites in this area would in themselves have an effect on flood peaks and, when taken in conjunction with a flood-warning system such as is operating in Shrewsbury, would reduce the dangers considerably.

One such reservoir, the Clywedog, has been in operation since April 1, 1968. Clywedog Dam (Fig. 6–1) is located 122 kilometers (76 miles) upstream of Shrewsbury on the Clywedog, which is a left-bank tributary of the Severn joining it near Llanidloes. The reservoir has a comparatively small catchment area, and should a major storm occur over other tributaries from the Clywedog, the dam would only have a minor influence on the resulting flood peaks. Clywedog no doubt reduces considerably the periodical but serious floods in the Newtown and Welshpool areas, but the effects on Shrewsbury must be much smaller.

The questionnaire survey showed that a great many floodplain inhabitants believed that Clywedog would have a major effect in actually preventing flooding in Shrewsbury in the future, and therefore did not expect to be flooded again. Various opinions were expressed by local authority officers and insurance agents in Shrewsbury. Local officers believed that Clywedog would have little or no effect on their problem. It seems that the views of many Shrewsbury floodplain inhabitants are overly optimistic and many have a false sense of security because of Clywedog Dam.

Elevation of newly built houses, shops, and shopping areas is a further approach adopted by the Local Authority. A riverside shopping area containing about 40 shops was recently completed in the once badly affected Smithfield Road area. The shopping area was constructed at an increased cost above the level of the highest recorded flood, 1946. Planning consent in the floodplain is now conditional upon floor levels of proposed buildings being above the 1946 flood-peak level. In Frankwell and Coleham, council flats and houses have been elevated in this manner, and several business premises in Abbey Foregate, Coleham, and Frankwell have similarly been raised. Although the 1946 flood is utilized to define the floodplain in Shrewsbury, flood frequency studies show this flood to have a recurrence interval of about 120 years. It is therefore quite possible for the 1946 flood to be exceeded at any time, placing these new elevated developments in some danger. Admittedly the allowed level of the floors is several inches above the 1946 flood level and elevation to this height would reduce damage even if the 120-year flood is exceeded. The degree of protection afforded by this means is obviously controlled to a large extent by increasing construction costs for successive increments of elevation.

Floodplain land-use control and change is also employed within the Shrewsbury redevelopment plans, often in conjunction with redevelopment at higher elevations. Clearance orders were made on flood-prone areas which have often seen high-density residential (slum) areas (150 per hectare, or 61 per acre) in the Coleham area immediately adjacent to the river. It is intended to redevelop this area bearing in mind the flood problem. In some parts owners retain their hold on land and compulsory purchase orders are difficult to obtain. Large areas of the floodplain are utilized as parkland, public car parks (with flood-warning procedures), allotment gardens, and sporting areas such as football fields, bowling greens, and tennis courts. To the west of Shrewsbury a large area of floodplain land is occupied by the Shropshire and West Midland Agricultural Showground—only in use during short periods during the year. Several buildings associated with these floodplain uses are found in these areas—e.g., an elevated bowling clubhouse and boathouses. The practice of consultation between river authorities and planning authorities to see that floodplain developments are not detrimental to the river regime and that developments themselves are not adversely affected by the behavior of the river is in operation on the Severn as on all rivers in England and Wales. Consequently, planning consent for a boating marina at Frankwell has recently been refused because it could restrict river floods already restricted by the Welsh Bridge.

No direct question on whether householders had taken out flood insurance was included in the questionnaire. However, a number of interviews were held with insurance companies in Shrewsbury. Before 1961, flood insurance on contents of buildings only was available and widely issued, but it was not until 1961 that insurance on damage to structures was offered on a wide scale (Porter, 1970). After the 1960 floods in Wales and the west of England, pressure was exerted to make flood insurance more widely available. Flood coverage was said not to be obtainable by many flood victims and the government was pressured to set up a National Disaster Fund. In August 1961, a statement was issued by the British Insurance Association and underwriters at Lloyds to the effect that in future flood coverage would be made available to the contents of all permanently occupied dwellings and coverage for dwelling buildings would be provided where reasonable. During October 1961, all offices offered this extended coverage. Certain companies now automatically include total flood coverage when quoting comprehensive rates, and the flood clauses may be optionally excluded with a corresponding reduction in premiums. Porter (1970) notes that following the 1961 announcement one insurance company reported a 25 percent immediate positive response of flood coverage for buildings in Shrewsbury.

In March 1965, following the 1964 floods on the Severn, particularly at Newtown, a Common Insurance Policy for the Severn was derived by insurance companies. This meant that companies adopted a common policy to ensure that if one company refused insurance another would not take advantage of this. Previous to 1965 very few people in Shrewsbury were insured for floods and the meetings of insurers were partly called to try to make inhabitants more aware of the availability and terms of flood insurance. By 1972, many more establishments had been insured. Nearly all residential property in Shrewsbury can be insured for both contents and buildings. However, some commercial properties in very high-risk areas such as Frankwell and Coleham are refused cover. Premiums vary according to the likelihood and severity of flooding, and each dwelling is assessed individually. Where there has been only one flood in recent times—e.g., Castle Foregate, flooded in 1964—premiums are lower. Mortgages on newly purchased houses are only granted if flood insurance is taken out. One insurance company said that it did not offer flood insurance in Shrewsbury but since Clywedog Dam was built attitudes have relaxed and insurance is now offered. It also seems likely that some insurance companies will lower premiums in the near future because of Clywedog.

Public relief

Two main forms of public relief are offered to floodplain inhabitants in Shrewsbury. The first consists of facilities and services such as the Shrewsbury Flood Welfare Scheme, Flood Warden Service, and supply of eventual food and coal for drying purposes. For this reason the Mayor's Charity Fund is set up and supplements finances expended by the local authority during a flood. If a flood is considered bad enough the mayor declares the fund open for subscriptions. Although this fund is still operating, costs during a period of flooding are borne increasingly by the ratepayer (taxpayer) through the local authority.

Second, if the flooding is serious, as in 1964 and 1965, the areas may be declared national emergency areas. This means that in the past Shrewsbury inhabitants have been able to claim for flood damages from the central government (the War Damage Commission) with the local authority acting as agents passing the claims on.

Flexibility of adjustments

All adjustments in Table 6–1 labeled as private adjustments, except permanent floodproofing (which is not widely adopted) require a minimum of preparations, very little monetary outlay, and no permanent alterations to buildings and are the type of adjustment that can be included in the broad category of "flood fighting." These are flexible adjustments and may be easily instigated or abandoned at short notice.

Those adjustments labeled as public are planned, permanent, and inflexible adjustments requiring high capital investments. The flood-warning and emergency evacuation scheme has been set up since the 1946 flood and procedures are followed automatically on preconceived lines, on receipt of the appropriate warnings. These procedures will not normally be abandoned unless more effective permanent measures are taken in the future rendering the scheme redundant. A flood fund for public relief is usually set up in the event of a major flood and is instigated depending on the severity of the event. In the past, inhabitants have been able to claim for damages from a War Damage fund but government policy has been to steadily reduce this type of relief in order to encourage inhabitants to insure against flooding. The degree of relief offered is generally a function of the severity of the hazard event and the degree of public pressure on the relevant authorities. Flood insurance is offered as part of household insurance policies in Shrewsbury and a householder could abandon this adjustment by not paying premiums if he so wished, but this is an unlikely action.

Conclusion

The Shrewsbury study has been concerned with an industrial society in which the degree of individual awareness of the hazard remains low, but in which community and government awareness, particularly in such matters as floodplain planning controls, has greatly increased in recent years. Generally, in the context of the United Kingdom, the strategy adopted in Shrewsbury represents an advanced stage of hazard adjustment. Other flood-prone settlements tend to have relied heavily upon structural measures. In contrast, Shrewsbury has developed nonstructural alternatives such as flood-warning and emergency plans, and development control combined with urban renewal programs, which reflect a high degree of community and government awareness.

Acknowledgments

Dr. Harding acknowledges the financial support of the United Kingdom Natural Environment Research Council in this study. The assistance of Mr. L. M. Punnett, Mrs. S. C. Punnett, and Mrs. S. Coggins is also gratefully acknowledged.

References

Howe, G. M., Slaymaker, H. O., and Harding, D. M. (1967) "Some aspects of the flood hydrology of the upper catchments of the Severn and Wye." *Transactions of the Institute of British Geographers* 41:33–58.
Porter, E. A. (1970) "The assessment of flood risk for land use planning and property management." Cambridge: Cambridge University, unpublished Ph.D. thesis.

7. Perception research and local planning: floods on the Rock River, Illinois

NORMAN T. MOLINE
Augustana College

Characteristics of the study area

The Rock River flows southwesterly through northern Illinois and joins the Mississippi River 770 kilometers (479.1 miles) upstream from the mouth of the Ohio River. The section of the Rock River floodplain considered in this study includes parts of the cities of Rock Island and Moline, the villages of Milan and Coal Valley, and unincorporated areas of Rock Island County (Fig. 7-1).

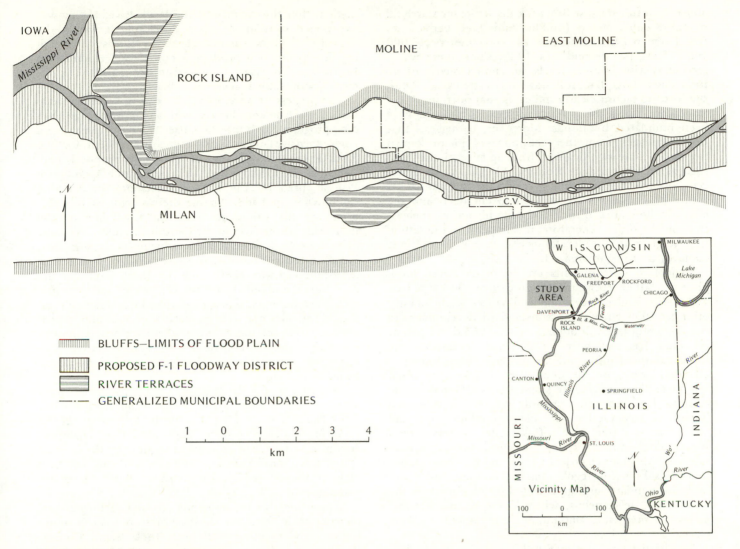

Fig. 7—1. Rock River study area

The floodplain varies in width from slightly less than 2 kilometers (over 1 mile) to approximately 5 kilometers (over 3 miles) at the mouth. Surface undulations are minimal, generally less than 3 meters (10 feet) except where glacially deposited sands and gravels form slight rises. Bluffs rise to heights of about 25–30 meters (80–100 feet) above the floodplain on both the north and south sides. The floodplain has poor natural drainage in many places with sloughs and swamps occupying the uncultivated and undeveloped areas. However, man-made drainage facilities have made many of the low areas cultivable and habitable. Soils generally are of the brown, yellow, and yellow-gray silt loam variety; in the more lowland areas the alluvial influence is reflected in a greater frequency of sandy loams. Most of the natural vegetation has been removed in favor of farming or residential development. Along tributary watercourses and in undrained areas, oak-hickory forests and rather dense brush still exist.

The main flood season for the Rock River is in late winter and early spring. Nine of the ten highest floods occurred in January through April. Ice cover and snowmelt are generally at their greatest during these months. This combination of conditions indicates the potential for ice jams at particularly constrictive parts of the channel and at bridge crossings during late winter and early spring period of high water. The effect of ice jams is basically the same as a dam in effectively blocking the flow and increasing the river's surface elevations upstream. Most of the larger floods have resulted from heavy rains in combination with snowmelt and/or ice jams.

The record flood occurred in February and March, 1948, when for 36 continuous days the river remained above bankfull stage with a peak discharge on March 22 of 1,308 cubic meters (46,200 cubic feet) per second. (Discharge at bankfull stage is 425 cubic meters (15,000 cubic feet) per second.) In 1960, since there was a secondary rise in river levels, an annual record of 60 total days above bankfull stage was established. Thus, the historic record clearly indicates that portions of the floodplain along this stretch of the Rock River have been inundated by floods. Moreover, it suggests that flood damages from the recurrence of major known floods or, more significantly, from an exceptional flood such as an Intermediate Regional Flood with its 1 percent occurrence probability would be substantial. If future developments are allowed to encroach into the effective flow area of the floodplain, higher water elevations might result. Therefore, it is imperative that attention be given to any developments on this floodplain if damages are to be minimized.

Generally, developments on the Rock River floodplain are rather recent, most dating since 1940. The lack of building codes and zoning ordinances in the county prior to 1955 encouraged substandard development in the unincorporated floodplain. Since 1955 development generally has not been substandard. Following the floods of the 1960s some families have been forced to abandon homes on the floodplain because they could not meet the reoccupance standards established by the county. Reoccupance standards require that flooded homes be structurally safe, have unpolluted hot and cold running water, have a flush toilet, be served by a functioning sewerage system or septic tank, and have safe and adequate electrical wiring and gas connections. Current subdivision regulations and building codes prohibit substandard and unprotected development on the floodplain. Nevertheless, within these regulations and codes, an increasing amount of development is taking place.

Residential land use in the floodplain is low in density and generally linear along both sides of single roads which run parallel to and at one-lot depth away from the shoreline. Aside from a few nonlinear clusters of homes and small businesses, the land "behind" this linear settlement pattern is vacant and thus is among the land most susceptible to future urban encroachment. Most of the residences along this part of the Rock River are in unincorporated areas of Rock Island County.

The incomes of the inhabitants of the floodplain are varied. Some families earn well over $10,000 a year; others are in poverty and on relief; the majority are characterized by middle-level incomes. Families with these varied incomes are mixed almost randomly along the river. In short, there is no distinct socioeconomic level or concentration among floodplain residents.

According to the comprehensive development plans of the Bi-State Metropolitan Planning Commission (the planning agency for the Davenport, Iowa—Rock Island and Moline, Illinois, Standard Metropolitan Statistical Area) as well as some of the individual municipalities,

areas within the vicinity of the Rock River Valley will experience substantial urban development within the next 15–20 years. History has shown that such growth on floodplains will produce a considerable increase in future flood damage if allowed to occur without sufficient floodplain management policies.

The process of developing effective floodplain regulations should be coordinated with a comprehensive planning program for future land use. The development of a comprehensive open space recreation system for the Rock River Valley is one of the major proposals contained in the Metropolitan Recreation and Open Space Plan adopted by the Bi-State Metropolitan Planning Commission in 1968. Among the elements of this proposal are the following: zone the floodplain on the north side of the river in the Central Urbanized Area for open uses to be developed gradually as a park area, incorporate the excess right-of-way situated between an interstate highway (280) and the south shore of the river into the planned open-space system, and zone the remainder for open space, conservation, and agricultural uses. (The pictorial representation of this plan which showed trees in places along the river where houses currently stand, has contributed to some misunderstandings about other proposed floodplain regulations, as will be noted below.)

Referring to the Rock River, the Corps of Engineers in their report *Rock River Flood Plain Information, Rock Island County* (1969) state the following: "If the flood plain is to become an asset rather than a liability, action must be taken as soon as possible by the County and affected municipalities to develop and enforce flood plain regulations which are coordinated with land use planning efforts."

Supporting this recommendation and seeking to implement their own, the Bi-State Metropolitan Planning Commission together with the Rock Island County Building and Zoning Department began to formulate more detailed plans and to develop new zoning ordinances. Ultimately, approval of these plans and ordinances rests with the elected county and city government units. Of course, these units are subject to the wishes of their constituents. Thus, since the decision-making on floodplain regulations ultimately rests with the citizens, some understanding of their perceptions of the flood hazard and of possible adjustments to floods is important to those charting the method by which final adoption might be achieved. In this context, our research efforts became involved with the residents of the unincorporated areas along the river, their governmental unit: the Rock Island County Board of Supervisors and its Building and Zoning Department, and the Bi-State Metropolitan Planning Commission.

Perceptions of floods and adjustments to floods

The first study was conducted in the spring of 1970. That year there was no threat of a flood and no reported flood damages. Information and data were derived from the reports, files, and staff persons of the local planning and zoning agencies and the Rock Island District of the U.S. Army Corps of Engineers, from field observations, and, most of all, from personal interviews using the basic questionnaire format developed by the natural hazards research group of the International Geographical Union's Commission on Man and Environment. Most of the interviewing was conducted by students in conjunction with a class project on man-environment issues. Parenthetically, involvement of students in public policy-oriented research has proven to be worthwhile for the students. Direct contact with floodplain residents and with public officials on a specific environmental issue has stimulated a greater interest in trying to understand the complexity of the problems involved and a sense of contribution in giving their findings to public decision-makers.

In 1971 a second study on this area, using the same research procedures, was conducted—since in late February and early March of that year the Rock River in this vicinity experienced one of its worst floods. At Moline the level of the highest reading had been surpassed only by the 1892 flood and the river was above bankfull stage for approximately 11 days. This river data is of somewhat limited value, however, since this flood was largely the result of ice jams rather than exceptionally large discharges. Accordingly, river stages fluctuated greatly from area to area and, likewise, flood damages were unevenly distributed along the floodplain. Persons living in the most affected sections clearly perceived this 1971 flood as being the worst they had experienced. Many persons even in the less affected areas perceived it as being the worst due to the damages or inconveniences they or their neighbors had experienced.

It was our expectation that the data from the interviews conducted in April and early May, 1971, would reveal some striking differences when compared with the 1970 data. However, available evidence indicated virtually no significant differences in the responses. Apparently, perceptions are so firmly established that an individual flood, regardless of its severity, is not likely to cause noticeable changes. Other findings from these two years of study are presented in succeeding paragraphs.

A clear majority (65 percent in 1970 and 69 percent in 1971) emphasized the advantages of their location, while a very small number (8 percent in 1971) emphasized the disadvantages. At the same time, the majority (75 percent) admitted that people in their neighborhood have trouble with floods. Fifty-seven percent admitted damages to their own household; of these, 36 percent described the damages as substantial or total. Yet, despite these troubles, residents stated a strong preference to remain even if damages had been substantial. Seventy-six percent intended to stay in their location many more years while only 7 percent stated a definite desire to move out. Of those persons who admitted to substan-

tial or total damages, 74 percent stated a definite intent to stay and of those who admitted to slight damages 73 percent planned to stay. Apparently, the level of damages experienced in this area is not a critical factor in the decision to stay or leave.

The people generally were aware that floods can happen in any year (76 percent in 1970 and 77 percent in 1971). At the same time, some people (13 percent in 1970 and 11 percent in 1971) believed that when a flood occurs, more floods will soon follow; others (6 percent in 1970 and 7 percent in 1971) believed that a certain number of years must lapse between major floods. Eighty-one percent believed that floods would come again in their lifetime, with 46 percent believing that a flood would come soon or in a few years. Only 3 percent believed that there would not be another flood in their lifetime.

Only when a flood approaches do most people start worrying about damages, inconveniences, emergency services, and the money needed to repair damages. Only 21 percent consciously tried to save money from year to year for possible flood damages. For some, there was little or no money to set aside while for others this habit merely reflected personal economic priorities.

A small percentage (13 in 1970 and 17 in 1971) believed that they personally could prevent damage from a flood by means other than sandbagging or lifting household possessions while a rather large percentage (55 in 1970 and 45 in 1971) believed with some certainty that there was little or nothing for them to do. The changing percentages from 1970 to 1971 may reveal that the people learned from the recent flood crisis that there are things which individuals and groups of residents can do to prevent or minimize damages.

Fifty-four percent of the residents turned to either government or, more often, to some special group other than family or friends for help in recovering losses from floods. This partial dependence on others was revealed, also, in the view held by 46 percent that public relief was a good adjustment to the flood problem. Regarding some of the other adjustments to floods, whether or not they were actually implemented, structural changes were perceived by 66 percent as being good while land use changes and flood insurance were perceived as good by only 48 and 34 percent respectively. The low regard for flood insurance is documented by the fact that as of July, 1972, there were only 48 policyholders in Rock Island County (which was declared eligible for flood insurance in May, 1971) and only 4 in the city of Rock Island (which was declared eligible in July, 1971). These figures are particularly significant when considering the fact that they represent residents along both the Mississippi and Rock rivers.

The people generally may not be aware of the extent to which various governmental agencies could provide and, in fact, do provide different kinds of assistance to residents during floods. Only 10 percent perceived that the government actually had helped anyone that they knew in time of flood. Furthermore, only 22 percent were inclined to talk over an important community problem with a government official and an even smaller number would talk to the Bi-State Metropolitan Planning Commission or one of the municipal planning commissions. Seemingly, this data suggests that "government" was not perceived by the residents as being in touch with the people. This conclusion is supported by the fact that a small brochure entitled "Rock River Flooding, Rock Island County, Illinois: How to Avoid Damage" apparently had never been circulated to most residents. (To remedy this latter situation, a copy of the brochure was left with each household after the perception interview was conducted.)

Unquestionably, there was a strong sense of "community" among Rock River residents. When asked for whom they feel some responsibility in a flood, less than 10 percent gave selfish or isolationist responses.

Some additional impressions of Rock River floodplain residents, though not statistically recorded, were derived from field observations and the interview experiences. Virtually all residents appreciated the atmosphere of "river living": the recreation opportunities, the lack of congestion, the cleaner air, the "more open" feeling, the cool river breezes in summer, etc. They believed that they are living in an area which is preferable to the more typically urbanized parts of the Bi-State metropolitan area.

The independence of the residents was clearly evident. They expected to bear their own losses and they took pride in their perceived self-reliance. It would be very difficult to convince these people that it would be in their own interest, as well as the community's interest, for the floodplain to be zoned purely for open space and recreation. Any hint of their possibly being asked to move away from the floodplain would be met with strong resistance.

While maintaining that they assume flood damages themselves, many persons expressed strong negative feelings when asked about government assistance in time of flood. Frequently, complaints were directed at the county and city governments for failure to repair road damages resulting from flooding conditions. A general dislike of the Corps of Engineers was evident. Many residents expressed the desire to have the river dredged to help prevent floods. They perceived this act to be a good way to reduce the potential of future floods and, since the Corps was believed to be the agency with the responsibility for doing such things, the lack of action by the Corps meant that the Corps was "responsible" for the floods.

Another contradiction to their perceived self-reliance was the sense of apathy and negative feelings toward flood insurance exemplified by reactions ranging from "don't know about it" to "it is not so much of a bargain" to "too expensive" to "worthless." According to flood veterans, experience has "shown" that it isn't necessary. Mandatory insurance for floodplain residents would bring out much opposition, especially if a risk factor was not built into the insurance premium schedule.

Residents expressed objection to the construction of the Interstate 74 bridge with its closely spaced support piers which encourage ice jams; they wondered how government agencies, whoever they were, allowed this construction to take place. It would seem that the residents have a legitimate objection on this point and that highway engineering should have taken this problem into consideration *before* the bridge was built.

Relationship to local planning and decision making

From the beginning this study was related to the Bi-State Metropolitan Planning Commission and the Rock Island County Building and Zoning Department which were developing a new zoning ordinance for the county including some regulations on land use along the Rock River. When in 1970 we proposed a perception study, they expressed interest in the information it might provide on the residents of the floodplain and their views on flood problems. With the completion of the 1970 study, the initial report, including some of the data presented on previous pages, was turned over to these agencies. From these findings, the agencies knew that in the future they should take these perceptions into greater account, increase their public relations and communications with these people, and share more information about the individual's potential role in reducing flood damages. They encouraged us to continue with more research in the same area.

In April, 1971, the County Board of Supervisors passed a resolution of intent to adopt the special article on the Floodway District from the pending new county zoning resolution in order that residents of the floodplain could be eligible for the national flood insurance program. (Selected paragraphs from the Floodway article are included at the end of this chapter.) Final action was postponed until the entire zoning resolution came up for approval. As part of the efforts to pass the complete new zoning ordinance in the fall and winter of 1971–72, we contributed the results of both the 1971 and 1970 studies.

The resolution was presented to various professional and private interest organizations and proceeded through the first fourteen public township hearings in the winter of 1971–72 with little opposition. However, at the last two township hearings, one of which included the major area of our study, the opposition surfaced. As described in a local newspaper (the Rock Island *Argus*): "A flood of anger apparently has swept through some of the river people of South Moline township. About 80 Rock River Valley residents . . . protested the proposed Rock Island County Zoning Ordinance." One of the major objections was that persons owning riverfront lots would not be allowed to develop them in the future. One citizen argued that he had made a sizable investment in a lot for development and if the pending ordinance were approved it would lose its value. In response to these objections, the proposed resolution was amended in two sections (63.01 and 64) which exempted existing "lots of record" from some of the new floodway provisions.

On March 16, 1972, the county Board of Appeals received final comments on the resolution. At this session, the results of our study as well as some additional background information on the nature of floodplain problems were presented and read into the official record. Since this presentation supported completely the new ordinance and since the study represented the most detailed analysis of Rock River floodplain residents' perceptions, we were encouraged to make an additional presentation at the meeting of the County Board of Supervisors the following week.

On March 20, the Building and Zoning Department and the Board of Appeals presented the final proposed zoning resolution to the County Board of Supervisors at a special meeting. The purpose of this meeting was to hear the new resolution, discuss it, and answer questions first from the supervisors and then from any citizens attending the meeting. Motions were not to be in order. Going into that meeting, county zoning and planning officials expected a difficult time; yet, they believed that the amendments would satisfy the complaints of earlier objectors and that the resolution would survive this evening of discussion and would be approved at the official meeting of the County Board on March 22. Because of the importance of this meeting and because of the supportive character of our data, we were encouraged to appear and to make a presentation at an appropriate time. Such a time never came. This meeting in itself was an incredible illustration of perception of floods and of decision making on adjustments to floods.

The Rock River residents turned out in force; various estimates of the standing-room-only audience ranged from 200 to 300. After politely listening to some of the opening remarks and explanations, members of this audience began to interrupt with questions and, more often, with strongly negative statements. Since this session was a meeting of the supervisors, these interruptions by the citizens technically were out of order. However, the chairman chose to ignore strict rules of order that night. As the citizens spoke, their perceptions of the flood problems on the Rock River came forth. For the most part, they echoed what the interviews had suggested; at the same time, some of their comments in this public forum seemed to be contrary to what had come from the private setting of the interviews. Some of the positions taken and the perceptions revealed that night are discussed in the following paragraphs.

Even though the proposed floodway district was based on the March, 1948, high-water levels rather than the Intermediate Regional Flood with its 1 percent occurrence probability, people expressed their dislike for the terminology and the concept of the "100-year flood." One spokesman for the group dogmatically refuted the fact that this phrase really means a 1 percent chance in any year for a flood of that magnitude; rather he stated firmly that it literally meant one such flood every 100 years. One citizen said: "The intermediate regional flood is a paper flood which never has occurred and probably never will during the lifetime of anyone present in this room tonight."

A number of persons blamed the floods on the construction of the interstate highway bridge which has supporting piers much closer together than they are on older bridges over the river. The citizens believed that this situation could have been avoided and held the government responsible for the mistake. Similarly, many persons emphasized that the flood problems would be reduced entirely or substantially if only the river were dredged. Among the specific comments were the following: "We think they should dredge it and get the mud out." "Is the Corps of Engineers doing anything?" to which the audience loudly answered "No!!" "How will you *zone* the mud out?" "They can put a man on the moon but they can't dredge Rock River." These sentiments coincided with those expressed in the interviews. While there has been a buildup of silt in the river, it certainly is not the major cause of the flood problem. Nevertheless, the views are widely held and intensely argued by floodplain residents. Accordingly, the agency which would do any dredging, the Corps of Engineers, was blamed for the floods through their neglect.

A presentation by the Soil Conservation Service agent showing that portions of the floodway have alluvial soil characteristics which clearly indicate that the area has been subject to flooding was emphatically denied by some citizens. (It was after the irrational, close-minded response to his data that the zoning administrator privately advised us that there would be little purpose in presenting our data on this occasion when an atmosphere of logic and rational thinking was not prevalent. Another session with the board was recommended as being more appropriate.)

Overwhelmingly, persons talked about private rights to live where they wish and do what they wish with their property. Any references to public rights, the public good, or ecological desirability were either nonexistent or negated by argument. Several specific comments are representative: "Let those of us who are willing to take the risk of living here take it." "We have taken care of ourselves. We knew the river flooded when we bought our property. We were given a choice and we expect a flood every year."

At the same time, and this contradiction seems significant, many people directly or indirectly voiced resentment toward government agencies; they felt they deserved more from the government than they received. "What would they spend on us? We've had water in our house but never got a penny yet." "We have men elected to office at all levels; there must be somebody who knows someone who cares to influence somebody." One person stated that the lack of action or positive response to his requests "is not what I fought for in World War II," which prompted applause from other citizens in the hearing room. Some persons were particularly resentful toward the Bi-State Metropolitan Planning Commission for its 1968 plan which showed only trees or open spaces where houses are now located. The main spokesman for the citizens sarcastically noted: "Obviously, the proposed zoning resolution would make that transition much easier."

Though this meeting was scheduled for information purposes and no actions were to be taken, one of the supervisors moved that sections related to special zoning in a floodway district be deleted. When informed that official actions were not in order at this meeting, he called for a "straw poll" of the supervisors to learn of their feelings. The parliamentary decision to allow this sample vote rather than to follow the policies set for the meeting was extremely important; in fact, perhaps it decided the final fate of the floodway provisions. The vote or "straw poll" was taken and *all* supervisors present favored the deletion of the provisions. On March 22, at the regular meeting of the Board, the floodway article was deleted officially. The work by the county's zoning department on this controversial and innovative article, the County Board's own 1971 resolution of intent to adopt, and the support for it from various planning agencies, other private interest groups, and our study were of little consequence in the final decision.

Two major questions were raised by these actions. Why had these supervisors who less than a year earlier officially had expressed their intent to adopt the floodway zoning provisions now unanimously changed their positions? One major reason was that this recent action was held less than a month before the elections for county board members and, thus, was particularly susceptible to political influences. Apparently, the pressures of a visible "vote" in front of 200 to 300 irate citizens prompted the supervisors to unanimously support the positions of those citizens. This fact leads to a second question: what happened to the opinions of other citizens in the 14 unobjecting townships and the citizens in the unaffected parts of the county who share through taxes in the financial support of the various flood services and post-flood clean-up and repair operations which the county provides when needed? (Incidentally, one estimate by the zoning department placed the 1971 costs for emergency flood services at $71,000.)

With the deletion of the floodway district, the territory which would have been in that district was rezoned. For the most part, sections of the floodplain included in this study were rezoned residential, while other sections of the Rock River floodplain in Rock Island County were zoned agricultural or residential. With this removal of special floodway provisions and replacement by more general zoning categories, the future of flood insurance became uncertain since special

floodplain zoning is a prerequisite for an area to qualify. Through a series of negotiations between the county, the Corps of Engineers, and the Federal Insurance Administration, individual properties protected by insurance purchased during the period when the resolution of intent to adopt was in effect were to be resurveyed and specific guidelines were to be presented. These actions have been taken and it again is time for the county to decide. If the county does not pass the necessary zoning resolution covering those particular properties or any others in the floodplain by July 1, 1973, existing flood insurance policies will be canceled and no new policies will be written.

As of April, 1973, there seems to be little likelihood that this flood insurance prerequisite alone will prompt the county to enact any floodplain zoning since there were then 47 policies in effect for all of Rock Island County. It appears that some additional motivation will be necessary for any special zoning to become a reality. However, in April, 1973, as citizens along the Mississippi and Rock rivers brace for the second major flooding by these rivers within a month, the highest floodwaters in the history of the Rock, and the second highest in the history of the Mississippi, there seems to be some renewed interest in any kind of adjustment to this persistent flood problem. Perhaps zoning again will be considered as a possibility.

Thus, though special floodplain zoning is temporarily a dead issue here, it is not likely to be permanently buried as long as people experience damages from floods and subsequently seek to find methods of minimizing such damages. Accordingly, our interest and involvement will continue until some coordinated, comprehensive program of adjustments to the floods is adopted.

Conclusion

Considering our involvement with governmental agencies up to this time, what are the results? Lines of communication and information sharing were established and strengthened with the appropriate agencies. From our study these agencies gained some new perspectives on the residents of the floodplain. Perhaps the most valuable impressions are: (1) the extremely close attachment of the residents to their river locations; (2) the residents' lack of knowledge about what they personally might be able to do or should do to prevent damages other than sandbagging and lifting possessions; (3) the lack of awareness by residents of the extent to which government agencies, in fact, do assist them in one way or another with their flood problems; (4) the perceived poor communications of government agencies with the residents; and (5) the low regard which many people have for various units of government. Each of these impressions seems to suggest the need for greater and more careful efforts at education and communication.

The fact that the research findings along with the

recommendations of planning and zoning and other resource agencies were of no influence when the final decision was made in itself suggests the need for better communications and education. The frequency with which floodplain residents revealed some erroneous impressions about and hostility toward the proposed floodway district in part can be attributed to the lack of significant involvement of and communications with floodplain residents throughout the development of the new resolution, particularly in the early stages. If steps can be taken to improve the amount of resident involvement in the planning process and the amount of communication with both persons living on the floodplain and with other residents of the county, long-term efforts to reduce flood damages in this area may be successful. Hopefully, our studies have not only called these needs to the agencies' attention but also pushed the agencies a bit into taking the necessary steps to meet these needs.

As an indirect outgrowth of these Rock River studies, an interdisciplinary faculty team at Augustana College has completed two and is working on two more environmental inventories on selected areas along the Mississippi and Rock rivers under contract with the Rock Island District of the Corps of Engineers. Perception studies of the type described in this paper are a recognized part of these comprehensive inventories and are included under the general heading of "humanistic elements of flood problems." Perception research is relevant to particular local environmental issues and can be of value to the public agencies who have responsibility for making decisions on those issues.

Appendix

Proposed Zoning Resolution—Rock Island County, Illinois Selected Paragraphs from Article VI, F-1 Floodway District

Section 60. *General Description.* This district is established for the purpose of meeting the needs of the rivers and streams to carry flood waters and protecting the rivers, creek channels and flood plains from encroachment so that flood heights and flood damage will not be appreciably increased; to provide the necessary regulations for the protection of the public health and safety in areas subject to flooding; and to reduce the financial burdens imposed upon the community by floods.

Section 61. *Uses Permitted.* Property in an F-1 Floodway District shall be used only for the following open-type purposes: 61.01. *In Areas Adjacent to Agricultural.* 61.011. Farm 61.02. *In Areas Adjacent to Residential, Commercial, or*

Industrial Districts. 61.021. Agricultural crops, but not the raising of farm animals or poultry within two hundred (200) feet of that property line adjacent to the residential, commercial, or industrial district.

Section 62. *Conditional Uses.* (A few selected open-type uses are permitted in the F-1 Floodway District subject to approval by the Board of Appeals. These uses and the appropriate areas are identified in this section.)

Section 63. *Board of Appeals Approval.*

63.01. *Approval of Plans.* No permit, except as provided in Section 64, shall be issued for the construction of any building or structure or for any open use within the F-1 Floodway District until the plans for such construction or use have been submitted to the Board of Appeals and approval is given in writing for such construction or use.

63.02. *Standards.* In their review of plans submitted the Board of Appeals shall be guided by the following standards, keeping in mind that the purpose of this district is to prevent encroachment into the floodway which will unduly increase flood heights and endanger life and property: 63.021. Any structure or filling of land permitted shall be of a type not appreciably damaged by flood waters, provided no structures for human habitation shall be permitted.

63.022. Any use permitted shall be in harmony with and not detrimental to the uses permitted in the adjoining districts;

63.023. Any permitted structures or the filling of land shall be designed, constructed, and placed on the lot so as to offer the minimum obstruction to and effect upon the flow of water;

63.024. Any structure, equipment or material permitted shall be firmly anchored to prevent it from floating away and thus damaging other structures and threatening to restrict bridge openings and other restricted sections of the stream;

63.025. Any water supply or sanitary sewage disposal facilities shall be constructed to avoid contamination or be contaminated by flood waters; and

63.026. Where in the opinion of the Board of Appeals topographic data, engineering and other studies are needed to determine the effects of flooding on a proposed structure or fill and/or the effect of the structure or fill on the flow of water, the Board of Appeals may require the applicant to submit such data or other studies.

Section 64. *Exceptions and Modifications.* In the F-1 Floodway District on a lot of record created prior to the adoption of this Resolution, a single-family dwelling and accessory buildings may be established or existing dwellings and accessory buildings may be enlarged, altered, or reconstructed provided that all other requirements of this Resolution are complied with.

8. Flood information, expectation, and protection on the Opotiki floodplain, New Zealand

NEIL J. ERICKSEN
University of Waikato

Introduction

This study of the floodplain at Opotiki, Bay of Plenty, is primarily concerned with the individual floodplain occupant's perceptions and attitudes toward flood protection for that town. Two basic hypotheses are tested:

1. Information of flood protection structures decreases expectations of future flooding.
2. Information of flood protection increases confidence in protected areas as places for human occupance.

In previous work on this theme, Roder (1961), in a study of floodplain occupance at Topeka, Kansas, was unable to discern a direct relationship between levels of knowledge of flood protection and expectation of future flooding. Nor was he able to find a direct relationship between public promise of protection structures and continued occupation of the floodplain. Indeed, he doubted whether dissemination of carefully prepared information on protection structures prevented uneconomic settlement within the protected areas. Burton (1961) came to broadly similar conclusions in a contemporary study of settlement along the Little Calumet River, northern Illinois, but in addition suggests that it is the type of past flood experience, rather than knowledge of protective works, which has a significant role to play in appraisals of flood hazard. These findings were largely verified by Kates (1962) in his study of hazard and choice perception in six United States towns with flood problems ranging from minor to severe. Perceptions of hazard, choice, and adoption of adjustments to hazards (including protection), and expectation of future flooding were found to increase as to the frequency and magnitude of flooding. In their studies of 17 United States cities, White et al. (1958) found that, in general, frequent flooding failed to inhibit the growth of settlement in urban floodplains. This finding and its implications also has considerable significance in the New Zealand situation (Ericksen, 1971).

Settlement and floods at Opotiki

The Opotiki floodplain is one of several pockets of similarly formed flatland found along the length of coastal Bay of Plenty. It was built up through sedimentation by the Otara and Waioeka rivers (Fig. 8–1). These rivers emerge abruptly from steep hill country and flow about 11 kilometers (7 miles) in a northerly direction across the plain to the coast. Some 6 kilometers (4 miles) apart when they first issue from their gorges, the two rivers converge to a common outlet 1.6 kilometers (1 mile) from the sea.

The floodplain of this prong-shaped river system is surrounded on three sides by uplands. To the south steep hill country rises quickly to heights of over 600 meters (2,000 feet), while to the east of the town is a terraced plateau of somewhat less than 90 meters (300 feet), matched in the west by rolling downlands. Opotiki lies completely within the floodplain between the angle of the Waioeka and Otara rivers immediately upstream of their conflux (Fig. 8–1).

Some elements of land use

Within the town, commercial establishments represent the whole range of business activities expected in a borough of Opotiki's size (population 2,590 in 1966). Most are located within three blocks on either side of the main business street (Church Street) which runs the length of the town (Fig. 8–2). The large number of food and clothing shops (20) is a distinctive feature for a town of this size and reflects the extent of rural hinterland that Opotiki relies on and serves for its economic well-being. The labor force of 824 (April 1966) relies largely on shop, office, and hotel work.

Although all commercial, manufacturing, and residential establishments within the borough are located on the floodplain, an increasingly important area of housing outside of the borough is developing on the low hills immediately west of the town. Of more than 100 houses in this area, known as Hospital Hill and Woodlands, over 20 percent have been built since the largest recorded flood in March 1964.

In contrast to rapid development on the "hill," many houses within the borough are old and deteriorating. For example, in February 1967 there were 658 houses in the borough, of which over half were more than 25 years old and almost 20 percent were built before 1920.

Frequency and magnitude of flooding

From 1860 when it was laid out until 1964, Opotiki was flooded by either river on no fewer than 25 occasions. Many more floods would have entered the town were it not for the construction of crude levees to the south and east of it in the 1930s. The erection of new levees, or stopbanks, following the record flood in 1964 had already prevented at least eight floods by 1967 when the field survey was conducted.

The pattern of flood spread within the town varied

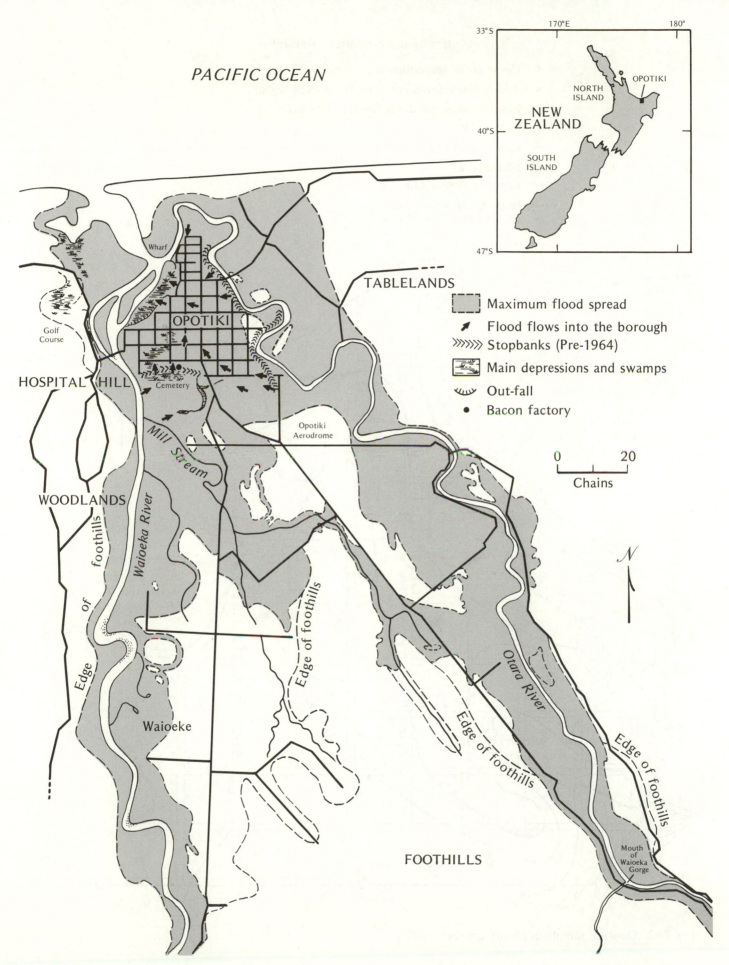

PACIFIC OCEAN

NEW ZEALAND

NORTH ISLAND

OPOTIKI

SOUTH ISLAND

TABLELANDS

OPOTIKI

Wharf

Golf Course

HOSPITAL HILL

Cemetery

Opotiki Aerodrome

WOODLANDS

Mill Stream

Waioeka River

Edge of foothills

Waioeke

Edge of foothills

Otara River

Edge of foothills

Edge of foothills

Mouth of Waioeka Gorge

FOOTHILLS

Maximum flood spread

Flood flows into the borough

Stopbanks (Pre-1964)

Main depressions and swamps

Out-fall

Bacon factory

0   20
Chains

N

Fig. 8–1. Maximum flood spread on the Opotiki floodplain, March 1964. Source: Poverty Bay Catchment Board (1964), Drawing 2035

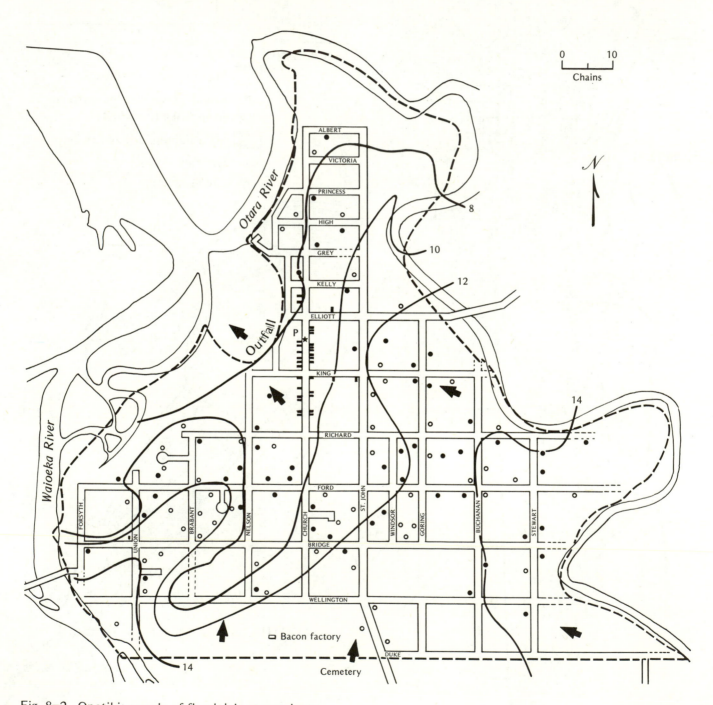

Fig. 8–2. Opotiki: sample of floodplain respondents

according to the location and magnitude of overflow, but in nearly all instances the business district of the town was flooded to varying depths.

The flow from the Waioeka River for the 1964 flood was augmented by water from the Otara River via Mill Stream. Following the receipt of some 65 centimeters (26 inches) of rain in 40 hours in its upper catchment, the Waioeka River burst through levees and inundated almost all of the borough's 772 acres to depths reaching over 1.5 meters (5 feet) in the business district. Only 24 houses escaped flooding. Within 24 hours of breaching the levees the floodwaters had receded, leaving behind gouged roads and pathways, smashed doors and windows, buckled walls, moved buildings, and extensive siltation which increased the burden of cleanup. At approximately N.Z. $600,000 (U.S. $700,000), this was the most expensive flood in the town's 100-year history.

Floods occurred with increasing frequency between 1957 and 1964, there being nine in 7 years. Thus, the great majority of the people who were to be interviewed had knowledge and experience of recent and increasingly severe flooding.

Flood protection

Early attempts to control flooding in Opotiki by the local government amounted to little more than the infilling of riverbank depressions. Frequent and severe flooding between 1957 and 1962 forced it to seek assistance from the Poverty Bay Catchment Board, a regional branch of the National Soil Conservation and Rivers Control Council. In 1962, this board put forward a plan which, in essence, required that Opotiki be encircled on three sides with levees protecting it from a design flood of 100 years (Poverty Bay Catchment Board, 1962). This design flood was exceeded by the record flood of March 1964, which had an estimated discharge of 2,200 cubic meters (77,000 cubic feet) per second and a recurrence interval of 250 years. The 1964 event was adopted as the new design flood (Poverty Bay Catchment Board, 1964).

One should perhaps note here that Catchment Authorities can only build to a project design which is within the economic capacity of the community concerned. While it might have been structurally desirable to have built levees of sufficient height to withstand a flood of, say, 2,500 cubic meters (90,000 cubic feet) per second in Opotiki's case, the final decision was primarily governed by the ability of the community to financially support the scheme, even though it received a 3:1 subsidy from central government.

The location, size, and state of construction of the levees proposed in the final scheme specified levees on both sides of the rivers with those on the town side having 15 centimeters (6 inches) more freeboard so that greater protection is provided for urban land. A number of supplementary measures designed to reduce levels of flood flow have been effected. These include the clearance from river channels of flow-impeding vegetation, river diversion cuts, and bank protection. An overflow channel between the Otara River and main outlet and a new river mouth were considered, but not acted upon because maintenance costs would be too high. Total cost of the adopted works was N.Z. $500,000 (U.S. $590,000).

The survey

Within what might then be considered a well-protected floodplain, freed from frequent and severe flooding recent enough to fall within the experience of most persons living there, a study was made of the relationship between flood frequency and those attributes of floodplain occupants which would help describe their perceptions of flood risk and their attitudes toward protection works and occupancy on the floodplain.

Two interview schedules were constructed: the first for commercial respondents (managers) in the center of town, and the second for residential respondents (managers). While the maximum probable flood and the regional flood for the area have not yet been computed, it is reasonable to assume that without levees all respondents would be potential flood victims, and that with the designed levees there someday will be a flood larger than 1964 that will overtop the system.

The interviews

The respondents were asked seven sets of questions. The first attempted to assess the information held by each manager on past flooding. The second assessed individual expectancy of future floods. The third and fourth sets of questions tested managerial perception of flood protection structures and sought attitudes on their effectiveness. Sets 5 and 6 were also concerned with expectation of future flooding, but in terms of the soundness of the levee structures and managerial assessment of the flood damage that could occur should the structures be breached or overtopped. Finally, a group of questions was aimed at attitudes toward continuing settlement on the floodplain.

The analysis

Does information of flood protection structures decrease expectation of future flooding? Past studies (Burton, 1961; Kates, 1962) indicate that, in the absence of reliable protection structures, information on past flooding increases expectation of future flooding. It could be assumed, therefore, that the erection of effective levees would disrupt the relationship.

Information on past flooding

There are many ways by which knowledge of flooding may reach a respondent. Perhaps the most important agent in Opotiki is the town's local newspaper, the *Opotiki News,* which reaches 80 percent of the residences. This paper takes a keen and responsible interest in many aspects of the town's flood problem. Reports range from soil and forest conservation practices through to plans for emergency actions by townsfolk at a time of evacuation. The costs of flood-control works and their expected degree of protection were given wide coverage. Various views on relocating the town onto higher ground were given full expression immediately after the major flood of 1964.

As shown in Table 8–1, all of the respondents interviewed had some knowledge of at least one flood occurrence. Several people were so well informed that they were able to relate, in detail, a history of the town's flooding.

Experience of flooding on a manager's property was also widespread. More than 80 percent of the people interviewed had experienced a flood. Of the 27 managers who had not, 5 were commercial managers who had begun business after the 1964 flood and 15 were residents who had moved to Opotiki after that flood. However, 3 of the 5 new commercial managers had experienced a flood as an owner of a house on the floodplain.

Expectation of future flooding

Replies to the question "Do you expect you will have, or there will be, another flood while you are living here?" were classed "Yes," "No," and "Uncertain." These may be viewed as pessimists, optimists, and neutralists, or, those who, given time, could become either optimists or pessimists. Of the 141 managers interviewed only 19 percent expect a future flood, 38 percent are uncertain about the future, and 43 percent reveal optimism by firmly denying the possibility of a future flood. Since eight floods entered the town between 1957 and 1964, well within the time span of most people who reside within the floodplain, this would appear to reflect their optimism toward the recent flood protection scheme. This inference is strengthened by the fact that, contrary to research findings elsewhere, no statistically significant relationship between expectation of future flooding and information on past flooding was apparent (Table 8–1).

The degree of optimism is greater for the commercial managers of the town. This could be due to their greater interest and assimilation of information on the technical aspects of flood protection since they as a group have suffered losses most consistently. Also, it could be be-

Table 8–1. Information and future flood expectancy of floodplain respondents[a]

| Information | Future flood expectancy | | | | | | | | | | | | | | |
| | Commercial | | | | | Residential | | | | | Total | | | | |
	Yes	No	Uncer-tain	Total No.	%	Yes	No	Uncer-tain	Total No.	%	Yes	No	Uncer-tain	Total No.	%
1. No knowledge, no experience	–	–	–	–	–	–	–	–	–	–	–	–	–	–	–
2. Knowledge, no experience	–	4	–	4	12.1	3	9	11	23	21.3	3	13	11	27	19.1
3. Knowledge with experience elsewhere	–	2	1	3	9.1	2	6	3	11	10.2	2	8	4	14	9.9
4. One experience on-site, on section only	–	–	–	–	–	1	5	6	12	11.1	1	5	6	12	8.5
5. One experience on-site, in home	3	6	1	10	30.3	2	2	5	9	8.3	5	8	6	19	13.5
6. Two to four experiences	2	–	2	4	12.1	5	7	13	25	23.2	7	7	15	29	20.6
7. More than four on-site experiences	2	7	3	12	36.4	7	12	9	28	25.9	9	19	12	40	28.4
Total	7	19	7	33	100.0	20	41	47	108	100.0	27	60	54	141	100.0

[a]H_0: that there is a significant relationship between information of past flooding and expectation of future floods. With information classes 2, 3–4, 5–7; $X^2 > 0.30$ with 4 degrees of freedom; not significant.

cause a high proportion of the town's officials are drawn from this section of the community and in their official capacity had closer contact than others with the technical aspects of flood protection.

Perception of protection works

Since it is postulated that flood protection works are an overriding factor influencing managerial perception of the flood hazard in Opotiki, some measure of their perceptions and attitudes toward flood control and abatement alternatives, including levees, is required. All those interviewed expressed some knowledge of levee works, and many were well enough informed to give details on the size of flood from which the town was protected. Again, the most detailed knowledge was that held by commercial managers.

Channel clearance and related measures, which as actions have less obvious environmental outcomes than levees, were generally not as well perceived. When they were, their use in reducing flood overflows, and therefore flood damages, was less obvious to the ordinary individual (Table 8–2).

Of the commercial managers interviewed, eight had contributed toward the adoption of these measures in their capacity as borough councilors. Most commercial managers and a few residential managers had taken part in meetings immediately after the major flood of 1964 which aimed either to gain a greater degree of protection or to have the town resited onto higher ground.

Attitudes toward public action in flood control and abatement for Opotiki are portrayed in Fig. 8–3. Only 9 percent felt uncertain or rejected the idea of levees.

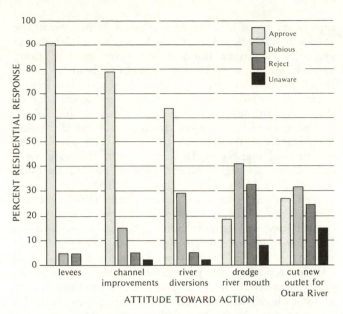

Fig. 8–3. Attitudes toward community action in flood control and abatement

Flood control and flood expectations

To further test the attitudes of respondents on the specific control measure of levees, managers were asked the following questions: "Do you think that the new levees will increase, decrease, or have no effect on the number of possible future floods?" and "Should flooding occur again, do you expect damages to be more, less, or about the same as before the levees were built?" The

Table 8–2. Attitudes toward community actions in flood control and abatement

Action	Attitude of respondents[a]							
	Approved		Dubious		Reject		Unaware	
	No.	%	No.	%	No.	%	No.	%
Levees	104	91.2	5	4.3	5	4.3	0	0.0
Channel improvements	90	78.9	17	14.9	6	5.2	1	0.9
River diversions	73	64.0	33	28.9	6	5.2	2	1.7
Dredge river mouth	21	18.4	47	41.2	37	32.4	9	7.9
Cut new outlet for Otara River	31	27.2	37	32.4	29	25.4	17	14.9

[a]N = 114; 27 residential managers gave no response.

replies are given in Table 8–3. Almost 91 percent of the respondents believed that the levees would lessen the certainty of future flooding. Whatever level of flood expectation may previously have been held by them, flood protection structures had lessened that expectancy. It would seem that those whose knowledge of flood control more closely approximated that of the technical expert were more optimistic about the future freedom from flooding in the town (100 percent for commercial managers). Dissenters from this view were only found among what appeared to be the less well-informed residential managers and from less than 5 percent of all those who responded. (Roder, 1961, found that levels of knowledge of levees were evenly divided among pessimists and optimists toward future flooding.) However, when *detailed* knowledge of flood control (taken as persons who had or had not read detailed reports of the protection scheme in the Opotiki *News*) was related to expectation of future flooding, no statistically significant difference among these variables could be found. This may reflect the inadequacy of the measure used for defining levels of knowledge of protection, since detailed knowledge can be obtained from sources other than newspaper media.

Opinion was far less concerted on the extent of possible future damages should flooding occur again. Respondents who expressed uncertainty (18 percent of the total) appeared to be better informed than their fellows. They held that damages would be less, providing the levees were not breached by floodwaters. They reasoned that overtopping of the levees would mean a smaller volume and lower velocity of water entering the borough than in 1964, and that the resultant flood would be relatively shallow except for ponding in the lower lying sections of the town. This reasoning closely approximates that given by the Poverty Bay Catchment Board and reported in the Opotiki *News*. However, they also recognized the possibility of massive breaching of the levees if a substantially larger flood than that of 1964 occurred.

Those who perceived damages as being greater in a future flood generally failed to understand the overtopping principle. Those who perceived damages as being less failed to appreciate the possibility of levees breaching and its consequences, and the risk of a flood occurring with a magnitude far in excess of that experienced in 1964. On the other hand, they may very well have perceived the possibility of greater destruction in a future flood, but denied that the event would happen to them (cf. Kates, 1962).

Comparisons with previous studies

Some measure of the influence that the Poverty Bay Catchment Board and their flood protection activity had on the expectation of future flooding by managers of the Opotiki floodplain can be seen in a much broader sense in Fig. 8–4. This compares several characteristics of the Opotiki flood situation with those of six sites studied by Roder (1961), Burton (1961), and Kates (1962). The extent of flood control and abatement works for all sites are as follows: Darlington, a watershed protection project; Aurora, a flood channel; El Cerrito-Richmond, some channel clearance; La Follette, some channel clearance, but not on a community scale; Topeka, considerable protection including upstream reservoirs and levees and proposals for levee extensions, floodwalls, and channel diversions; Desert Hot Springs, no measures taken; Munster-Hammond, small-scale and piecemeal protection works including levees, channel improvements, and pumping plants; Watkins Glen, an inadequate overflow channel dating back to 1937; and Opotiki, a comprehensive levee protection, channel clearance, bank protection, and river channel diversion scheme.

With the exception of Darlington, Opotiki has had the greatest frequency of flooding. According to both Burton and Kates, therefore, it should have a high level of information on flooding and expectation of future floods. In fact, Opotiki has 100 percent knowledge and the third highest percentage of managerial experience of flooding, but its expectation of future floods ranks third to last before Munster-Hammond and Watkins Glen. Since, with the exception of Topeka, Opotiki is the only town of the nine that has anything like large-scale river-control works, its low ranking for expectation of future floods may be attributable to the levee protection scheme. (This assumes, of course, that the cultural differences between Opotiki and the other eight sites is no greater than that among all sites.) It is interesting to

Table 8–3. Attitudes toward levees and their effects on expectation of future flooding

| | Respondents | | | | | |
| | Commercial | | Residential | | Total | |
Attitudes	No.	%	No.	%	No.	%
Know of levees and expect future flooding to . . .						
1. increase	–	–	3	2.8	3	2.1
2. decrease	33	100.0	95	87.9	128	90.8
3. no effect	–	–	4	3.7	4	2.8
Total	33	100.0	102	94.4	135	95.7
Know of levees and expect size of future flood to be . . .						
1. more	16	48.5	47	43.5	63	44.7
2. less	7	21.2	29	26.9	36	25.5
3. no effect	4	12.1	6	5.5	10	7.1
4. uncertain	6	18.2	19	17.6	25	17.7
Total	33	100.0	101	93.5	134	95.0

Study sites

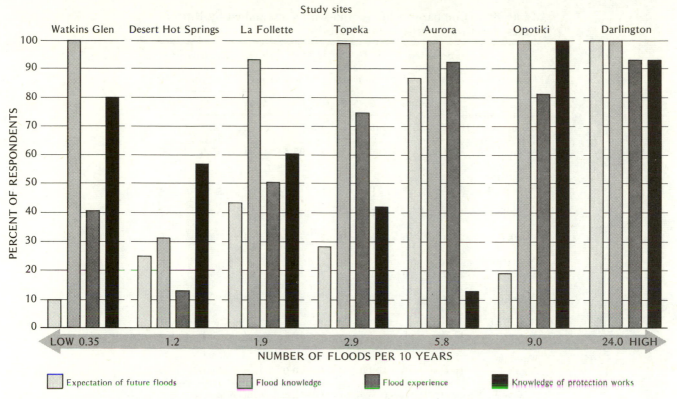

Fig. 8–4. Comparison of Opotiki with sites studied in the United States

note that Topeka also has a lower flood expectation value than one might expect from its hazard experience. However, as indicated earlier, Roder (1961) did not find a significant relationship between attitudes toward future flood and varying degrees of protection knowledge. Although highly perceived, it appears that the nature of protection at Darlington (watershed treatment) fails to depress flood expectancy as in Opotiki (Table 8–4).

Levee protection: implications for settlement

Since knowledge of levee protection in Opotiki appears to have decreased expectation of future flooding, it seems reasonable to assume that this knowledge will inspire confidence in the protected area, discouraging those who fear floods from moving off the floodplain, and increasing settlement.

This second hypothesis was tested in the following way. First, respondents were asked these questions: (1) Did you possess knowledge of the flood situation when you first decided to move into this house? (Residents only) (2) Knowing what you now know about flooding, would you still choose to locate here? (3) Did you consider leaving the borough (floodplain) after the 1964 flood? (4) Where would you like to live, given free choice?

Responses show that nearly 80 percent of householders knew of flooding when first they located in their homes. With hindsight, 55 percent of all residents and 94 percent of commercial managers believed that they would still locate in their same premises today. It is difficult to tell to what extent this response is colored by their later knowledge of new levees, especially since approximately one-third of them (35 percent, shopkeepers; 30 percent, householders) had seriously considered selling out and leaving the floodplain immediately after the 1964 flood. Had they done so, it would have swelled considerably the 30–40 families (approximately 200 people) which actually did vacate the floodplain between 1964 and the time this survey was made. That they finally decided against moving may be attributable not only to levee protection, but to a number of other socioeconomic constraints as well. For example, of the 16 commercial managers who reported considering moving after the record flood, 2 decided against it upon receipt of adequate insurance compensation; 3 because of their strong social links with the area; 6 because such a move would be too costly; 1 because he did not believe that a flood as large as that in 1964 would occur again; and 4 when they learned of the new protection scheme.

Householders were asked, therefore, to reconsider the question of leaving the borough assuming that all socioeconomic constraints were magically removed. Only 30 percent then preferred to remain within the borough. A small group (7 percent) would leave the district entirely

Table 8–4. Comparison of Opotiki with sites studied by Burton, Kates, and Roder in the United States

Study site	Expectation of future flood		Flood knowledge		Flood experience		Knowledge of protective works	
	%	Rank	%	Rank	%	Rank	%	Rank
Darlington	100.0	1	100.0	1	92.3	2	92.3	2
Aurora	86.7	2	100.0	1	93.3	1	13.3	9
El Cerrito–Richmond	45.4	3	90.9	4	72.7	5	81.8	3
La Follette	43.3	4	92.7	3	49.5	6	59.6	7
Topeka	28.0	5	99.9	2	74.7	4	42.0	6
Desert Hot Springs	25.0	6	31.2	5	12.5	9	56.2	8
Opotiki	19.1	7	100.0	1	80.7	3	100.0	1
Hammond-Munster	11.2	8	?		49.3	7	78.8	5
Watkins Glen	10.0	9	100.0	1	40.0	8	80.0	4

Sources: Burton (1961), Kates (1962), Roder (1961), and Opotiki field data.

because of its isolation and lack of employment opportunities. The remainder (63 percent) considered the flood-free hills west of the town as their first choice for relocation and almost 80 percent of these people (that is, 48 percent of all householders) gave *flooding* as their main reason for wanting to do so.

This result is somewhat surprising since only 19 percent of the householders had previously said that they expected a future flood (Table 8–1). When a question with essentially the same meaning was posed in different form, the pessimistic grouping swelled to 48 percent. These were the persons who, given freedom of choice, said they would relocate beyond the floodplain *because of the flood factor.* Their response suggests that they had remained on the floodplain, not because of the flood protection scheme, but out of socioeconomic necessity.

Why this different response to two sets of questions with similar meanings posed in different form? One likely reason is questionnaire design. It is well known that the interview itself provides an educational situation to the extent that responses may change as questioning proceeds. A related reason is the neglect of personal and/or situational factors that may impinge upon responses to specific questions. An opportunity for consciously expressing some of these factors was obviously provided in the questions that referred to socioeconomic constraints (Wicker, 1971). A psychological reason, but one somewhat different to that just

referred to, might well be that of a cognitive dissonance situation where, in order to reduce conflict between verbal attitudes and overt behavior, respondents answer differently from what their deeply held beliefs might otherwise suggest. Thus, if a respondent feared floods, but for socioeconomic reasons was not able to relocate, he might well conceal his true expectations of future flooding so that his behavior, occupance on the floodplain, appears consistent (Festinger, 1957).

In order to examine the relocation response in more detail, information on free choice in residential location was compared with expectation of future flooding, expectation of flood losses, and knowledge of levee construction. At this stage of the analysis, residential respondents were subdivided into two groups to determine whether past flood experience had any bearing on results. In the first group were residents who had experienced two or more floods; in the second those who had one or no flood experience. No significant distinction in results on this basis was found.

As can be seen from Table 8–5, a relationship of moderate strength (Pearson's coefficient = 0.3184) exists between free choice in relocation on or off the floodplain and attitude toward future flooding and flood losses. The pessimists form a fairly homogeneous group with almost 75 percent opting for homes outside the floodplain, mostly because they do not wish to experience the flooding they still believe could occur. The inconvenience of a possible future flood is the price

Table 8–5. Free choice of relocation by attitude toward expectation of future flooding and flood losses[a]

Expectation of future flooding and future flood losses	Choice of residence		
	Remain on floodplain	Move off floodplain	Total
Optimistic	22	17	39
Uncertain	9	32	41
Pessimistic	5	13	18
Total	36	62	98

[a]H_0: that there is a significant difference between expectation of future flooding and losses and relocation on or off the floodplain. X^2 0.995 with 2 degrees of freedom; significant.

that the remaining pessimists are willing to risk paying for continued occupance of the floodplain. On the other hand, the optimists are divided, with 45 percent favoring relocation beyond the hazard area, but, unlike the pessimists, for reasons other than flooding. Of the 40 percent of people who felt uncertain of the future most (78 percent) preferred to live outside the floodplain, primarily because of the flood factor.

Since it was demonstrated that knowledge of levees markedly depressed future flood expectations, we might reasonably expect that such knowledge would deter movement off the floodplain additional to the socioeconomic factors referred to above. However, a statistically significant relationship between free choice of location on or off the floodplain and varying degrees of knowledge of the protection scheme reports was not apparent.

Conclusion

For persons who wish to settle or to continue settlement in the Opotiki borough, the topographical and hydrological nature of the area is such that there is no alternative to locating in an area which, until 1964, had undergone frequent and severe flooding. Low-lying hills immediately west of the borough do offer some flood-free sites for residential development.

Although relocating the town off the floodplain was dropped in favor of a "cure-all" flood protection scheme, the 1967 survey demonstrates that a large section of the community is far from confident about future prospects on the floodplain. An analysis of reported perceptions and attitudes of individual floodplain managers (commercial and residential) toward flood hazard and protection at Opotiki shows that knowledge of levees decreases managerial expectation of future floods. In this, the survey provides more conclusive results than the American studies. The extent to which knowledge of levees increases confidence, and therefore settlement within the hazard zone, remains

unclear. It must be remembered that the protective structures were barely 2 years old when this survey was made and this may be insufficient time for confidence to have been restored to many living within the flood area. Given time, will the large group of people who felt uncertain about the future swell the ranks of the optimists? In 1967 about half of those sampled remained on the floodplain not so much because levees abate flood fear, but because socioeconomic constraints place relocation to a flood-free area beyond their reach.

Speculation may turn in another direction. As the socioeconomic constraints are removed, will individuals move from the floodplain, leaving the costly protection works to protect a stagnant and deteriorating town, thus compounding a problem, common to many other small service centers in New Zealand, of a town whose economy relies largely upon a farming hinterland?

As one of almost 100 urban communities in New Zealand that suffer damaging floods, what implications does the Opotiki experience have for national policy on flood hazard?

The flood hazard in Opotiki is relatively easy to assess, yet technical expertise was not focused on the problem until 1962. Until then, management of the floodplain was based primarily on layman perception of risk and resulted in minimal protection and continued invasion by land uses with high damage potential. While application of technical expertise to the risks involved is welcome, the Opotiki experience suggests that construction of large-scale protection structures need not always solve the human and economic problems associated with flood hazard. It seems essential, therefore, that the rapid increase in hydrological knowledge evident in New Zealand in recent years be coupled with a national policy on flood damage reduction which recognizes and adopts a range of measures additional to that of conventional engineering (Ericksen, 1971). For example, regulating floodplain development in urban areas through zoning ordinances and building codes is one measure which may lead to less costly and more efficient uses of urban floodplains. Although these measures have been loosely defined in the Town and Country Planning Act, 1953, communities have, by and large, failed to utilize them to the extent warrantable (Ericksen, 1970).

Acknowledgments

This paper is based on part of a University of Canterbury Master's Thesis for which research grants were provided by the Department of Geography, the Lester Fund, and the Opotiki Borough Council.

I wish to thank the many people and organizations in Opotiki and officers of the Poverty Bay Catchment Board who provided information and material for this study.

References

Burton, Ian. (1961) "Invasion and escape on the Little Calumet." In G. F. White, ed., Papers on Flood Problems. Chicago: University of Chicago, Department of Geography, Research Paper No. 70, pp. 84–94.

Ericksen, Neil J. (1970) "Regulating urban flood plain develop-
    ment in New Zealand." *Proceedings: Sixth New Zealand
    Geographical Conference*, Christchurch, pp. 150–57.
_____. (1971) "Human adjustment to floods in New Zealand."
    New Zealand Geographer 27 (2):105–29.
Festinger, Leon. (1957) *A Theory of Cognitive Dissonance*.
    Evanston, Ill.: Ross, Peterson.
Kates, Robert W. (1962) *Hazard and Choice Perception in Flood
    Plain Management*. Chicago: University of Chicago, De-
    partment of Geography, Research Paper No. 78.
Poverty Bay Catchment Board. (1962) *Opotiki Flood Control
    Scheme No. 1*. Report No. 92 (Gisborne).
_____. (1964) *Opotiki Flood Control Scheme No. 2*. Report No.
    1451 (Gisborne).

Roder, Wolf. (1961) "Attitudes and knowledge on the Topeka
    flood plain." In G. F. White, ed., *Papers on Flood Prob-
    lems*. Chicago: University of Chicago, Department of Ge-
    ography, Research Paper No. 70, pp. 62–83.
Town and Country Planning Act. (1953) Wellington, New Zea-
    land: Government Printer. Reprinted with amendments,
    January 1, 1967.
White, Gilbert F., et. al. (1958) *Changes in Urban Occupance of
    Flood Plains in the United States*. Chicago: University of
    Chicago, Department of Geography, Research Paper No.
    57.
Wicker, Allan. (1971) "An examination of the 'other variables'
    explanation of attitude-behavior inconsistency." *Journal
    of Personality and Social Psychology*, 19 (1):18–30.

9. Coastal erosion: the meaning of a natural hazard in the cultural and ecological context

ROWAN A. ROWNTREE
Syracuse University

Coastal erosion, when compared with other natural hazards, is a predictable and continuous process of nature
which manifests itself in human damages of relatively
low magnitude. The characteristics of this phenomenon
pale among such environmental threats as hurricanes,
tornadoes, floods, and earthquakes. Contrasted with the
hundreds of thousands of impoverished Asians wiped
out in the tropical cyclones of coastal Bangladesh, it is
often the wealthy few, who have chosen to live on
ocean bluffs because of the local amenities, who are
endangered by coastal erosion. Nevertheless, as an everyday part of their environment, the retreat of shorelines
must be dealt with by many communities in the United
States and, presumably, in other coastal countries as
well. The manner in which it is dealt with depends on
the meaning that coastal erosion holds for each resident
and, more importantly, for the community as a whole.

This study was conducted in the small town of Bolinas, situated on the bluffs overlooking the Pacific Ocean
on the coast near San Francisco. In its physical setting,
Bolinas (population approximately 2,000) is typical of a
handful of communities coping with the problem of
coastal retreat in Central California. Among them are
Pacifica, Santa Cruz, and Capitola (Fig. 9–1). Culturally,
Bolinas is somewhat special for two reasons. First of all,
the people of this village have a keen sense of place;
life-style and environment are fused into an awareness

of just what the community is, and should be, in space
and time. Secondly, the town is undergoing a rapid, and
rather agonizing, cultural transition, not so much in the
basic values held by the people but in the way these
values are socially and intellectually articulated. Most
important from our point of view, several environmental
issues are at the root of this cultural transition. Coastal
erosion is one of them.

The purpose of this paper is to describe how a natural
hazard takes on meaning at the individual and community levels of cognition. Simply, in the mind of the
individual coastal erosion is not of great concern nor are
its characteristics clearly discernible. It has little meaning except for those persons whose homes are directly
threatened by the ocean, and these are in the minority.
In the community mind, however, coastal erosion has
great meaning and it is at this level of organization that
the hazard is articulated, and given symbolic value.

The following section describes the character of the
hazard, its occurrence, damages, and past adjustments,
together with a discussion of what appear to be the
alternatives open to the community for dealing with the
phenomenon of coastal erosion. The final section elab-
orates on the differences between individual and community perception of coastal erosion focusing on the
interaction of culture groups and their attitudes, and the
debate over environmental issues, both of which give

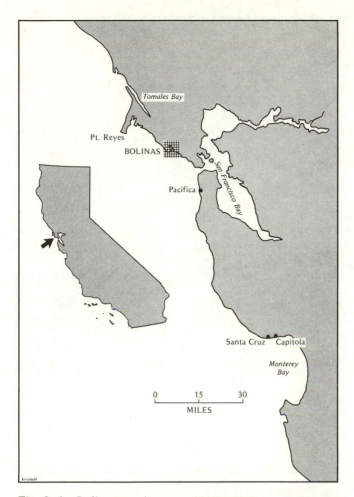

Fig. 9–1. Bolinas study area, Marin County, California

coastal erosion sufficient meaning for a community choice to be made, a choice involving important trade-offs about the future of Bolinas.

Characteristics of the hazard

Physical factors

Fifteen miles north of the entrance to San Francisco Bay the coastline makes an abrupt westward turn at Bolinas to form the massive Point Reyes Peninsula. Geologically, the peninsula is severed from the "mainland" by the San Andreas Fault, which can be traced along its north-south axis by following a series of depressions, sag ponds, and fissures that run between Bolinas Lagoon and Tomales Bay. Extensive damage occurred in this area during the disastrous San Francisco earthquake of 1906. Thus, the village of Bolinas is no stranger to environmental hazards, lying with one flank on the San Andreas Fault and the other on the rapidly eroding cliffs of the Pacific (Fig. 9–2). The town itself rests on two separate geologic formations. Just west of the fault, and underlying points A and B on Fig. 9–2, is

a wedge of Plio-Pleistocene sandstone approximately 1,800 meters (2,000 yards) wide at the inlet to Bolinas Lagoon, and diminishing in width north of the estuary. This rock is referred to as the Merced Formation and forms two small mesas near the inlet to the lagoon. The mesa closest to the ocean (point B) supports a number of homes in the high-hazard zone.

West of the Merced Formation the peninsula is comprised of Miocene shale and sandstone referred to as the Monterey Formation. The grid development called Bolinas Mesa (point C) lies on an extensive marine terrace underlain mostly by beds of Monterey shale dipping to the west. The Merced and Monterey rocks differ slightly in their microerosional behavior but retreat from the ocean at about the same rate, the major factor being exposure. The west-facing cliffs between Bolinas Point and Duxbury Point are exposed to the prevailing wave action from the northwest throughout the year. The shale beds forming these cliffs dip on the average of 40 degrees toward the sea and during the winter the bedding planes become lubricated so that when the cliffs are undercut by wave action, massive slides occur. The average rate of erosion appears to be higher in this region than along the more densely populated south-facing cliffs of the Bolinas Mesa and Little Mesa, which are sheltered by Duxbury Point and Duxbury Reef. However, winter storms rising out of the southwest bring massive erosion to this area, often removing up to 2 vertical meters (2.2 yards) of beach material at the base of the cliffs. Then, with no sand for protection, the cliffs are undercut by the waves, resulting in severe slumps and landslides.

The long barrier spit separating the lagoon from the ocean is presently about 75 percent developed with expensive new homes. Because the spit is well stabilized and high enough to escape severe flooding, the hazard from storm waves is minimal. (Lying on unconsolidated sands, however, makes these homes most susceptible to seismic hazards.) The town of Stinson Beach at the base of the spit sits on an unstable bedrock of graywacke, shale, chert, and serpentine of Mesozoic age. This material comprises the Franciscan Formation, which underlies much of the central California coastal region and is responsible for a large number of residential landslides in the San Francisco Bay Area. Between Stinson Beach and the entrance to San Francisco Bay the cliffs reach heights of over 100 meters (110 yards) and display the results of a long history of undercutting and sliding. However, virtually no homes are located on these cliffs. The same holds for the coast running north of Bolinas for many miles. Therefore, this study restricts itself to the inhabited south-facing cliffs between Duxbury Point and the inlet to Bolinas Lagoon.

Variations in rate of erosion

Cliff erosion in this area manifests itself in the form of discrete events that vary widely in space and time. One

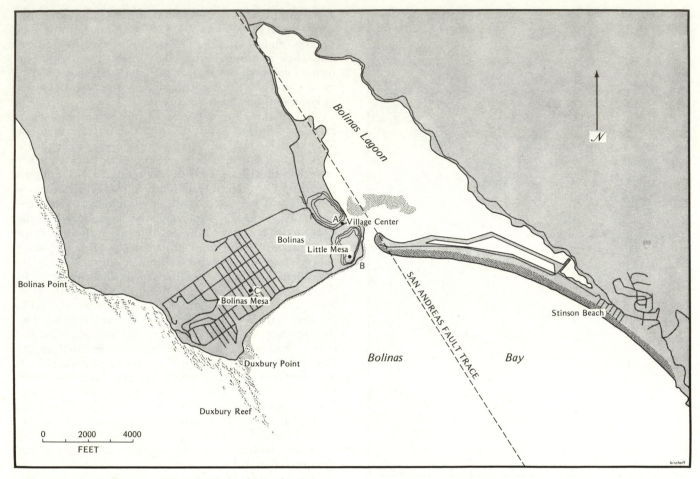

Fig. 9–2. Bolinas, California, showing the San Andreas fault; trace of 1906 earthquake

year may see a large slump or landslide while other years may pass with no appreciable retreat of the cliffs. Thus, to speak of *average* erosion rates over large areas is to couch the hazard in terms different from those which would perhaps best describe its occurrence. However, a widely circulated U.S. Geological Survey report of sedimentation in the Bolinas vicinity suggests that "the cliff between the inlet (to the lagoon) and Duxbury Point is eroding at the phenomenal rate of about 2.3 feet (0.7 meter) per year" (Ritter, 1969). Mostly due to this statement, the figure of "2 feet [0.6 meter] a year" is used commonly within the community to indicate the average rate of erosion.

There are only two sources for documented measurements of cliff retreat in the Bolinas area. One is the 1929 U.S. Coast and Geodetic Survey report, which estimates cliff erosion for the period 1859 to 1929 from actual surveys (U.S. Coast and Geodetic Survey, 1929). The other is A. J. Galloway, geologist with the California Academy of Sciences, who measured the distance between a portion of the cliffs and a relict seawall constructed about 1890 and broken up in the winter of 1913 (Galloway, 1972). This measurement was made in the vicinity of the Little Mesa and gives us an estimated rate for the period 1913–62.

Along the west-facing cliffs, USC & GS rates for the period 1859–29 vary from 0.54 meter/year (1.8 feet/year) to 0.99 meter/year (3.3 feet/year) with an average of about 0.75 meter/year (2.5 feet/year). For the south-facing cliffs (Duxbury Point to the Lagoon inlet) the USC & GS estimates a rate of 0.42 meter/year (1.4 feet/year) for the period 1859–1929. Galloway's measurement provides a rate of 0.45 meter/year (1.5 feet/year) that corresponds remarkably with that of the USC & GS. These rates are, of course, for cliff retreat under natural conditions. Seawalls and wooden bulwarks erected near the Little Mesa have acted to reduce erosion somewhat in that vicinity.

There is no record, other than the memories of local residents, which might reveal the variation in intensity of the hazard from year to year. However, most of the interview respondents (see below) noted 1955–56 and 1969 as being the worst years for coastal erosion at Bolinas. It should be mentioned, however, that these interviews were conducted prior to the winter of 1971–72, which was, according to experienced observers, one of the most devastating seasons in their memory. Intense storm waves ate under piling-and-plank bulwarks which had stood for many years. The entire beach fronting the south-facing cliffs was removed down to bare bedrock and several major slides occurred.

Summary of damages and expenditures

An estimate of total damage to property, structures, and life must be based on an analysis of historical photographs and the reports of certain residents. To date, no estimate of cumulative damage has been made by any public agency. Table 9–1 is a summary estimate of property damage and a rough estimate of the amount of money spent over the years to cope with the hazard of coastal erosion in this vicinity. (As far as is known, no lives have been lost as a direct result of coastal erosion in Bolinas.)

With the exception of public-utility repairs, expenditures for protection and reconstruction have been borne mainly by individual residents or small groups of residents. Until the last several years there has been no government involvement beyond remedial work done by the Bolinas Public Utilities District and road maintenance performed by the Marin County Public Works Department.

Table 9–1. Coastal erosion damages and expenditures at Bolinas, California

	Value ($)[a]
Damages	
Homes (3)	75,000
Real property (the equivalent of 15 lots @ $5,000 each)	75,000
Public utilities (roads, pipelines)	50,000
Total	200,000
Expenditures	
Seawall construction	75,000
Cliff drainage and planting	25,000
Rip-rap of road area at top of cliffs	5,000
Other road repairs	10,000
Moving of homes (2)	10,000
Problem studies[b]	10,000
Total	135,000

[a]1971 dollar equivalent.
[b]U.S. Army Corps of Engineers, Bolinas Beach and Erosion Study, $5,000 per year for 2 years.

Coping with the hazard

Past adjustments

For the most part, residents of Bolinas who have experienced coastal erosion have had to fend for themselves. When asked to whom one goes for help after experiencing damage to property or structures, the majority of those queried answered, "A friend." Past adjustments to the hazard range from doing nothing and hoping for the best to making a substantial investment in a concrete seawall or even moving one's house. Between these extremes, protective measures include (1) planting the surface of the cliff with perennial grass, shrubs, or trees, (2) installing drainpipes in the cliff face to prevent saturation and slippage, (3) placing sand and gravel at the cliff base to retard undercutting, and (4) spreading plastic over the face of the cliff to prevent infiltration of rainfall and gully erosion from surface runoff. To be truly effective, these measures must be employed in conjunction with a wood or cement bulwark, or seawall, at the cliff base which retards or prevents undercutting. In cases where this has not been done, attention to the cliff face alone has been of little help in preventing slumps or slides.

Though half of the total number of respondents thought that insurance would be a preferable adjustment, it is, in fact, unavailable to homeowners endangered by coastal erosion.

Two types of bulwark, or seawall, are employed: wood plank-and-piling, and formed concrete. The high cost of concrete walls makes the plank-and-piling bulwarks the most common at Bolinas. These are found in the vicinity of the Little Mesa Cliffs and for a distance of some 100 meters (110 yards) to the west.

Two homes have been moved. One new house recently constructed on the cliff was designed with this in mind. Of course, evacuation has been necessary in several cases.

Future alternatives

Discussion with community leaders, public officials at the local and county levels, and coastal stabilization technicians in the U.S. Army Corps of Engineers suggests that four alternatives exist for coping with the hazard at Bolinas.

1. Zoning to prevent hazard-zone development. Zoning would be adopted to prevent development along the top of the cliffs. No structures to retard erosion would be provided and natural recession of the cliffs would continue. This approach would be effective in varying degrees only in the vicinity of the Bolinas Mesa tract, where there is undeveloped land along the top of the cliffs. The width of this land behind the cliff varies considerably, and assuming natural recession continues, it will be only a matter of time before some homes are affected. At the east end of the tract, the first line of

existing homes lies about 25 meters (80 feet) from the edge of the cliff. Moving west from this point, the distance between developed land and cliff edge increases and, presumably, the rate of cliff retreat decreases in the lee of Duxbury Point. Zoning of this kind would, of course, be most effective in the undeveloped region between Duxbury Point and Bolinas Point.

2. Public land acquisition of hazard zone. The area subject to erosion would be acquired for public recreation and open space. Natural recession of the cliffs would continue and damages would be restricted to low-cost recreation facilities. The costs of acquisition would range widely depending on whether the program was limited to undeveloped private land or included the acquisition of homes that presently exist on the edge of the cliff in the high- and medium-hazard zones.

3. Rubble-mound seawall along base of cliffs. Of the two alternatives involving extensive construction, this is the less expensive. It consists of a long seawall of broken rock located at the toe of the cliffs extending from the Little Mesa to Duxbury Point. The seawall would be designed and constructed by the U.S. Army Corps of Engineers at an initial cost of $3–$4 million and an average annual cost of $200–$270 thousand (annual cost plus initial cost discounted at 5-1/8 percent). Once the base of the cliff is stabilized, the top will continue to retreat until a natural angle of repose is reached. This would result in continued recession of 10–20 meters (30–65 feet) along the top of the cliff unless other engineering structures were erected to prevent this.

4. Groins, beachfill, and energy dissipator. The most elaborate and expensive alternative for dealing with the hazard consists of three rubble-mound groins placed at right angles to the cliffs and a rubble-mound energy dissipator doglegged out from Duxbury Point (Fig. 9–3). The groins would impound about 300,000 cubic meters (392,000 cubic yards) of imported beachfill, extending the beach area at least 200 meters (218 yards) seaward from the base of the cliffs. The initial cost for this alternative is estimated by the Corps of Engineers at $4–$6 million, the average annual cost at $250–$390 thousand.

Wave energy coming from the SW–NW quarter would be reduced effectively by the Duxbury Point dissipator. Large storm waves from the south, however, could remove an undetermined but significant amount of beachfill, exposing for a time the south-facing bluffs to high-energy storm conditions. Adding a rubble-mound seawall along the base of these cliffs, as in alternative 3, would nearly double the cost, though this would be the most effective combination of structures for arresting cliff erosion. Again, even when the base of the cliffs is stabilized, the top will continue to retreat until a stable angle of repose is reached.

In addition to increased recreational benefits from the expanded beach, this alternative would provide a safe anchorage for small boats in the lee of Duxbury Point. On the other hand, there would be severe destruction of resident biota on Duxbury Reef, which is presently a marine ecological reserve for intertidal organisms. Similarly, there are extensive clam beds located along the existing beach which may be destroyed by large amounts of imported beachfill.

The cost of alternatives (3) and (4) would be divided among federal (50 percent), state (25 percent), and local (25 percent) interests. Because the town of Bolinas is unincorporated, a special assessment district would have to be created to generate local moneys for the project. The boundaries of this district have so far not been discussed.

The perception and meaning of coastal erosion

During the summer of 1971, 120 interviews were conducted with heads of households located in three zones of hazard severity: (1) *high* (13 percent of the interviews): homes immediately adjacent to the cliffs with damage to land or structures likely to result from erosion and landsliding within the next 3–5 years, (2) *medium* (23 percent): homes located a distance of one or two city lots, 50–100 meters (55–109 yards) from the cliffs, and (3) *low* (64 percent): homes located inland of zones (1) and (2).

The purpose was not to restrict our sample to just those who were directly affected by the natural hazard. We sought also to understand what coastal erosion meant to persons who contribute to community decision making but on whom the hazard does not impinge. It was felt that in a community where only a minority of residents live under direct threat of the hazard, but where decisions are made by the community as a whole, it was most important to follow this procedure.

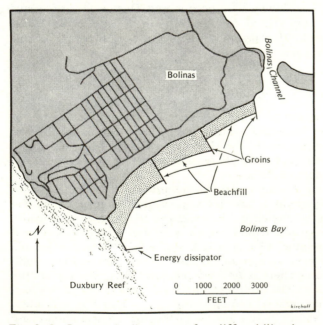

Fig. 9–3. Proposed adjustments for cliff stabilization

Individual perception

Residents of the high-hazard zone, whether renters or owners, saw retreat of the cliffs as a direct and serious threat to their presence there and were willing to spend, or advocate spending, large sums of money for preventive measures. Not unexpectedly, the level of individual understanding about the characteristics of the hazard (frequency, magnitude) was greater in this zone, generally speaking, than anywhere else. Renters were almost as knowledgeable about the hazard as property owners, but were less concerned about the long-term implications. Often, a sense of comradeship was discernible among residents of this zone, arising perhaps from the fact of their having to deal actively with cliff erosion in small groups. Without question, all residents sampled in the high-hazard zone thought that outside (federal, state, and/or county) help would solve their problem. However, there was a certain reticence about the impact that a large cliff stabilization project would have on the amenities of their personal environment. Too, it seemed that there was an underlying feeling of reward, never articulated, which derived from living under limited environmental stress. In any event, the respondents agreed that the amenities were worth the risk. (Homes in this zone have an unmatched view across the Gulf of the Farallones to the Golden Gate, entrance to San Francisco Bay, and to the more distant hills of San Francisco itself.)

The residents of zone 2 shared an anxiety about the retreat of the cliffs which apparently grew out of their inability to do much about the problem (they were deprived of the sense of purpose and group activity which engaged their seaward neighbors) even though they were still only a stone's throw from the ocean. Many of them, particularly renters and new owners, were basically aware of the hazard but knew little about it in terms of its frequency of occurrence, severity, and predictability.

In zone 3 respondents were aware that coastal erosion existed, but generally had difficulty articulating their thoughts and feelings about the phenomenon as a natural hazard. Many felt that the interview questions were couched in such a way as to be unanswerable or ambiguous. Part of this stemmed from the standard format and cross-cultural nature of the questionnaire. For the most part, though, the problem appeared to be that the concept of coastal erosion had little meaning for the respondents outside of the context in which they were used to thinking. Only when the questionnaire portion of the interview was completed and open discussion was generated was it possible to glimpse the context in which the concept "coastal erosion" resided in the respondent's mind. Because the concept was not related to their everyday experience, as it was to those living in hazard zones 1 and 2, the structure of the meaning was different. This is not to suggest that the concept was not as important to them; it simply existed at a different

level of abstraction and in a different context. (Exceptions to this were respondents who had close friends or relatives living in zone 1 and who, consequently, viewed coastal erosion more along the lines of those directly affected by it.)

In many cases, the importance of the concept was of a symbolic or allegorical nature. For example, among members of a group of ecologically concerned, communally oriented young people, coastal erosion was seen primarily as a revered natural process relentlessly consuming private property. This view can be better understood when one grasps the meaning of the two concepts "natural process" and "private property" in the context of a community debate over certain deep-seated issues about the town's future. Therefore, in the next section we will try to describe the ways in which the concept of coastal erosion was used at the community level. A note should be made here that to clarify much of this we had to turn to certain aspects of symbolic interaction theory (Blumer, 1962) and methods of participant observation (Bruyn, 1970; and especially Bogdan, 1972). Also, a number of additional interviews were conducted to provide a basis for the following discussion.

The community context

Bolinas was settled in the first half of the last century, late in the Spanish period before statehood. Its history includes hay and potato farming, heavy lumbering, shipbuilding, and cattle grazing. Today, the hills surrounding the lagoon comprise a pleasant mosaic of grassland and forest. The pastoral atmosphere is enhanced by a few remaining cattle kept on the grassy interfluves and around the outlying homes on Bolinas Mesa.

In the minds of many, both young and old, external pressures to put Bolinas on the tourist map fly in the face of its working history. Too, the town was clearly not built for tourism. The beach fronting the bluffs is often under water at high tides, the lagoon is a fragile ecosystem incapable of supporting heavy recreation, and, most importantly, the village center is not designed for heavy automobile traffic. Parking is totally inadequate. Two large state parks near Stinson Beach draw big crowds during weekends and summer months. To these travelers, metropolitan Bay Area newspapers and tourist guides tout Bolinas as the unspoiled village "just a few more miles" out the coastal road. Tranquil Bolinas Lagoon has long been coveted by the thousands of powerboat owners of San Francisco Bay as a pleasant location for an outer coast harbor. On top of all this is the recent federal acquisition of Point Reyes National Seashore to the north and the Golden Gate National Recreation Area to the south. Bolinas lies, rather uneasily, squeezed in between the two federal recreation centers, anticipating even greater pressure on its already strained resources.

Without a doubt, the community feels a threat to its present existence. Its past is an unpretentious history of farms and cattle and coastal schooners. Its place is one of bucolic isolation, Pacific storms, and a contentment with its geography. The basic concern, shared by almost everyone, is to preserve the identity of their community in the face of outside pressures for residential and recreational development.

Milton Rokeach, in his distinction between public attitudes and values, suggests that values "transcend specific objects and situations" and concern *"modes of conduct* and *end-states of existence"* (Rokeach, 1968, p. 550). It can be said, then, that the people of Bolinas share a common *value* concerning their "end-state of existence"; that is, they want to preserve their sense of place and character. On the other hand, opinions differ sharply within the community on the appropriate "modes of conduct" of its members. This stems primarily from the introduction and recent dominance of a new culture group in the community. With a different idea of what constitutes appropriate conduct, this group has created a climate of controversy within the community; but, importantly, a climate that has allowed basic questions to be raised about what in fact constitutes a desirable end-state of existence for the town. Thus, it is this very process of cultural transition which has sharpened and given added definition and meaning to the concepts touched on below, including, of course, coastal erosion.

For purposes of this discussion, we are making the assumption that the meaning(s) given to coastal erosion, or to any other concept for that matter, vary from one culture group to another (the notion of reference group is appropriate here: Shibutani, 1955; Kuhn, 1964). Thus, it becomes necessary to divide the community of Bolinas into culture groups and briefly note what meanings each group assigns to the concept of coastal erosion. Unquestionably, it is pretentious and often naive to divide into simple groups a community undergoing active cultural transition. Nevertheless, for this discussion, we can identify (1) the descendants of early settlers and longtime agricultural residents, (2) weekend and summer residents, (3) retired permanent residents, and (4) activist environmentalists. It must be emphasized that this simple division of the Bolinas community into four groups does not imply that each group is homogeneous in its value system or in its attitudes toward all environmental and social issues. It suggests only that the meanings given to the concept of coastal erosion are shared to a reasonably high degree among the members of each group.

There is no basis for making a separate group of the business people and entrepreneurs simply because of their small numbers. Only recently have absentee property owners formed an association to protect their voice in the community. They, also, are not considered here as a separate group.

1. Descendants of early settlers and longtime agricultural residents. Many of these people have understandably strong ties to the past and can document with varying degrees of accuracy changing conditions in the environment and in the community's life-style. One of the most important environmental changes that concerns them is the filling (siltation) of Bolinas Lagoon. They recall the days when crabs could be forked out of the deep channels and coastal schooners stopped in the lagoon for passengers and produce. Their collections of historical photographs are drawn upon to flesh out the community's sense of history and to document the cumulative effects of cliff erosion and lagoon siltation. Today, they see the lagoon as existing precariously on the edge of biological and hydrological senescence due to the large amounts of sediment deposited annually. There appears to be a near consensus within this group that this sediment is derived predominantly from the Bolinas cliffs and brought into the lagoon on the flood tides. Because few of them live on the cliffs themselves, there is no deep concern for the property of those in hazard zones 1 and 2. But there *is* concern for the role that cliff erosion is playing in the destruction of the lagoon. (This view is shared by many in other groups as well.)

2. Weekend and summer residents. These families reside permanently in communities within about two hours' drive of Bolinas: San Francisco, Berkeley, Santa Rosa, and the larger metropolitan Bay Area. Many of them have homes in the high- and medium-hazard zones and are concerned about cliff erosion as a direct threat to their property. While they are articulate and well educated, their absence from the day-to-day social interaction within the community ultimately lessens their political influence when it comes time to make important decisions.

Because of their life-style and due to the location of their homes, they are very much aware of the impact of weekend and summer tourism on Bolinas. Sightseers stream past their front doors in continuous lines of cars and beachcombers scramble up and down the unstable cliffs under their homes. In an earlier day, the excitement of weekend village life was the experience of seeing old friends and catching up on the week's news in front of the post office or general store. Now it is a rare Saturday or Sunday when a person can make his way through the stalled traffic to the village center. The increased leisure time and affluence of central California seem to be focused on their very doorstep.

Thus, while on the one hand this group might benefit greatly from a cliff stabilization program, the likelihood that it would bring additional people to Bolinas is distasteful. This seemingly ambivalent view becomes very important in the ultimate community choice about coastal erosion, summarized in the last section.

3. Retired permanent residents. This group comprises people who have recently moved to Bolinas for retirement after years of spending summers and weekends there as part of group 2, and people who have been permanent residents of the community before retirement, having strong ties with group 1. For the most part, these retired residents are property owners rather than renters, with fixed incomes and limited mobility.

While perhaps not the most vocal element in the community, they are certainly one of the most deeply concerned about the future of Bolinas, for they see themselves locked into that future. Unlike their friends in group 1, the retired people think less about the details of Bolinas's rich history than about the threats that future change may pose for them in their personal lives. Stability is their most precious amenity. This includes stability in cost of living and stability in community life-style, though there are some retired persons who feel that recent changes in the latter have added quality and variety to the village. Coastal erosion is important to them primarily in the way it affects their future taxes. That is, if taxes in Bolinas must be raised to support a cliff stabilization program such as that described earlier, they are generally against it. So, too, with assessments for a regional sewage treatment plant, an issue that is closely related to the problem of coastal erosion.

4. Activist environmentalists. Bolinas, like many small towns on the outer California coast, has always attracted intellectuals, artisans, and members of one or more countercultures. And, although it is admitted today only in private circles, it is a fact that these people have been an important element in forging the attractive character of Bolinas over the last half century. But, in the late 1960s (coincident with but not necessarily related to the "hip" exodus from San Francisco's Haight-Ashbury district) increasing numbers of foot-loose, long-haired young people began drifting in and out of the town, mostly on the waves of surfers who come and go with the tides. The majority of them brought little to the community in those first days, for they were for the most part transient and temporary. Still, the town became very sensitive to these new people because of their large numbers and increasingly divergent life-style.

Among them, but virtually inseparable from the crowd, came a small number of well-educated, intelligent persons (not exclusively young) who wanted to remain in Bolinas and become members of this small community which they saw existing in an incredibly beautiful setting. The Bolinas Lagoon, however, offered more than just beauty; it appealed to those with an interest in ecology along both intellectual and esthetic dimensions. For these people, the lagoon was a changing and complex system, but at the same time it was a symbol that focused the vague rhetoric of the turn-of-the-decade ecological revolution into meaning for their day-to-day life. (The role of Bolinas Lagoon as a powerful environmental icon was further elaborated during the furious defense of the lagoon against an oil spill from the collision of two Standard Oil Company tankers in the straits of the Golden Gate.) The values of this group involved a commitment to an ecological way of life that was to eventually antagonize the larger community.

The transformation from small culture group to politically powerful community coalition began for these people with the controversy over Bolinas Lagoon and culminated with the divisive issue of the regional sewage disposal system. The question of coastal erosion was intimately involved in both of these debates. And, the meaning given to coastal erosion in this context was not just a function of its ecological relationships to such processes as lagoon siltation. It was part and parcel of the way the activists perceived the values of each of the other groups. And vice versa of course.

The Bolinas Lagoon controversy was over a plan for the development of a marina in the lagoon. This was part of a larger plan to erect jetties at the lagoon inlet and to stabilize the cliffs with seawalls and groins. To have a marina, said the plan, sediment transport from the cliffs into the lagoon must be halted. Many who were concerned about the rapid siltation of the lagoon thought that the proposed marina dredging and cliff stabilization programs were appropriate measures to be taken. Furthermore, residents who viewed Bolinas historically and geographically as a coastal harbor with an already rich maritime history thought that a marina was consistent with the town's real character. For others, there was also the profit motive.

The elaborate argument in support of the marina drew heavily upon the idea of coastal erosion as an environmental hazard which posed a threat to the whole community even though only a minority were directly affected by erosion of the cliffs. However, the environmentalists rebutted that coastal erosion was not a natural hazard but a natural process, and so was lagoon sedimentation. Neither should be tampered with. Of great importance to our understanding of context, here, is the fact that the argument for preservation of natural processes came into direct conflict with the concept of preserving private property (on the cliffs) and of enhancing private profits (from the marina), two ideas that were rejected as quite alien to the environmentalist value system.

The controversy over a regional sewage treatment plant allowed leaders of the environmentalist group to take a majority of the seats on the Bolinas Public Utilities Board in a single election (1971). This board is the one most important agency with respect to environmental control in Bolinas. Again, coastal erosion was an issue: a portion of Bolinas, in the vicinity of the village center (point A, Fig. 9–2), is sewered directly into the lagoon. This produces an average discharge of untreated waste to the estuary on the order of 40,000 gallons per day. The town had been under a cease-and-desist order from the state of California for several years, and, on the heels of the marina controversy, a plan was generated that called for an $8 million treatment plant to serve the whole community. An important element in the argument for this plan was that sewering Bolinas Mesa (point C, Fig. 9–2) would eliminate existing septic tank seepage to the cliffs, thereby retarding erosion.

The environmentalists, many of whom lived in conventional and communal families on Bolinas Mesa, saw this as an unsupportable contention and worked avidly to rebut it. But hard data were not available to prove or disprove that septic tank seepage was contributing to erosion of the cliffs. Thus, the point of attack became, instead, the assumption that cliff erosion should be considered in the sewage treatment debate.

One of the prime concerns of the environmentalists was the potential effect of the treatment system's ocean outfall on the nearshore biota, particularly the intertidal community on Duxbury Reef. (The outfall was planned for a location just south of Bolinas Point.) An alternative plan offered by this group involved the complete recycling of waste and the ultimate use of reclaimed fecal matter for fertilizer. This fit nicely into their conception of an ecologically sound subsistence agricultural community. Important to the whole debate, however, was a growing awareness that in the end a large sewage treatment system would only encourage growth of the community beyond limits which many thought were appropriate. One had to acknowledge that there would be a strong tendency to spread assessments for the treatment system over a larger and larger future population in order to minimize per capita costs.

The environmentalists appealed to the community at large on this last point and came into political power by gaining control of the Public Utilities District. Their appeal was aimed at the underlying desire, shared by almost everyone, for preserving the size and character of the community. Upon this shared value, they built an argument that was ecologically sound (no ocean outfall, recycling of waste) and economically palatable (smaller treatment facility, leaving the Bolinas Mesa on septic tanks). Their conception placed coastal erosion, which had heretofore been considered primarily as a "hazard" in the context of the marina debate, into the perspective of a natural process to which man must accommodate himself. This is not to say that each of the culture groups mentioned above now sees coastal erosion only as a natural process. It is simply that this view has been institutionalized in the community political process. We now look at the major policy decision which resulted from this new community view of coastal erosion.

The ultimate choice

In 1967, a reconnaissance study of the Bolinas cliff erosion problem was initiated by the U.S. Army Corps of Engineers (San Francisco District) at the request of local government. The conclusions of the study were simply that, while it was technically feasible to arrest cliff erosion, project costs were too high relative to the public benefits to be derived. Public benefits, in this case, were necessarily restricted to beach recreation because most of the property endangered by erosion was, in fact, private. In addition to an unfavorable benefit/cost ratio, the financial contribution required of the local residents was in excess of that normally thought to be appropriate by the federal government for this type of project.

In 1970, the study was reinitiated under section 110 of the River and Harbor Act for "beach erosion control and *related purposes*" (U.S. Army Corps of Engineers, 1970, p. 1, italics added). In that year, there was intense community pressure for reducing siltation rates in Bolinas Lagoon and it was thought that an important "related purpose" would be to cut off sediment supply by stabilizing the cliffs. The Corps was willing to consider reduction of sedimentation in the lagoon as an additional benefit and went so far as to agree to tie the beach erosion study to a study of remedial dredging in the lagoon for the sole purpose of increasing flushing action and retarding siltation. (This, interestingly, had to be considered by the Corps as an experimental dredging program because it had no continuing authority to conduct dredging for any purpose other than navigation.) In the early stages of this study, the Corps generated the costs and plans summarized on page 74 above. It did not take long, however, for the community to arrive at a major policy position with respect to coastal erosion. It decided to opt out of the federal program.

An early sign that this position was forming was apparent at the first public meeting on erosion protection held by the Corps of Engineers in Bolinas. This occurred shortly after the environmentalists had made their successful bid for political power. Recall that this bid was based strongly on the interpretation of cliff erosion and lagoon sedimentation not as natural hazards but as natural processes which should be allowed to continue unmolested; then it is no surprise to learn that only a dozen members of the community attended the hearing (U.S. Army Corps of Engineers, 1971). Public meetings on environmental issues in Bolinas traditionally draw 75 or more citizens, and Corps officials acknowledge that this was one of the lowest recorded attendances for a community of this size. Finally, as of this writing, the author is informed by the Corps of Engineers (Sustar, 1973) that Bolinas had decided to formally end consideration of federal assistance.

The community of Bolinas sees a federal cliff stabilization program as having two fundamental disadvantages: First, in order to achieve its objectives of cliff protection and to create, at the same time, public benefits, the effects of the project would be to bring to bear much greater recreational pressure on the community. The larger and more stable beach (Fig. 9–3) could very easily stimulate an order-of-magnitude increase in tourism at Bolinas. Second, there would be, again, a tendency to spread local assessments for the project out over a larger base, acting as an incentive for community growth; a point which was very important in the argument over sewage treatment. Thus, the community decided to place the burden for coastal erosion solely on its own shoulders in order to preserve the essential size and character of the community.

Recently, the town formed a community-wide planning group which is presently in the process of deciding specifically how to deal with coastal erosion as an element in the regional master plan. It is doubtful whether the plan will ultimately call for the acquisition of endangered property. Costs would be very high. Zoning, on the other hand, is viable only where lots are vacant. Realistically, the near future sees two holding actions:

1. In the vicinity of the Little Mesa where existing bulwarks and seawalls have been successful, in combination with cliff plantings and drainage pipes to preserve the present high angle of repose, small groups of residents directly affected by erosion may get widening support from the community in their continuing effort to maintain these protective devices. The degree of support will depend on the extent to which the community feels their larger policy decision has disadvantaged these cliff-dwelling residents. One who knows Bolinas is optimistic that the community will take over the responsibility for maintaining these bulwarks in spite of the fact that they serve to protect only private property.

2. Along the face of the Bolinas Mesa, a county-maintained road runs between the first row of exposed homes and the retreating cliff. The county of Marin is responsible for road maintenance and, one supposes, this agency will continue to take remedial action to retard recession of the cliffs. Eventually, though, the cliffs will recede to the first row of homes in spite of interim measures.

For decades, industrial societies have viewed natural phenomena that impinge upon our lives as hazards to our well-being. The traditional response has been to try to eliminate the hazards with the technology at hand. Dams are built and levees line the rivers inviting future generations to settle in the hazard zone where apparently all has been made safe. The justification of project costs is the land reclaimed from the hazard for human use. Thus settlement occurs and the scene is set for an even greater disaster.

The Bolinas decision is one of a few cases where a community has chosen to accommodate itself to the hazard, rejecting the large-scale technological fix. The people of this village saw elimination of the natural hazard as not only being too expensive for the community, but, furthermore, as an inducement to growth and tourism. In the end, these phenomena were the real hazards.

Acknowledgments

The author wishes to express his appreciation to Mrs. Frances Stewart and Mr. Peter Warshall of Bolinas for assisting in this study. The initial interviews were conducted by Lore Dobler, Linda Griffin, and Bruce Hamilton.

References

Blumer, Herbert. (1962) "Society as symbolic interaction." In A. M. Rose, ed., *Human Behavior and Social Processes.* Boston: Houghton Mifflin, pp. 179–92.

Bogdan, Robert. (1972) *Participant Observation in Organizational Settings.* Syracuse: Syracuse University Press, p. 106.

Bruyn, Severyn T. (1970) "The methodology of participant observation." In W. J. Filstead, ed., *Qualitative Methodology*, pp. 305–27.

Galloway, Alan. (1972), California Academy of Sciences, San Francisco. Personal communication.

Kuhn, Manford H. (1964) "The reference group reconsidered." *Sociological Quarterly* 5:6–21.

Ritter, John. (1969) "Preliminary studies of sedimentation and hydrology in Bolinas Lagoon." Menlo Park, Calif. U.S. Geological Survey Open File Report.

Rokeach, Milton. (1968) "The role of values in public opinion research." *Public Opinion Quarterly* 32:547–59.

Shibutani, Tamotsu. (1955) "Reference groups as perspectives." *American Journal of Sociology* 60:562–69.

Sustar, John. (1973) Project Engineer, Bolinas Beach erosion study., San Francisco: U.S. Army Corps of Engineers. Personal communication.

U.S. Army Corps of Engineers. (1970) Public brochure, "Beach erosion study, Bolinas, California." San Francisco.

——. (1971) Transcript of public hearing, "Bolinas Beach erosion study." San Francisco.

U.S. Coast and Geodetic Survey. (1929) Topographic Map No. 4520, "General vicinity of Duxbury Point." Washington, D.C.

10. Windstorms: a case study of wind hazards for Boulder, Colorado

D. J. MILLER, W. A. R. BRINKMANN
AND R. G. BARRY
University of Colorado

Introduction

Severe windstorms have occurred, on average, more than once per year in Boulder since observations began (Julian and Julian, 1969; Whiteman, 1973). The winds, which are downslope on the eastern side of the Rockies, are frequently referred to as chinooks (or foehns) but detailed analysis of their characteristics shows that in many cases they do not fit the classical definitions of such winds (Brinkmann, 1971, 1973). However, since their velocity characteristics are not unlike those of winds in the lee of mountain ranges in many parts of the world, a study of the wind hazard and responses to it in Boulder is of more than local interest. We begin by outlining the characteristics of these windstorms and then proceed to examine the damage caused in the severe storms of January 1969 and, finally, the responses of the population and city management to this hazard.

Physical characteristics

To determine the detailed physical characteristics of Boulder's windstorms, data from nine stations in and around the city for the three winters 1968–71 have been analyzed. The period of data coverage varied slightly depending on the instrument availability. A windstorm period is defined as one during which maximum speeds at the Boulder stations exceed 22 meters per second (50 mph), with at least one station recording a gust of hurricane force—over 33 meters per second (73 mph)—at a height of 3.4 meters (11 feet) above ground.

Using this criterion, a total of 20 windstorms occurred over the three winters; 40 percent of the cases occurred in January (Table 10–1), in line with the findings of Julian and Julian (1969) and Whiteman (1973) that January is the prime month for windstorms in Boulder. Strong winds occur infrequently in the summer, and these may be of a different origin. It is also interesting to note that the frequency in the individual winters was 5, 3, and 12. The occurrence of large interannual variations is borne out by the other two studies.

Diurnal variability was determined from the number of occurrences of 5-minute mean speeds of 15 meters per second (34 mph) or more and gusts of 25 meters per second (56 mph) or more summed over 15-minute intervals to smooth the histograms. An example of the results, for the most exposed station in Boulder (Fig. 10–1), clearly shows that strong winds and high gusts in

Table 10–1. Annual variation of windstorms in Boulder, Colorado

Month	Distribution of the cases examined by Brinkmann (1973)		Distribution 1906–69 according to Julian and Julian (1969)	
	No.	%	No.	%
September	3	4		
October	4	5		
November	16	21	3	15
December	10	13	2	10
January	22	29	8	40
February	5	7	3	15
March	11	15	3	15
April	4	5	1	5
May	1	1		

Boulder occur most frequently between about midnight and 10:00 A.M.

An outstanding characteristic of Boulder's windstorms is not only the hurricane force speeds but also their extreme gustiness. Fig. 10–2 shows anemometer

Fig. 10–1. Frequency of wind gusts ≥25m s^{-1} (top) and 5-minute mean speeds ≥15m s^{-1} (bottom) plotted for 15-minute intervals for twenty windstorms at Southern Hills, Boulder

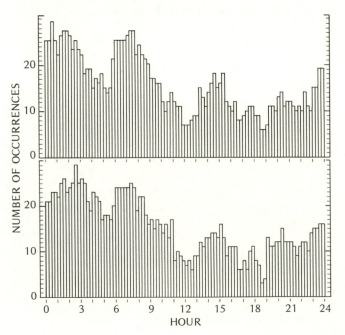

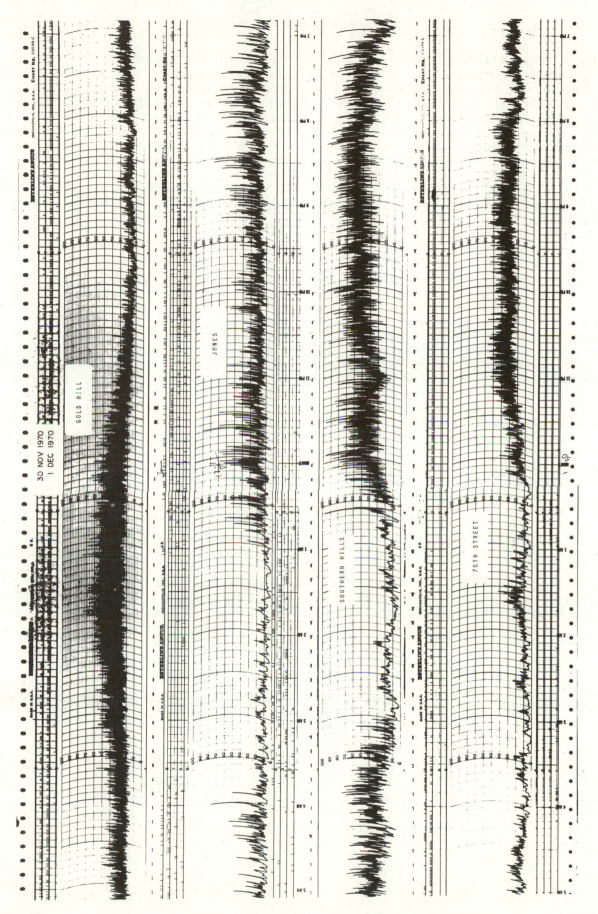

Fig. 10–2. Anemometer traces from stations on the east slope of the Front Range, Colorado, November 30–December 1, 1970. Time is read from right to left on the abscissa; the ordinate scale of wind is mph

traces for November 30–December 1, 1970, from four stations, arranged with respect to location of the instruments: the uppermost one, Gold Hill, is from a mountain station 980 meters (3,200 feet) above and 13 kilometers (8 miles) west of Boulder; Jones and Southern Hills are located in Boulder (the former below a steep escarpment, the latter having a more open exposure and thus being more representative of conditions in Boulder); and the bottom trace, 76th Street, is from a station 9 kilometers (6 miles) east of Boulder. Examination of these traces shows that gustiness is especially marked in Boulder, at the foot of the mountains, where the mean speeds are also highest. The most striking feature is the rapid and frequent fluctuation of the wind speed from 7–10 meters per second (16–22 mph) to 35–40 meters per second (78–90 mph) or more during the storm periods. It is also apparent that there is an out-of-phase relationship between the occurrence of the windstorm in Boulder and at the mountain stations. This is a manifestation of "moving" maxima related to the dynamics of the windstorms (Brinkmann, 1973).

Gustiness is commonly assessed by the gust factor, which is defined as the ratio of the peak gust to the mean speed for a given averaging period. The peak gust duration is typically 3–5 seconds. Lettau and Haugen (1960) calculate mean and maximum gust factors using data from a number of sources, assuming that differences of instrumentation and exposure heights do not affect the representativeness in terms of average gustiness conditions. Their results are shown in Fig. 10–3 together with mean and maximum gust factors for seven stations in and around Boulder, averaged over all 20 windstorm cases. The curves show the usual decrease in gust factor with increasing mean wind speed, which is due to a slower increase in gust as compared with the

mean speed, and the small variations of the gust factor with high mean speeds (Davis and Newstein, 1968). The large differences between the curves in Fig. 10–3 clearly demonstrate the much higher than average gustiness of the Boulder windstorms. Comparison with wind-speed traces during hurricane passage at coastal stations also indicates the greater gustiness of Boulder's winds, because of higher mean wind speeds but similar maxima in hurricanes.

Many of the downslope winds occurring in different parts of the world are described as extremely gusty. For comparative purposes, anemometer traces were obtained for downslope winds in the lee of the Pennines, Great Britain, and on Hokkaido, Japan, as well as for a foehn (warm wind) at Altdorf, Switzerland, and a bora (cold wind) at Dubrovnik, Yugoslavia. These traces permitted the abstraction of gust and mean speeds for 1-hour intervals. The gust factors for these cases are shown in Fig. 10–4 together with mean gust factors for the Boulder area (recalculated for the same averaging period). Considering the differences in instrumentation and anemometer heights the gustiness characteristics of Boulder's windstorms are apparently very similar to those of other warm and cold storms around the world.

Another feature of Boulder's storms, which is common to most downslope winds, is the extremely low relative humidity. Drops to 20 percent or 10 percent are a common occurrence in Boulder in winter, setting the stage for fires which are difficult to control during strong and gusty winds.

Wind effects on buildings

While the action of wind is of obvious concern to the structural engineer, there is no simple basis for assessing

Fig. 10–3. Mean and maximum gust factors plotted against 5-minute mean speeds for Boulder and corresponding averages given by Lettau and Haugen (1960)

Fig. 10–4. Gust factors plotted against hourly mean speeds for Boulder and other stations around the world during windstorms

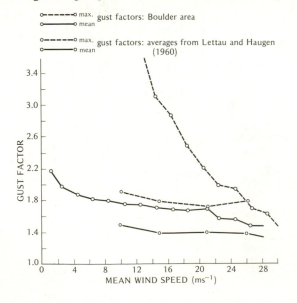

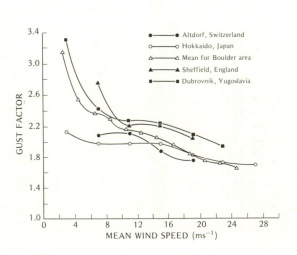

its likely impact. The pressure (P) in millibars exerted by wind on a building is proportional to the square of the wind speed (V^2) in meters per second (Peyronnin, 1962):

$$P = 0.5\rho(kV^2) \times 10^{-2}$$

where k = the gust factor and ρ = air density in kilograms per cubic meter. Since Boulder is situated at 1,600 meters (5,250 ft.) elevation, the 15 percent reduction in air density reduces this wind pressure by about 8 percent. Building regulations in the United States generally allow for wind speeds of 40–45 meters per second (90–100 mph) by specifying the ability to withstand pressures of 12–14 millibars (25–30 pounds per square foot); Boulder's code specifies 12 millibars. However, the total pressure in the lee of an object perpendicular to the wind is increased by the added suction pressure. For a signboard this may amount to a further 10 percent and for a building a further 60 percent (Thomas, 1930, cited by Brooks, 1950). Moreover, the pressure differential between the walls of a building is unaffected by the offsetting reduction of air density.

The design working loading of a material is usually well below its collapse loading, but the arrangement of buildings may significantly affect wind loads and stresses so that leeward buildings experience greater loads and stresses than windward or isolated ones. This effect was dramatically demonstrated in the collapse of cooling towers in a gale in Yorkshire in 1965 (Shellard, 1967).

Damage caused by storms

Losses

A windstorm began in Boulder in the late afternoon of January 7, 1969, and continued until about 11:00 P.M. There were at least three gusts exceeding 45 meters per second (100 miles per hour) at 3.4 meters, with many instances of a wind increase from 5 meters per second to over 40 meters per second in a matter of seconds. It is estimated that almost half of the residences in Boulder suffered damage and that the losses were of the order of $1–$1.5 million. In addition to residential property damage, buildings and scientific equipment installations of NOAA in the area suffered losses of about $0.5 million, 6 light aircraft worth $250,000 were destroyed by wind or fire at Boulder Municipal Airport and a total of 14 others there and at Jefferson County Airport were damaged, and more than 100 trees and 20 main powerline poles were brought down. Two deaths were attributed to the winds and 21 injury cases required hospital treatment. The winds also triggered four brush fires in and around the city, one of which was brought under control only a few meters from cabin residences.

A further severe, but shorter windstorm occurred on the morning of January 31, 1969. Following these two storms, a committee set up by the City Manager's Office carried out a survey of residential property damage (City Manager's Office, 1970) for a random 10 percent sample of the 10,640 single-family residences in Boulder. A 71 percent return of questionnaires was achieved and the results were analyzed by the TAXIR information retrieval system (Brill, 1971).

Table 10–2 summarizes the information on damage and losses to property. On this basis, it can be estimated that approximately 4,700 homes in the city received damage in the first storm and 1,575 in the second. Such instances are apparently not rare; windstorms causing severe storm on January 11, 1972, resulted in an estimated $2.5 million property damage (Lilly and Zipser, 1972), primarily to mobile trailer homes in Boulder; wind damage was also estimated at $0.5 million on January 15–16, 1967, and at $0.5–$0.75 million over a 2-week period in December 1964. These figures may be compared with an estimated current range of annual losses on insured property in the United States, due to extratropical windstorms, of $25–$50 million (Hendrick and Friedman, 1966).

In an attempt to uncover evidence of any spatial patterns of damage in Boulder, approximately 20 percent (665) of all insurance claims exceeding $25 for single-dwelling residences filed with respect to the January 7 storm were examined. (The 20 percent estimate is substantiated by the almost exact agreement between the number of claims exceeding $1,500, multiplied by 5, with that estimated from the city's 10 percent random survey.) Although maximum losses were sustained in the southern part of the city, it is not considered that this reflects any topographically determined characteristic of the windstorms. Reports in the Boulder *Daily Camera* from the turn of the century demonstrate the occurrence of damage in the town center. The comparative absence of shelter may be the only common factor; some evidence for this is discussed below.

The 665 insurance claims totaled $243,600, giving an average of $366 and an estimated total for the city of $1.2 million. An analysis by districts, mainly in the southern half of the city, where there were enough cases to obtain meaningful averages, indicates that the average claim ranged from $120 to $1,777 for 20 districts. These figures represent 0.6 to 9.6 percent of the improvement

Table 10–2. Damage to single-family residences in Boulder (based on 754 questionnaires)

Damage	January 7, 1969 (%)	January 31, 1969 (%)
0	55.5	85.2
1–100	26.7	12.3
101–500	13.2	2.0
501–1,500	3.8	0.4
> 1,500	0.8	0.1
Total	100.0	100.0

value of the lot (i.e., the house and other constructions). For most of the districts the figure was between 1 and 2 percent.

Types of property damage

For the January 7 storm structural damage (loss of a portion of the roof, walls or ceiling moved or cracked, exterior wall siding damage) was reported in just over 5 percent of the houses surveyed by the city. As might be expected, 71 percent of this (excluding roofs) affected the westerly side of the houses. "Minor" damage was widespread: 30 percent of houses with shingle roofs, which represent 90 percent of all houses in the survey, lost some shingles. Moreover, where the gable pitch was shallow (less than 1:3) the proportion of houses so affected was almost twice as great as for houses with steep gables, in conformity with design expectations. The survey showed that the orientation of the house was insignificant (although as we shall see this is not the case with trailer homes), but the proximity of another building on the west side, between southwest and northwest, apparently affords some protection from loss of shingles. 35 percent of houses without such protection lost shingles whereas only 21 percent of those with protection did so.

24 percent of the houses in the survey received window damage and, significantly, less than 4 percent of the respondents indicated that their windows were covered by shutters! Although the direct cost of window damage is usually below $100, associated damage to furnishings and the risk of personal injury make it a significant item. The survey showed that a considerable amount of debris (shingles, sand and gravel, fencing material, and loose items stored outside including trash cans and lids, lumber, etc.) was airborne during the storm. Particular hazards include the existence to the west of the property of open ground, buildings under construction, and building with gravel roofs, although it is usually impossible to isolate the exact cause of window or other damage in any specific instance.

The main conclusions of the report were that much storm damage can be avoided by installing shutters on west-facing windows and by the regulation of strength standards for wooden fences.

Hazard perception

Perception of the wind hazard in Boulder was analyzed on the basis of a survey of 120 households conducted in 1970 (Miller, 1972) and subsequent discussions with local officials. The survey provided a systematic sample of 40 houses in each of four geographical areas of the city (north, south, east-central, west-central). The areas were stratified into three income-based categories (high, middle, and low) with 10 houses in each group. The 120 respondents were all female, primarily housewives. Fifty-one percent of the sample reported some damage

in 1969 (67.5 percent less than $100, 9 percent greater than $500), so it may be considered representative of the city as a whole. All of those sampled perceived the winds as a hazard, and all but two enacted adjustments designed to reduce this hazard.

Two separate assessments of hazard were made. The individual was asked to rate the hazard at his home as being high, moderate, or low, and the investigator, using previous damage reports as a guide, ranked each household also. Interestingly, the respondents felt that they were more subject to high velocities, and less to moderate velocities, than the investigator. Perhaps, this is attributable to the fact that the individual homeowner has more personal and direct experience of the hazard. In addition, the investigator was limited to structural damages as an indicator of severity, whereas the homeowner based his assessment on several variables, as seen in Table 10–3.

Only a fifth of the sample considered structural damages as being the most common disruption, indicating that the nonphysical manifestations of the windstorms determined the perception of the severity of the hazard. Examining this further, we find that most of those ranking the disruption as substantial reported anxiety as the dominant effect (Table 10–4), while those ranking the disruption as slight regarded structural damage as the major factor. It is apparent that the structural damages (or ones noticeable to an outsider) may not be reliable indicators of the overall effects. Those noticed only by the individuals being affected tend to serve as more accurate indicators of actual disruption and, in turn, severity.

Table 10–3. Types of disruption reported by 120 respondents

Type of disruption	% of Sample
Nothing	26.7
Structural damage	20.0
Activity	14.2
Anxiety	31.7
Other	7.5

Table 10–4. Type of disruption versus its perceived extent

Type of disruption	% of total reporting an effect[a]	Sub-stantial	Slight	None
Anxiety	43.2	19.3	17.0	6.8
Structural damage	27.5	4.5	21.6	1.1
Activity	19.3	5.7	5.7	8.0
Other	10.1	5.7	6.5	—

[a]The 26.7 percent of the sample who reported no effect are excluded.

Adjustments

Various adjustments reported in the survey were peculiar to one household. For purposes of generalization, therefore, the adjustments enacted have been grouped into the following four classes:

1. Actions taken to prevent structural damages or bodily injury during a windstorm.
2. Structural modifications made prior to a specific windstorm.
3. Actions designed to lessen wind-derived tension.
4. Relocation of inhabitants both within and outside the home.

Insurance was excluded since, first, it is not initiated solely as a means of alleviating wind damages and, second, 94 percent of the sample were certain they had insurance in force that would cover wind damages, with the remainder stating they did not know.

Two households, possessing no peculiar traits and initiating no specific type of adjustments, are excluded from the analysis. The remainder of the sample (118 households) initiated category 1, either alone or in conjunction with other types of adjustments. Category 2 is of special significance since prior planning is inherent in its use, whereas the others are immediate adjustments requiring no prior action. Eight combinations of the four adjustment categories can be derived from the data, and further grouped into three distinct classes, as shown in Table 10–5—using adjustment category 2 as the distinguishing factor: (I) the initiation of 2 with 1, (II) 1 alone, and (III) 1 with others excluding 2. About 27 percent of the households rely only on taking emergency action to prevent damage; more than 41 percent in addition try to lessen tension and move within or outside the house. The remaining 32 percent take steps to reduce damages before a windstorm is predicted.

The relationship between the adjustment classes and other characteristics of the households is summarized in Fig. 10–5. Six variables proved to be statistically significant as determinants of the type of adjustments enacted: location within the city; severity; the number of years lived in a specific home; the acceptance of a means

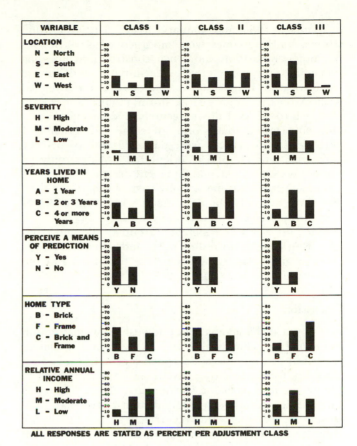

Fig. 10–5. Adjustment classes (see Table 10–5) and percentage responses for six significantly related variables

of prediction; the type of home; and relative annual income. The first two are apparently the most important although this may only be a reflection of the more severe damage occurring in the southern part of the city in January 1969. The survey divided the city into sectors based upon income data and four of these were selected for examination. The area with the highest severity is characterized by adjustment class III, the most complex, while the area with the lowest severity shows no dominant adjustment class. It appears that as perceived severity increases, so does the complexity of adjustments enacted. With respect to income, 90 percent of the high-income-group homes have been structurally modified compared with only 50 percent of those in the medium-income group. This may account for the finding that the latter view the hazard as more acute.

Hazard abatement measures

Apart from individual adjustments, efforts have been made by local government to minimize the hazard. Action to reinforce the building regulations seems to have been prompted by the highly damaging windstorms

Table 10–5. Adjustment classes

Adjustment class	Combinations of Adjustment Categories	% of households practicing
I	1 Alone	27.1
II	1 and 3 1 and 4 1, 3, and 4	41.5
III	1 and 2 1, 2, and 3 1, 2, and 4 1, 2, 3, and 4	31.4

of January 1969 and January 1972. After the 1969 storms, building codes were modified with restrictions on roof composition and fence construction; licensing procedures for contractors were initiated and provide for revocation if laws and regulations are not complied with. After the January 1972 storm, mobile homes were required to be tied down securely and in many cases blocked. During that storm, of some 182 orientated north-south 36 percent were damaged and 10 percent destroyed, whereas of 138 orientated east-west only 17 percent were damaged and 3 percent destroyed.

The control measures are obviously a step in the right direction, but only the next major windstorm will provide a proper test of their effectiveness.

Since there is a considerable hazard to life and property from broken powerline poles and associated power outages, appropriate actions such as the burying of the lines could be a next step.

Discussion

The physical characteristics of downslope windstorms in Boulder (and climatically and topographically similar localities) can now be enumerated succinctly and a firm basis for forecasting these events more reliably than in the past is now available (Brinkmann, 1973; Klemp and Lilly, 1973). It has been amply demonstrated that windstorms are a frequent and recurring feature of Boulder's climate. Informed awareness of the hazard, coupled with improved forecasting, should help to reduce the evident anxiety which a windstorm evokes. Simple preventive measures, such as the use of shutters on windows, can eliminate much minor damage while adequate building regulations and code enforcement should minimize structural damage, except perhaps during the most extreme events.

References

Brill, R. C. (1971) "The TAXIR Primer." Boulder: University of Colorado, INSTAAR Occasional Paper No. 1

Brinkmann, W. A. R. (1971) "What is a foehn?" *Weather* 26: 230–39.

———. (1973) "A climatological study of strong downslope winds in the Boulder area." Boulder: University of Colorado, INSTAAR Occasional Paper No. 7/NCAR Cooperative Thesis No. 27.

Brooks, C. E. P. (1950) *Climate in everyday life.* London: Ernest Benn.

City Manager's Office. (1970) "Analysis of damage to residential properties from high winds occurring on January 7 and January 31, 1969 in Boulder, Colorado." City of Boulder, mimeo.

Davis, F. K., and Newstein, H. (1968) "The variation of gust factors with mean wind speed and with height." *Journal of Applied Meteorology* 7 (3):372–78.

Hendrick, R. L., and Friedman, D. G. (1966) "Potential impacts of storm modification on the insurance industry." In W. R. D. Sewell, ed. *Human Dimensions of Weather Modification.* Chicago: University of Chicago, Department of Geography, Research Paper No. 105, pp. 227–46.

Julian, L. T., and Julian, P. R. (1969) "Boulder's winds." *Weatherwise* 22:108–12, 126.

Klemp, J. B., and Lilly, D. K. (1973) "A mechanism for the generation of downslope windstorms." Boulder, Colo.: National Center for Atmospheric Research, unpublished. (submitted to *Quart. J. Roy. Met. Soc.*).

Lettau, H. H., and Haugen, D. A. (1960) "Wind." In *Handbook of Geophysics.* New York: Macmillan, pp. 1–8.

Lilly, D. K., and Zipser, E. J. (1972) "The Front Range windstorm of 11 January 1972: a meteorological narrative." *Weatherwise* 25:56–63.

Miller, D. J. (1972) "Human perception of and adjustments to the high wind hazard in Boulder, Colorado." Boulder: University of Colorado, Department of Geography, unpublished M.A. thesis.

Peyronnin, C. A., Jr. (1962) "Designing for hurricanes." *Mechanical Engineering* 84 (9):40–45.

Shellard, H. C. (1967) "Collapse of cooling towers in a gale, Ferrybridge, November 1965." *Weather* 22(6):232–40.

Thomas, A. M. (1930) "Memorandum on high wind pressures on tall structures." British Electrical Allied Industries Research Association, Technical Report F/T 42.

Whiteman, C. D. (1973) Personal communication.

11. Drought in Eastern Kenya: nutritional status and farmer activity

BENJAMIN WISNER
University of Dar es Salaam

PHILIP M. MBITHI
University of Nairobi

The reality of drought

Twice during 1970–71 the rains in many parts of Kenya reached only 50–75 percent of the long-term average. Crop failure, livestock death, or serious production losses directly affected the livelihood of about 3 million Kenyans. At the peak of the ensuing famine nearly a tenth of these persons were supported by famine relief. For the nation as a whole, the impact of drought was mediated by the presence of a highly developed highland economy where the direct effects of drought are seldom felt. Indeed, Kenya's generous mixture of ex-

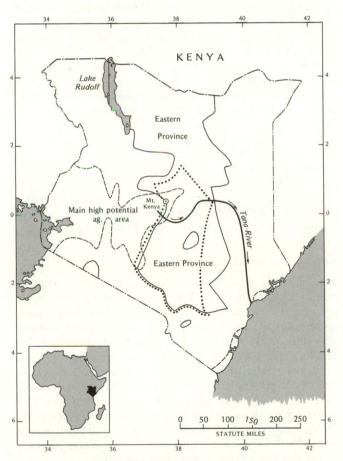

Fig. 11–1. Eastern Province and study area within it. Areas of high-potential agricultural land are delimited by dashes. Study area enlarged in Figure 11–2 is delimited by dots

tremes of environment and degrees of rural development make the national assessment of drought particularly difficult. Recognizing the difficulty of national assessment of drought's significance, this paper reviews the 1970–71 lessons of drought in the light of the history of drought and drought policy in Kenya over the past 20 years.

The focus is on the individual farm family and small community in drought-affected areas. Because it is difficult to grasp the concrete reality of drought at such a level of disaggregation, the analysis begins with a description of what it means to the peasant farmer and his family and how their lives are subtly altered in some ways, violently wrenched in others. The 17 study sites in eastern Kenya were chosen to contain nearly the full range of environmental and economic variations found in Kenya (Fig. 11–1). They range from "high-potential" coffee-tea zones on the slopes of Mount Kenya, through "medium-potential" cotton-maize zones, and onto the semiarid plains of the upper Tana River Basin where millet, goats, and thornbush dominate the landscape (Pratt, Greenway, and Gwynne, 1966). Further details will be found in other papers by the authors (Mbithi and Wisner, 1972; Wisner, 1972a, 1972b).

For the sake of simplicity portraits are taken from three sites inhabited by farmers of the Kamba tribe, one of the six tribal groups considered in the study as a whole. This Kamba gradient contains examples of both extremes of environment presented by the range of study sites and one intermediate site. The three sites shown in Fig. 11–2 have the characteristics shown in Table 11–1.

Katse: dry-river highways leading through the Thorn Bush

The Elliot's Bread truck grinds its way along the stony road 64.4 kilometers (40 miles) north of the administrative center of the Far Northern Division of Kitui District, a favorite rest stop for the army convoys going to and from Garissa in the northeast. Drier as you draw nearer the Tana River, acacia gives way to dry thorn scrub and huge, lonely outcrops of rock standing in the sandy soil of the ancient basement complex. Now and then the truck passes a small boy herding goats and a few cattle. It is difficult to drive livestock through the bush, so boys prefer to follow the dry riverbeds in search of the remaining grass. The truck shudders into second gear as it plummets into a drift across a small river. When the rains are heavy even larger trucks have been washed away while fording such streams. Now

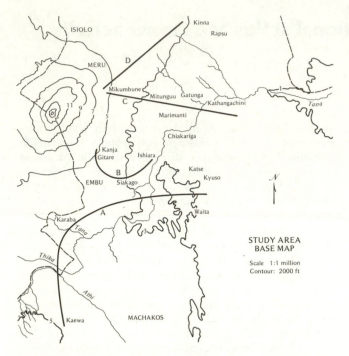

Fig. 11–2. Study area base map showing sites and major environmental gradients. Gradients: A-Kamba; B-Embu; C-Meru; D-Northern Frontier.

Table 11–1. Three sites on the Kamba gradient

Site name	District	Elevation above sea level (m)	Rainfall (mm per annum)	Standard deviation of rainfall in %
Katse	Kitui	750	565	45
Karaba	Embu	1,200	762	35
Kaewa	Machakos	1,650	1,270	25

women dig 1.8 meters (6 feet) deep in the sand and lift out water. Donkeys and children with small hand carts stand waiting.

Many people in Katse have been on famine relief for months. The rains have failed for two seasons. Munyasia knows the routine well by now. In 1961, 1965, and again this year he went to Runyenges, high on the Embu side of Mount Kenya, to work on farms to buy food for his family. He finds he has to pay as much as 70 shillings for a bag of maize, more than twice the normal price. For this reason some of his neighbors prefer to be paid in kind for their casual labor. Munyasia has no cattle left, and has only planted 0.6 hectare (1½ acres) of cowpeas and bulrush millet this season. The routine of distant farm labor for the head of household and famine relief for his dependents has little to recommend it. If he walks to Runyenges, it is 55 kilometers (35 miles), 240 kilometers (150 miles) by bus.

For generations the Kamba have moved about ex-

changing goods and services with kinsmen and people of allied tribes during hard times. Marriages still take place, social visits are paid, small civil cases are heard, beer is still brewed. Generations of Kamba have learned well the unwritten lessons of dry farming in the area—intercropping, drought-resistant crops, staggered planting, large shifting fields. And there is still land available.

Sometimes extreme drought challenges even the unwritten teaching. This season Kaugi managed to plant only 1.2 hectares (3 acres)—sorghum, bulrush millet, and local maize, since he was unable to get early-maturing seed from the government. In 1965 he had to go find work high on the Embu side of Mount Kenya, about as far away as Runyenges. In 1961 and this year he was able to get enough money from sales of livestock and cattle hides to make ends meet. He has a 14-month-old daughter who is fortunate still to be on the breast. She weighs 90 percent of the standard weight for her age. "Standard weight" refers to Harvard Standard Weight-for-Age (Jelliffe, 1966). Had she been weaned during this drought, one would have had to expect the worst, especially since cholera swept close by northern Kitui midway through the drought.

Simeon with no livestock, has interplanted 2.4 hectares (6 acres) of the more drought-resistant crops: bulrush millet, sorghum, cowpeas, pigeon peas, and green grams. In 1961 and 1965 he sold livestock, like Kaugi. Now he has a job as a night watchman in Nairobi, so he is able to send money back to his wife. His 17-month-old son has been weaned and is 90 percent of his standard weight. This is an accomplishment considering that more than a third of the children in this area are near or below the 70 percent mark, where clinical signs of malnutrition are quite obvious. However, this accomplishment costs him long absences from his family.

Karaba: black cotton soil and spice

The smell is everywhere. Women, children threshing oceans of spice for sale: for purchase of grain this year, for school fees, school uniforms, and taxes next season if the rains are good. The drought has not hurt Mutuku very much. He has bred his own variety of maize by crossing the government's early-maturing variety with local maize. He interplanted this with coriander, yellow and black grams, and beans on two 2-hectare (5-acre) plots which he cultivated with a rented tractor. The maize harvest was not good, but there was money from the sale of spice and grams to buy flour. He has sent some food to relatives who live on the Yatta plateau, across the Tana River in northern Machakos.

He is fortunate. The black cotton soil holds moisture well; the market for spice and grams is well established. Of the eight people living in his household, five are strong adult workers. His eight cattle are safely off near the river and some are at the grazing scheme 16 kilometers (10 miles) away, where there is still enough grass. All three children under 5 in the household are above 80 percent of the standard weight.

There have been worse times. In 1965 Mutuku had to work as a laborer on the huge famous Mwea-Tebere rice irrigation scheme just a few miles north of Karaba. Now he sits outside the New High Life beer club in the smell of spice and evening light and talks it over with other Kamba who have immigrated to Karaba since the early 1940s. Goats and ancient tractors pass by slowly in the dust. Water sellers wheel steel drums on unsteady handcarts. A queue (line) has formed in front of Old Man Kisoa's butchery. Around the bumpy dusty square of half-finished shops and beer clubs, women thresh spice. The sun glints off the windows of the teacher's house behind a "living fence" made of euphorbia; the drinkers chat: "You can usually tell if the rains are going to be poor. If August and September are cloudy . . . " "And if the lightning in October is weak . . . " "Or if the Masolo birds do not appear early."

Most of these men have had to work as casual laborers in the past when the rains have been unusual. They don't go far away to work; usually down to the rice scheme. As soon as the rains break, there is a flurry of activity as they work the heavy black soil and plant as much as quickly as they can. Many have a few cattle which they keep in camps as far away as Yatta. And then, of course, there is the spice.

Kaewa: women in the valley bottoms

The map sings like a blind elder, sweet half-truths drowned in honey wine, whine of a stringed instrument. Try to tell Mrs. Imara that she lives in Ministry of Agriculture inspected, measured, mapped, and certified high ecological potential, agroeconomic zone, Iveti hills, central Machakos!

Many factors in the highland increase the vulnerability of the people to drought despite the favorable soil and annual average rainfall. Iveti is very densely peopled (200–300 per square kilometer, 500–750 per square mile). The farm plots are small and fragmented. Quite a few men have left the area to take up jobs in Nairobi, which is only 80 kilometers (50 miles) away (Redlich, 1971). In Kombo sublocation, bordering the Kaewa study site, from 48 to 72 percent of the families have a member employed outside the sublocation, depending on the "village" within the sublocation. Others have migrated down into the plains to the east in an attempt to find more land. As a result, women in central Iveti have begun to take over many of the agricultural tasks traditionally allocated to men.

Mrs. Imara buys 2 shillings' (28 cents U.S.) worth of sugarcane, presses out the juice with a wooden press at home, spends 50 cents (01 cent U.S.) to get the tin of juice to Machakos market, and makes 2 shillings 50 cents (35 cents U.S.) profit when it is purchased for the manufacture of local beer. She also buys single trees and sells firewood to her neighbors.

Such income is important to her family's well-being during a drought, but it all takes time. Her husband has migrated down to the plains, where in 1970–71 things were worse than in Iveti, nearly as bad as in Katse, so no financial help can be expected from him. At home in Kaewa 12 persons are dependent on Mrs. Imara's 2 hectares (5 acres). Two of these are over 50 years old and six are under 15. There is one adolescent boy and two young women to help Mrs. Imara, who is 60. The household owns no livestock, but fortunately owns a small plot of valley bottom land which continues to produce cabbages, some sugarcane, cassava, sweet potatoes, arrowroot, and kidney beans. These crops are sold during good seasons, but become increasingly important famine reserves depending on the severity and duration of a drought. She doesn't own any coffee, which plays a role in the famine economy of other households in Kaewa.

Her children are very important to Mrs. Imara, and she feels a heavy responsibility for all the children in the household. Three-fourths of the women in this area engage from time to time in local self-help projects to build and support schools for their children (Roberts, 1962). However, with no milk cows and such a small amount of cash income, the children seem inevitably to be subject to heightened nutritional risk during droughts.

There are two children under 3 in the household. One child is still on the breast at 17 months and measures 80 percent of the standard weight. She may be undernourished, but is not in clear danger. The 25-month-old has been weaned and is typical of this highest-risk group—weanlings during a drought. He has attained only 60 percent of normal weight. He suffers from eye trouble and diarrhea. His hair has changed color and texture, one sign of protein-calorie malnutrition. To make matters worse, he was recently burned while playing near the cooking fire.

Kaewa is normally a reasonably prosperous and progressive yeoman farming area where government development programs have been active since the early 1950s. From this final portrait it should be clear how difficult it is to separate the "drought problem" from the general problem of rural development.

Drought as a national problem

The cost of drought to the nation can be divided into the direct monetary costs which the government incurs in spreading the burden of drought over more than the affected population, primarily through *famine relief.* Other costs arise from *production losses,* value not added to the economy because activities in which farmers have invested time, money, and labor fail: cattle lose weight, die, do not bear calves; plants wither or bear a fraction of their normal harvest. There are other *social costs* to the nation measured by increases in nutritional problems and nutritionally related disease. Finally, drought has an important overall impact on the pace of technological change and *rural development* in the affected areas.

Famine relief

We calculate that during 1961 Kenya spent for internally purchased maize and transportation alone K. shs. 12 1/2 million (U.S. $1.8 M): K. shs. 5.5 million for transportation to railheads for distribution and about K. shs. 7 million for maize purchased from the Maize and Produce Board (Roberts, 1962). That year about 40 percent of the famine relief maize came free from the United States, and in accordance with the national goal of self-reliance, the value of this free maize should be added to the 1961 cost. A similar estimate for famine relief cost during January 1970–January 1971 reaches about K. shs. 20 million (U.S. $2.8M). For that period we estimate an average of 50,000 persons on famine relief with peaks as high as 250,000, at a cost of 1 shilling (14 cents U.S.) per head per day for a year. These figures give an idea of the range of magnitude only. Further credibility is lent these estimates when it is noted that Tanzania spent T. shs. 20 million (U.S. $2.8 M) on famine relief during 1969.

Production losses

A total Kenyan estimate cannot be made with confidence; however, it has been calculated that Tanzania (whose general rainfall reliability pattern is similar to Kenya's) loses on the average about 10 percent of primary production (less minerals) a year. This is about 4 percent of its GNP (Kates and Wisner, 1971). Maize production losses were estimated to be about T. shs. 45 million (U.S. $6.3 M) a year (0.6 percent of GNP). Working with 1962 figures, it is estimated that 1961 losses to the Kenyan livestock industry alone could have been as high as K. shs. 140 million (U.S. $19.6 M).

It has been estimated conservatively that the total cost of the 1961–62 drought and floods was 10 million K. pounds (U.S. $24 M) (E. A. Standard, 1971). If this is so, then a ratio of total costs to famine relief cost of 10:1 seems a safe estimate.

Social costs

Drought alone does not account for large-scale starvation except where a population is highly vulnerable. A single season drought can be usually met by late-planted cash crops, sales of animals, loans, and "bush" foods, such as roots and wild fruits. A series of such droughts (two or more seasons) or combinations of drought, flood, or pest invasion (in 1961–62 Kenya suffered all three) intensifies food shortage to the point where the danger of some deaths is present. The one-season drought may be contrasted with the extreme famine condition often present in civil war.

Children of weaning age experience a considerably higher risk during drought. Decrease in caloric intake and milk supply can precipitate clinical protein-calorie malnutrition, a risk that may also be increased by lower domestic water use, hence poorer hygiene and greater danger of diarrhea. A controversial aspect of protein-calorie malnutrition is its possible retarding effect on the mental development of surviving children. The possibility exists that drought and resulting famine may contribute to poor school performance and mental/emotional development of thousands of Kenyan children.

Rural development may in some cases actually be speeded up by drought. Drought may hasten the process of innovation and the adoption of new economic ideas, including widespread though low-level involvement in the cash economy or the accelerated cropping of cattle by nomadic tribes (Mbithi, 1971). It is difficult to say whether such positive effects balance the clear losses to the nation. The sequence of orderly rural development may be thrown out of phase by a drought. Resources are diverted into investments like famine relief where they are only marginally productive and have very low rates of return (Mbithi and Barnes, 1973).

Types of drought problems and famine potential in Kenya

In Kenya as a whole over the past 30 years, three kinds of drought appear.

National drought directly affects the production of more than 10 percent of Kenya's population, lasts two or more growing seasons, and generally involves serious loss of production in most ecological zones and usually two or more provinces. This type of drought seems to occur about once each decade. Farmers in eastern Kenya mention very serious droughts in 1913–18, 1925, 1936, 1946, 1954, and, of course, the droughts of 1961 and 1970. Further insight into the use of oral history to establish famine chronologies can be gained from the following: in our Tharaka sample (lower Moru), taking all responses together (not relying solely on one man's memory), 47 percent of the years since the turn of the century had poor enough harvests to be remembered. There were 15 "worst droughts recalled" since the turn of the century. Dr. Ndeti (University of Nairobi) has dated the following famines in Kamba oral history: 1836, 1850, 1861, 1880, 1899–1901, and 1910. Heavy livestock losses are usually involved, and can amount to 40–50 percent or more of the herds (e.g., Kajiado Masai herds in 1961 and Samburu herds in 1970). Rehabilitation of herds takes longer than reseeding of farm lands. Beans will give a catch crop in 3 months. With loss of condition, reduced calving rates, increased mortality among calves, and sales of the breeding nucleus of a herd a serious undersupply of milk can remain a problem in a pastoral area for 6 to 12 months after the meteorological "end" of the drought.

Most ecological zones suffer loss of production during national drought. Interviews with farmers in high, medium, and low agricultural potential zones of eastern Kenya reveal food crop shortages in even the high potential coffee/tea zone near the Mount Kenya forest. In sum, these characteristics present a unique challenge to government policy which is different from the challenges presented by regional and local droughts.

Regional drought directly affects the production of less than 10 percent of the population of Kenya, lasts one or two growing seasons, and is generally confined to the medium- and low-potential areas, especially the semiarid dry-farming zone and the arid and very arid rangelands. The occurrence of regional drought varies according to the kind of crops grown and densities of livestock and grazing patterns. With local maize such a drought might be expected once every 3–4 years. With full adoption of drought-resistant Katumani maize it might occur only once in 8 years. Millet in northern Kitui and southeastern Tharaka seems to fail on an average of once in 5 years. Thus, on the average, two or three such regional occurrences might come each decade. Within living memory of the farmers interviewed in Tharaka (lower Moru) there were such droughts in 1951, 1954, 1961, 1965, and 1970. If stocking densities are low enough, pastoralists seem well adapted to getting through a single season failure of the rains by increasing their range of movement with their herds. If the cost of famine relief in a national drought is K. shs. 20 million (U.S. $2.8 M), such a regional drought probably costs around K. shs. 10 million (U.S. $1.4 M).

Local drought probably occurs every year somewhere in Kenya. Especially in the marginal agricultural zones of the eastern plateau foreland (Machakos, Kitui, Tana River, Kwale, Lamu, and Kilifi districts), the variability of rainfall is such that individual ridges and sublocations can experience crop failure or serious shortfalls in harvest because of localized combinations of slope, soil, and rainfall conditions. This type of drought is usually handled well by traditional gift and loan relations among the farmers and their kinsmen and friends and by the normal social welfare allotments made to district authorities. The Kenyan budget contained K. shs. 50,000 (U.S. $7,000) as the "normal" fund for local drought. This probably should be multiplied by at least 10 to give the actual average annual cost of such droughts, e.g., K. shs. 500,000 (U.S. $70,000).

Famine potential

Drought in Kenya can also be classified by the pattern of agricultural potential: high, medium, or low. However, in considering the impact of a given drought, it is very important to look at the pattern of population movement and growth in various farming regions of Kenya (Ominde, 1968).

Comparative 1962 and 1969 data shows:

1. Substantial emigration of populations from Central, Nyanza, and Western provinces into the Rift Valley related to the Million Acre Settlement Scheme.
2. Substantial emigration of populations from the east-

ern plateau foreland and the coastal hinterland in the coastal plateau.
3. Extensive intraregional population flux tending to push more people into the hitherto less densely populated medium and marginal potential lands of West Pohot, Elgeyo Marakaret, eastern and northern Meru, Embu, Yatta, Makueni, Darajani, North Kilifi.

It is the movements into these poor agricultural zones, which in the absence of drastic technological improvements pose the greatest risk of famine, environmental deterioration, and total economic dislocation.

Theoretical framework

Agricultural drought should be viewed as a natural, though extreme outcome of the interaction of man and nature. When rainfall exceeds the upper limit of a range of physical conditions, the farmer must use practices not normally employed to protect his crop from mildew, pests, and flood. Likewise when rainfall falls below the lower limit of the normal range, farmers employ practices, or adjustments, which reduce the damage caused by drought.

As Kates noted (1972, p. 12) depending on the rainfall range which normal farming practices allow, "drought" would be declared at different times by different farming systems. The farmer who plants early-maturing maize, plants early, and weeds early would not define a season which brings 180 millimeters (7 inches) of rain as a drought season. His neighbor, who normally plants local maize, plants late, and weeds late would call the season a serious drought season, and he would set in motion a series of adjustments, nonnormal practices, in order to feed his family. He might even leave off farming temporarily and go to seek wage work.

In these circumstances, a relatively low-cost and high-benefit approach for the government in dealing with drought problems is to build upon local patterns of adjustment to drought which have grown up in the different ecological zones, fostering those which seem to be effective, discouraging some which seem wasteful, and introducing new ones, like early-maturing maize, where appropriate.

This emphasis on the complexity, variety, and flexibility of farm and small community-level adjustments rests within the general framework of human ecology (Burton, Kates, and White, 1968), and microsociological theory (Mbithi, 1967, 1970). Such an approach focuses initially on the environmental experience and cognition of small groups of farmers and on the range of nondeviant resource exploitation within the group. Although classical innovation theory is appropriate, the starting point is "spontaneous, localized innovation or adjustment" rather than innovation originating outside of the group. Man is a creative actor, attempting to cope with an environment which is both physical and sociological, which is constantly changing and which exhibits unreliable patterns. Through choice of technology, ritual means, and functional social linkages, man "adjusts" to and improves upon his environment to increase its productive capacity and reduce risk.

Study strategy in eastern Kenya

Figure 11–2 shows study sites chosen to give information about the farm level impact of drought and response of farmers in a range of ecological zones nearly representative of Kenya. These sites can be arranged along several "gradients" from high-altitude, high-potential lands of higher, more reliable rainfall down to low-altitude, marginal agricultural lands of lower, less reliable rainfall. Figure 11–3 shows one such gradient, which begins near the Mount Kenya forest on the Embu side and falls through the "medium"-potential, sub-humid and semiarid zones of Mbeere Division, across the

Tana River to the "marginal" thornbush lands of Far Northern division, Kitui. Other possible gradients are indicated by arrows on Fig. 11–2. To complete the eastern Kenya picture interviews were conducted in Isiolo District and background studies in Marsabit in the extreme north. In all, 610 randomly selected farmers were interviewed and about 120 children weighed and measured for nutritional assessment.

Drought impacts and patterns of adjustment

A first impression of the overall suffering due to drought can be obtained from Table 11–2. Farmers' responses to several of the questionnaire items were scored to give a comparable index of the amount of drought suffering they themselves perceive. The score is based on the number of serious famines they remember, the level of hunger, crop and animal losses, and deaths

Fig. 11–3. Section of typical gradient NW-SE

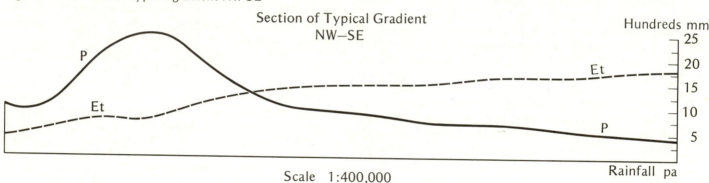

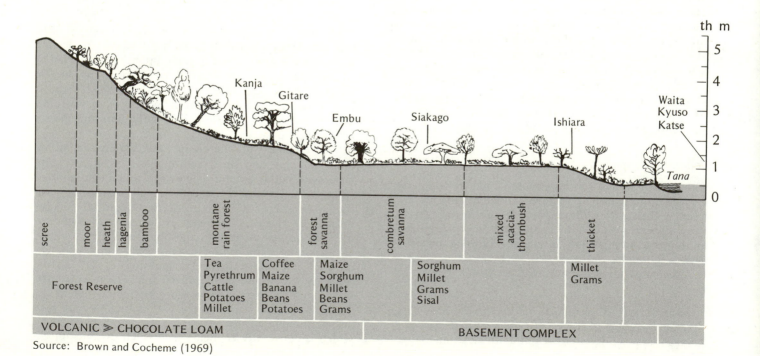

Source: Brown and Cocheme (1969)

Table 11–2. History of drought suffering along Menu-Tharaka gradient (percent of all farmers in a site)

Site	No Suffering	Negligible Suffering	Mild Suffering	Moderate Suffering	Severe Suffering
Mikumbune	0	5	90	5	0
Mitunguu	0	0	75	25	0
Chiakariga	0	0	45	50	5
Kathangachini	0	0	25	50	25

remembered and reported. Site 115 is about 1,800 meters (6,000 feet) on the Meru side of Mount Kenya, and the gradient falls toward site 116, just 10 kilometers (6 miles) from the Tana River in the Sansevieria-bush zone. Overall suffering due to drought increases along the gradient.

Malnutrition is central to the government concern with famine relief. A group's normal level of nutrition determines to a great extent the vulnerability of persons to the stress of drought, how long they can remain healthy and productive on reduced diets, and the extent to which it is associated with the incidence of such diseases as diarrhea and pneumonia and increases in morbidity/mortality from such diseases as measles and tuberculosis.

Table 11–3 summarizes preliminary observations of the nutritional status of children under 3 years and the quality of adult diet along the Katse-Mikumbune gradient.

The picture here is not as clear as in Table 11–2 on overall suffering. Many local factors influence the quality of diet and the condition of children, and further analysis is required to make sense out of the site-to-site differences observed. In the two driest sites a significant and disturbingly high percentage of children are in danger. Both these sites lie in the "eastern plateau foreland-marginal zone" mentioned earlier as the zone of highest famine potential in Kenya.

Animal losses are concentrated in the low-potential zones. An indication of the magnitude of drought im-

pact on livestock can be seen in the Kitui sites. In the Kitui sample of 120 farmers, only 53 percent owned cattle because, as many explained, they had just sold the last of their cattle to make ends meet. Fifty-nine percent owned goats or sheep; 71 percent said they sell more cattle during a drought period than they do during good times. Seventy-six percent of the sample had seen cattle die in droughts before 1970 and the same percentage had cattle die during the 1970–71 drought. The best estimate is that from 20 to 33 percent of the cattle in Far Northern Kitui died in the 1970–71 drought. 1965 was the latest previous time serious cattle death had been experienced in the area, and the level of death was only slightly higher during 1970–71.

Farmers reported that while one in four calves born during a good time die while young, the death rate increased to two out of three during the recent drought. Twenty-nine percent of the farmers had been forced to sell milk cows. This is a drastic decision because it means less milk for the children and a slower rate of herd replenishment once grazing is restored and calving rates are normal.

Crop losses must be seen in the light of the normal food-crop reserves and cropping pattern. In the high-potential areas it was found that farmers did not normally keep a large reserve of grain, but relied on purchased maize bought with coffee or tea earnings. They reported they had grown some maize, but had eaten it green in the fields. They found it difficult to purchase maize because of the nationwide shortage. However, Irish potatoes, cassava, yams, and vegetables (especially cabbage) were still plentiful.

In the medium-potential areas less than 50 percent of the households kept a significant reserve of grains. Here the effect of diversification and extensive dry-land cultivation was obvious. In Karaba, even though the maize harvest had been very bad, they still had cowpeas, green, yellow, and black grams, and cardamom to sell. These farmers cultivated large acreages with rented tractors, and their children seemed not much worse off than the children in higher areas. By contrast the farms in Siakago, just 40 kilometers (25 miles) away, were not

Table 11–3. Child nutrition and adult diet in six drought study sites

Sites	High Potential	Medium Potential		Low Potential		
	Mikumbune	Karaba	Siakago	Chiakariga	Kathunga-Chini	Katse
Adult diet[a]	16	15	17	14	10	14
Child nutrition[b]	17	13	26	15	25	38

[a]This scores the quality of the food eaten the previous day in terms of vitamins, type of starch, number of meals and expensive extras: sugar and oil.

[b]This is the percentage of the children under 3 years who measured 70 percent of the standard weight for their age or below. This is the level below which one generally expects to find clinical signs of malnutrition, and it should be taken as an indication that the child is probably highly vulnerable to measles, pneumonia, etc., and is, in short, in considerable danger.

mechanized and were much smaller with fewer crops. Farmers there were in more trouble, and twice as many children of Siakago were found to be malnourished than in Karaba.

In the low-potential areas normal reserves were highest. Eighty-seven percent of the farmers in Katse normally store large amounts of grain, and a large assortment of crops were grown. These include bulrush millet, sorghum, cowpeas, pigeon peas, and many varieties of grams; the millet harvest had failed in many areas.

Short-term migration to the nearest upland area in search of wage employment was very common in the marginal zones of Meru, Embu, and Kitui. Tharakans (lower Meru) tended to go to the Nyambeni range and up toward Meru to pick coffee and work at other agricultural tasks. Mbeere (lower Embu) traveled to the lower slopes of Mount Kenya of the Embu side and to Chuka. The Kamba in Kitui sometimes crossed the Tana River and traveled to farms on the Embu side of the mountain; others went toward Mombasa. These wage migrants are usually paid in kind and carry food back to their families at intervals. This has the benefit of providing some food for the people of the area and a large, unreliable, seasonal labor supply. However, such migration also tends to disturb family life and drains the affected areas of labor which could plant catch crops when the rains break and engage in other local anti-drought measures of the self-help variety.

Individual adjustment to drought

Many ways of coping with drought have been institutionalized over the years, and are now permanent features of the agricultural and social systems. Examples are the characteristic mixture of crops and livestock and the widely spaced network of farm fields, cattle camps, and fields belonging to kinsmen which spread over large areas and which ensure at least *some* little affected or unaffected economic activity during a local drought. However, during regional and national droughts which affect the sites studied, these built-in features of the systems are not adequate to meet the threat of famine.

Then other practices appear which are not often encountered during normal rainfall seasons.

Table 11–4 summarizes the percentage of farmers naming a given adjustment among their three most preferred and most frequently practiced. The sites from which data is taken are fairly representative, and are the same three described in the introductory portraits.

Agronomic adjustments (planting in wet valleys, planting early) predominate in the primarily agricultural/horticultural site in highland Machakos. Even though such adjustments do not rank among the most preferred in the two drier zones, there is much that can be done to improve dry-farming technique. There would be scope for extending knowledge of these adjustments among the farmers in such places as Karaba and Katse.

Community adjustment to drought

In Kenya and Tanzania the call for Harambee and self-reliance is exposing unique potentialities. Officials are no longer blind to the prospect that alleviation of suffering in these areas may lie with the local people! How many people can really be fed by a government relief program in an area which has experienced a total food-crop failure? Figures on relief recipients show that the proportion per village is rarely high. *How do the others cope?*

Local communities as well as individual farmers are continuously adjusting to environmental stress. These community adjustments are critical to the success of any new program introduced by government and they are tangible bases for innovation.

The main difference between colonial era Kenya and modern Kenya is the spirit of Harambee and the startling transformation of rural people from docile apathetic people to active ones. The studies show that among the colonial dry-land projects which failed because they were rejected, such as afforestation, cattle destocking, peasant irrigation projects, the same projects as community Harambee projects are achieving a surprising degree of success. In fact community initiative and activity during the past few years actually burdened the Kenyan

Table 11–4. Preferred adjustments in three sites (percent)

High Potential: Kaewa (Iveti, Machakos)		Medium Potential: Karaba (Embu)		Low Potential: Katse (Kitui)	
Plant in wet valleys	63	Wage labor	68	Buy food	81
Plant early	63	Buy food	56	Sell livestock	61
Buy food	50	Help from kin	42	Wage labor	58
Pray	38	Sell livestock	37	Long-distance wage labor	29
		Help from government	26		
				Help from kin	26

government: numerous health centers and schools were built by communities which government cannot staff. Communities turn to government for aid in completion of locally initiated water schemes.

In famine-prone areas there are "oases" of unique development. Taita-Taveta has unique local production of high-quality bananas. Karaba-Embu demonstrates locally developed peasant tractorized maize and coriander production. Kimutwa, Machakos, initiated vegetable production for export. Mwingi-Waita, Kitui, developed a bushland livestock industry for the beef market in Nairobi and Mombasa.

This unique localized development needs careful study. Who supplies the initiative? Who identifies uniquely advantageous circumstances for self help innovations? These questions have not been asked because all initiative has been assumed to come from the government. But the dry expanses of Marsabit, Moyale, and Kitui show that government initiative has been very limited. Even if government initiative were available, without local cooperation little is practicable. This reinforces the importance of thorough understanding of local farmer and community response to drought and the relevance of microgeographical and sociological approaches.

Present and possible government approaches to drought

Kenya's agricultural development strategy reveals a skewed concentration of technology in high- and medium-potential farming areas. Reuthenberg attributes the rapid growth in agricultural production to a multiplicity of approaches (1966). At the national level the main approaches have been through land reform, increased efficiency and intensity in agricultural administration and extension (dependent upon technological packages), smallholder tea development, development of coffee, pyrethrum, mixed dairy farming, farmer training, and the introduction of viable cash crops.

Agricultural economists have developed serious myopia in failing to appreciate that existing dry-land farming technology and adjustment patterns in Kenya, Israel, Mexico, and Australia can revolutionaize extension effectiveness in these areas. This myopia has shown itself in four ways:

1. There is still costly repetition of ill-designed projects, such as the Ishiara irrigation scheme, the Samburu grazing rotation scheme, the Machakos soil and water conservation program.

2. Selection and training of agricultural research experts is based on the assumption that Kenya's agricultural systems lie exclusively in the highlands, the Lake basin, and the coastal strip. This goes together with the assumption that the only other possible land-use system in the rest of Kenya can be subsumed under the term "range management."

3. Dry-land extension systems are developed with limited technological packages to be exploited through the concept of "crash programs." The dearth of dry-land technology leads to the indiscriminate importation of intensive, wet-land farming practices such as fencing, ley farming, heavy mineral fertilizer use, and the extension of medium-potential field crops, so that the risk of crop failure is increased. Within the extension service, transfer to northern or southern districts often is seen by the officers concerned as a disciplinary measure, leading to low officer morale and productivity.

4. Periodic and costly famine crises continue without solution.

If the governmental approach to 1972 was inadequate to meet the drought problem, what changes might improve the approach? Three opportunities are suggested.

1. Government can foster the application of existing effective adjustments to drought on the local level. These adjustments can be made more effective through cooperative, self-help exploitation of local resources.

2. Government can provide inputs needed for increased effectiveness of local and community adjustments. Often constraints seem to be *credit* and *markets*, but less commonly knowledge. Individuals and potentially small groups of individuals in the driest areas are aware of highly localized resource potential (a seasonal swamp, stream bottomland, small but highly fertile and irrigable areas). However, the key to government involvement must be flexibility and small-scale planning. This requires highly motivated extension workers at the lowest echelon. There is a need for a cadre cum agricultural assistant who will act as advocate for groups of farmers in the marginal zone, cutting across ministerial boundaries to work out the details of credit, land tenure, association registration, and technical problems arising out of the people's own identification of a unique community resource.

3. Government can introduce new adjustments to complement existing ones. The prime case in point is the Katumani early-maturing maize program. The most effective government response since 1958 to crop failure has been breeding of synthetic and composite maize varieties which extension activity introduced widely in the dry areas after 1964 under the name Katumani. Adoption of Katumani would ideally reduce moisture requirement of maize from about 300 millimeters (12 inches) per season to about 180 millimeters (7 inches).

In 1967, 80 percent of the farmers in several dry-land sample villages were using Katumani (Mbithi, 1967). Why hasn't there been a significant reduction in the frequency of food shortages in these areas? The findings offer two clues. Although the average family of five adults requires over 1.2 hectares (3 acres) of land for subsistence at a productivity level of 7–10 bags of maize per hectare (3–4 per acre), the average acreage of Katumani maize was only about 0.6 hectare (1½ acres). In addition, the ratio of Katumani to total maize acreage including high-moisture-demanding varieties was less than 40 percent. The acceptance rate of Katumani in terms of effective acreage was extremely low, and explains the lack of significant improvements in food supply. This is not a *technical* failure. The main problems seem to be in bulking and distributing the new seed at the proper time, in appropriate quantity, and at low cost together with the problem of farm level competition between early-maturing maize and alternative crops.

4. Government probably must work within the framework of overall patterns of risk and adjustment. A good example is the problem of drought-caused migration. It could well be that migration is a waste of manpower and energy and contributes to regional inequalities. However, even if this is true, there seems little that can be done about it in the short term. At a minimum the government could provide a mobile labor exchange service to provide information in the famine area about where farm labor is required and provide transport to labor deficit areas and return.

Another example is the problem of settlement. The Kwale calls for close cooperation between settlement officers and the agencies in charge of famine relief and antifamine planning. Over the years just before 1972 the Makueni settlement scheme in eastern Machakos effectively tripled in size (unofficially) with no corresponding increase in control over settlement or advanced planning for such an increase.

5. Government can coordinate a famine warning system. The East African Meteorological Department should be persuaded to site some of its new agrometeorological stations in the eastern marginal lands. That would provide technical support for any future program of experimentation with dry-land farming techniques and also contribute to a national famine warning system. Other elements in such a warning system would be improved agricultural reporting and reporting of the nutritional vulnerability and reserves of food among high-risk populations.

6. Government can take either a defensive or offensive posture while pursuing the above approaches. The fact that in much of Kenya drought is the dominant environmental factor implies that drought adjustment must be one focus of development efforts. It by no means implies that government must constantly fight a defensive battle, in which development in the marginal and pastoral areas never progresses as far as in the better-endowed zones. The idea of famine prevention or drought adjustment need not be negative.

A positive approach calls for a strongly increased commitment to applied *nutrition work.* Emphasis would be on mobile dispensaries and rehabilitation centers, together with nutrition education and family planning. It also calls for increasing *cash incomes* in the marginal and pastoral areas through improved marketing and industrialization. The "cash-crop fixation" which characterized agricultural policy in Kenya was a fallacy. The immediate need is to find markets for what drought-prone areas are already producing and to increase their productivity, not to invent new cash crops.

There are indications that government is beginning to take the offensive role. It is now generally accepted that all famine relief should be replaced by antifamine programs while residual relief cases (widows, the rural blind and lame, orphans) are absorbed under the normal welfare functions of local government.

7. The government approach in the pastoral areas will have different emphasis. A recent report by one of the authors on the small irrigation scheme at Kinna, Isiolo District (Wisner, 1972a, 1972b), where Borana pastoralists were resettled, contains the core of our thinking on the rangelands. All projects in this area should be small-scale at first. The nutritional, agricultural, and ecological as well as social problems present on the 80-hectare (200-acre) scheme demonstrate vividly how little is known about the improvement of pastoral systems and their integration with arable agricultural and horticultural systems. Small-scale irrigation should be planned simultaneously with group ranching, perhaps as small nuclei or growth centers in the middle of larger ranching schemes with which the vegetable cultivators would have reciprocal exchange relations. Development efforts in the pastoral areas should be phased so that in the early phases obvious improvements in the standard of living of the masses takes top priority. To accomplish this the first phase must be focused on such things as mobile dispensaries, improved marketing, livestock improvement, and water resource development, not initially on education, which has a longer-range benefit.

A final, and most important, suggestion is stated well in a recent report on the Kaputiei Maasai group ranches. The report concludes as follows (Halderman, 1972):

> a. It should be realized by development planners that the settling of nomadic and seminomadic pastoralists requires more than the physical development of the area (water development, cattle dips, etc.)—it requires that the ecological conditions *permit* permanent settlement (as the reversion to seminomadism by the Poka [group ranch pilot scheme] members in 1971 demonstrated).
>
> b. Development plans for arid and semi-arid areas must take into full consideration: the traditional system of resource utilization of the indigenous pastoralists—so that development proposals will be based on the existing structure and be acceptable to the owners of the land and livestock.
>
> c. The solutions to the problems of development in Maasailand require dialogue between planners, implementers, *and* Maasai in order to determine policy—and then cooperation by all those involved aimed at effectively implementing the proposals.

A future

Camels and ox plows race the communal lorry (truck) in slow-motion floodplain choreography. Receding water from the Ewaso Ngiro River will supplement the rain and fill again this year, 1995, the Society's store and the smaller kin-group silos. Bwana Uhuru pauses over the sun-drier he has been filling with vegetables. He watches the camels rock and sway into the thornbush haze beyond the millet blocks. He looks up at the sun, whose fierce pull at evaporating soil, and transpiring green things still puts, and probably will always put, an edge of risk on the life of his people. However, since the first advocate-cadre arrived, nearly 20 years ago, stress has become a spice, like coriander, to his people's life. New and old ways, common life and mutual help have stabilized, in part, that life. . . . The camels move further into the haze, through the lens-shaped pits which harvest run off, through the date palms, burdened with loads of dried vegetables and the newest edition of the Jomo Kenyatta memorial literacy manual. From behind the dispensary Uhuru hears craftsmen pounding on low-cost oxcarts. Laughter. Small boys chuckling, chasing Sodom apples near the calves and milk cows they are tending. He recalls the identical game. He recalls the hunger. His gaze unconsciously shifts to the dispensary where a group of mothers with their babies gossip outside the weekly well-baby clinic.

Acknowledgments

The authors wish to thank their research assistants for their energetic help under difficult field conditions: Germano Mwabu, Daniel Ndonye, Fred Katule, John Kirimi, Mutua Muriera, Paul Goto, and Halake Dido—students at the University of Nairobi; Moses Wesonga, research assistant at I.D.S.; Fred Kiraithi, Peter Sekundu, Albert Ndigi, and Erastus Ndwiga—students at the Embu Institute of Agriculture.

We also express our thanks to Frank Schofield and Johannis van Luijk of the Department of Community Health, University of Nairobi, and to the farmers and government officials in our study area, and to the National Christian Council of Kenya.

References

Burton, Ian, Kates, R. W., and White, G. F. (1968) *The Human Ecology of Extreme Geophysical Events.* Toronto: Natural Hazards Research Working Paper No. 1.

Halderman, John M. (1972) "An analysis of continued semi-nomadism on the Kaputiei Maasai group ranches: sociological and ecological factors." University of Nairobi, I.D.S. Working Paper No. 28.

Jelliffe, D. B. (1966) *The Assessment of the Nutritional Status of the Community.* General WHO Monograph No. 53.

Kates, R. W. (1972) "Foresight, hindsight, and insight: problems of measuring drought adjustment." Worcester, Mass.: Clark University, mimeo.

——, and Wisner, Benjamin. (1971) "The role of agricultural drought in a developing economy: examples from Tanzania and Kenya." Gödölö, Hungary: UNESCO Seminar on Natural Hazards.

Mbithi, Philip M. (1967) "Famine crises and innovation: physical and social factors affecting new crop adoptions in the marginal arming areas of eastern Kenya." Makerere University College, R.D.R. Paper No. 52.

——. (1970) "Self-help as a strategy for rural development: a case study." Dar es Salaam: Universities of East Africa Social Science Conference.

——. (1971) "Non farm occupation and farm innovation in marginal, medium, and high potential regions of eastern Kenya and Buganda." University of Nairobi, I.D.S. Staff Paper No. 114.

——, and Barnes, Carolyn, (1973) *The Squatter Problem: Landlessness, Famine and Land Carrying Capacity.* Report prepared for IBRD.

——, and Wisner, B. (1972) "Drought and famine in Kenya: magnitude and attempted solutions." University of Nairobi, I.D.S. Discussion Paper No. 144.

Ominde, S. H. (1968) *Land and Population Movements in Kenya.* London: Heinemann, p. 159.

Pratt, D. J., Greenway, P. J., and Gwynne, M. D. (1966) "A classification of East African rangeland." *Journal of Applied Ecology* 3:369–382.

Redlich, L. C. (1971) "The role of women in the Kamba household." University of Nairobi, Department of Sociology Occasional Paper, mimeo, p. 7.

Reuthenberg, Hans. (1966) *Agricultural Production Development Policy in Kenya, 1945–1965.* Berlin: Springer-Verlag.

Roberts, M. J. (1962) *Famine and Floods in Kenya, 1961.* Nairobi: Government Printer.

E.A. Standard. (1971) "Relief plan for drought province." The author of this item seems to claim that "relief" cost £10 million (U.S. $28 million) in 1960–61. We can get nowhere near this figure in our own estimates, so take it as a journalistic error. He probably means the *total* cost of drought and not just relief.

Wisner, Benjamin. (1972a) "Kinna site subproject, preliminary report no. 1." Report commissioned by the National Christian Council of Kenya.

——. (1972b) "Dryland farming settlement by Rendille pastoralists on the southern slopes of Mt. Marsabit: notes on the viability of the Songa scheme." Report to the National Christian Council of Kenya.

12. Response to drought in Sukumaland, Tanzania

THOMAS D. HANKINS
West Virginia College of Graduate Studies

How do Sukuma farmers respond to drought? The relationship of moisture availability to the drought-hazard experience of the farmers, to the adaptive capacities of their farming systems, and to adjustments they employ to reduce the effects of drought are the three aspects of this broad topic considered in detail. As background, some pertinent characteristics of the Sukuma environment and farming methods are discussed.

The field studies

The analysis is based on data collected as part of two projects. Most of the basic farm data were obtained and prepared by the Sukumaland Interdisciplinary Research Project (SIRP) of the University of Dar es Salaam during 1969 for the crop year 1968–69. Robert H. Hulls, James R. Finucane, Arne Larsen, and I were the members of that research team. The farmers were chosen as a random, cluster sample stratified by districts (SIRP, 1971).

A total of 219 farmers in 11 areas were interviewed four times by SIRP field assistants. Included in these interviews were careful measurements of farm plots. The sample was not limited to full-time farmers. Small businessmen and persons employed by government, schools, and missions were also included. One of the study areas is dominated by a mission hospital (SIRP, 1971) and one lies adjacent to Mwanza town (SIRP, 1971). After some months of work with this data the SIRP group was satisfied that this was a quite representative sample of the four districts. Unfortunately Maswa District was not included. This is especially unfortunate for the drought study because study sites in Maswa would have provided drier areas than any included in the present group.

The drought-hazard questionnaire was administered to the farmers of the SIRP sample in January-March 1971. One of the 11 SIRP sites in the wetter zone was not revisited because of transport difficulties. In the other 10 areas a total of 166 farmers out of the original 194 were interviewed (68 in the wetter zone, 98 in the drier). Of those missing some had died, some had moved, and some were away from home. No check has been made to determine how either of these changes has biased the sample.

The administration of the drought study questionnaire was not closely supervised, but the two interviewers hired for this project always worked with experienced assistants from the SIRP. In cross-checking the drought questionnaire with 1969 interviews, similar results were found for common questions. Nevertheless, the reader is cautioned to draw conclusions only from the cumulative evidence of several variables rather than from single ones.

Drought in Sukumaland

Drought has been a persistent hazard to Sukuma farmers over the years. Either alone or in combination with other factors (poor soils, cattle disease, crop parasites) it has produced recurrent famines in the region. Brooke (1967), using records from district offices and mission stations, reported 15 years of famine in at least some portions of the region during the period 1890–1964, an average of 1 famine year in every 5, one half of these being serious food shortages. McLaughlin (1971), without specifying his sources, stated that a shortage of water and grazing over several successive years produces extra-severe drought conditions on an average of every 6–9 years, these being so severe that up to one-third of the cattle may die.

A 1969 SIRP survey adds support to these suggestions. At that time 61 percent of the farmers reported drought had damaged their crops, and over 50 percent of these reported damage as frequently as 1 year in 3 (Berry et al., 1971).

Rainmakers should also be mentioned in this context. Cory (1951) describes how they were traditionally hired by chiefs or agents of chiefs and provided an important means of dealing with the problem of drought. Although the chiefs have largely passed from the scene, drought has not, and so the rainmakers remain. Twenty-seven practicing rainmakers were located within a radius of a dozen kilometers or so (7 miles) in central Shinyanga in 1970.

Some Sukuma proverbs also reflect concern with the possibility of poor rains. The Sukuma say: "The hoe handle has died, killed by the clouds," meaning that without the rain there is no work for the hoe (Cotter, c. 1967, p. 48). Similarly they may describe an area that did not yield a crop by referring to it as "the place where the rain stopped" (Ibid., p. 97).

Perhaps the most telling of all the evidence of the impact of drought on Sukuma life is that in Cory's small book *The Ntemi* (1951). Much of the description of traditional Sukuma chiefs relates to the many activities performed to ensure adequate rainfall for the chiefdom. So important was the chief's responsibility to supply rain that failure to do so adequately was sufficient reason to select a new chief.

Sukumaland rainfall characteristics

Annual rainfall in Sukumaland averages between 750 and 1,000 millimeters (30–40 inches) throughout most of the region. Only near the shore of Lake Victoria near Mwanza and westward into Geita District do averages exceed 1,000 millimeters.

Average annual rainfall tends to be higher to the north and west, and decreases to the south and east. The five stations listed in Table 12–1 demonstrate this trend. Stations 1–4 lie on a roughly northwest-southeast line (Fig. 12–1) along which mean annual rainfall decreases. Station 5 at Geita illustrates the higher precipitation common further west.

In general, the rains begin in October-December with a first peak in November or December, a lessening of rainfall in January-February, and a second peak in March-April. The rains begin earlier and last longer in the higher rainfall areas to the north and west. In the lower rainfall areas of the southeast the bimodal nature of the rains is less marked both because of lower peak values and because of a less distinct break in the rains.

Within a single season there is great variation in rainfall over space both on regional and local scales. Stations in the northwest may record less precipitation than stations in the southwest, and neighboring areas (2–20 kilometers or 1–12 miles) may differ greatly (up to 200 millimeters or more or up to 8 inches or more) in the amount of rain they receive (SIRP, 1971).

The data shown in Table 12–1 together with rainfall records from the unevenly distributed stations in Sukumaland were used to group study sites into areas with similar crop moisture availability characteristics. The higher rainfall and available moisture index of areas next to Lake Victoria and in the northwest of the region appeared to provide justifiable grounds for dividing the region into two zones. Although other more detailed divisions might be preferable, neither the availability of data nor the flexibility of the sample permitted this. A general line demarcating the two moisture availability zones would run generally north/south along a line east of Sanjo and Nyanza study sites.

Other characteristics of Sukumaland and Sukuma farmers

In 1970 over 1 million Sukuma lived in Mwanza, Kwimba, Geita, Shinyanga, and Maswa districts. Much of this territory was claimed by them only in recent decades as they moved outward from core areas in the central part of the region. Geita District and eastern Maswa and Shinyanga districts were the major areas of expansion for the Sukuma, and they are now moving into western Geita, northern Nzega, and eastern Kahama districts.

The relatively dense settlement and cultivation in the region has resulted in the removal of most of the natural vegetation. Only in the newer, more sparsely settled areas such as western Geita do any large areas of the natural woodlands remain. This condition of the region led early writers to refer to it as the "cultivation steppe" (Rounce, 1949).

The landscape in the region varies from flat to low relief, with only a few areas of larger hills interrupting the rolling to gently rolling character of the countryside. In general the local relief is greater in the north and east of the region than in the south.

Table 12–1. Rainfall stations

Nearest study site[a]	No. on map	Station name and No.[b]	Years of record	Mean annual rainfall (mm)	Available moisture index[c]	No. of months with: $P > E_0$ 2	$P > E_0$ 10
12, 13	1	Mwanza 92:32000	41	1,022	49	7	11
12	2	Ukiriguru 92:33004	29	854	38	6	9
43	3	Ngudu 92:33005	36	828	39	6	8
51	4	Shanwa 93:33005	39	804	38	5	8
23, 24	5	Geita Gold Mine 92:32003	28	1,007	51	7	10

[a]Only in the case of site 12 at Mwanza is the station close enough to be more than suggestive of conditions at the site.

[b]East African Meteorological Department registration number.

[c]An index of moisture availability calculated after Woodhead (1970). Essentially this is the proportion of water available to satisfy annual evapotranspiration demand. For each site, soil moisture storage capacity was estimated at 230 mm. Walter Charles Murphy of Clark University wrote the program used to calculate the indexes.

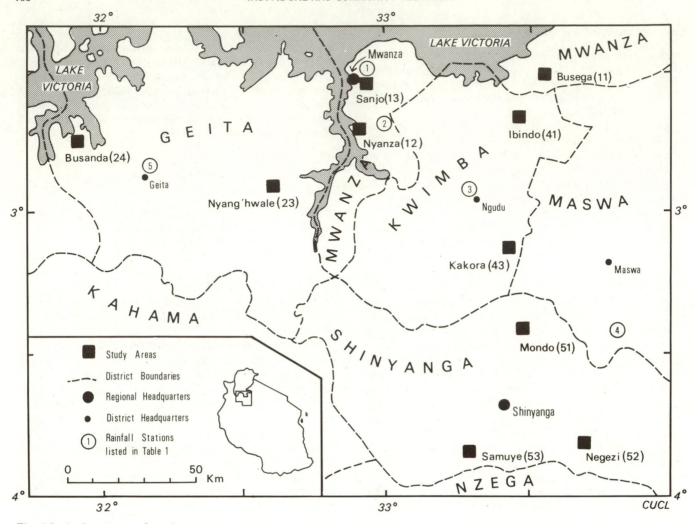

Fig. 12-1. Location of study areas

Soils vary with location on the slope. Upper soils are sandier and more easily worked. Downslope the soils become heavier, the increasing clay content providing capacity for greater moisture storage. Soils also vary with their parent material of which granite is the most common with ironstone and others also being important in some areas. The Sukuma pay close attention to soils and have strong preferences for certain crop-soil combinations (Malcolm, 1953).

Sukuma farmers plant a variety of crops. Maize, millet, rice, sorghum, sweet potatoes, and cassava are the primary food crops but chick peas, groundnuts, bambarra nuts, grains, several types of beans or peas, and bananas and other fruits are also grown and are of importance locally. Cotton is the main cash crop, but rice, chick peas, cassava, and other crops are also sold by farmers with surpluses.

Traditionally the Sukuma have cultivated on ridges constructed with short-handled hoes. Although these ridges, ranging from 1 to 3 meters (3 to 10 feet) from crest to crest, are laborious to construct, they provide effective weed control in the early weeks of crop growth as well as increase the moisture-holding capacity of the soil and slow the rate of runoff. Ox plowing has now become popular in the south and east of the region and in some parts of Geita District because it provides a considerable saving in cultivation labor. Fields cultivated by this method do not have ridges and the advantages they provide, however.

Almost all labor used in crop production is household labor. Nonhousehold sources are sometimes used; in fact, many households make use of them at times, but the total nonhousehold labor used amounts to only about 5 percent of the total labor input (SIRP, 1971). Tractors are sometimes hired; in 1968–69, 7 percent of the area cultivated was plowed by tractor (SIRP, 1971).

Capital inputs into Sukuma farming are very slight. For most farms they consist of a few hand tools, buildings, and baskets for storage. Insecticides are rarely applied and fertilizers are used even more rarely. Oxen and ox plows are the only large capital investments to have become a major input into the crop-farming operations in any large portions of the region.

Table 12–2 illustrates the similarity in households in the two hazard zones. The average farm household has five or six persons, about the equivalent, from the standpoint of labor supply, to two and a half adult men. The ratio of men to women is about 1, and about 10 percent of the men have more than one wife.

Other variations are evident in comparing cultivation and livestock characteristics. Farms in the drier zone are larger; they have a larger proportion of their land in maize and other grains; they are more apt to have livestock, especially cattle, but they do not exhibit the diversity of crops that wetter-zone farms do.

The median length of farm residence at the sites indicates little difference among the zones. There is, however, considerable variation within each zone, as each has "new" and "old" areas of settlement.

While Sukumaland contains the largest number of livestock of any region of the country, nearly all Sukuma families depend primarily upon crops for their livelihood. Livestock and crop farming are not closely integrated except for the use of oxen to plow; nevertheless the animals are very important to their owners. They are a major component of most bride-price payments; ownership of a large herd brings prestige; they provide milk and meat; and the animals or their hides can be sold for cash. The portion of farmers owning livestock varies from place to place; for the SIRP sample 48 percent of the farms had livestock and 40 percent owned cattle.

Sukuma livestock have so far contributed only a fraction of their potential to the national livestock markets. The region is home for 36 percent of the nation's cattle, but it supplies only 15 percent of the marketed number. Corresponding figures for Shinyanga Region alone are 30 percent and 10 percent (Ministry of Economic Affairs and Development Planning, 1968).

Sukumaland's importance to the national economy is due mainly to its role as a cotton producer. Ninety to 95 percent of the cotton produced in Tanzania comes from Sukumaland or adjacent districts to the north or south (Tanzania, 1971). Cotton has consistently been among the leading Tanzanian exports in terms of value over the past decade, accounting for an average of roughly 13 percent of total export revenue (Tanzania, 1971).

Table 12–2. Characteristics of farm units

Characteristics	Available moisture zones	
	Moderate High	Moderate Low
Household size		
Number	5.68	5.69
Man-equivalent units (MEQ)[a]	2.34	2.41
M:F ratio	1.07	1.00
% with > 1 wife	9.7	9.5
Cultivation		
Plots per farm	2.8	3.4
Ha cultivated per person	0.26	0.38
Ha cultivated per MEQ	0.64	0.90
Ha food crop per MEQ	0.28	0.52
Food crop:nonfood crop ratio	0.98	2.25
% cultivated land in maize	27	42
Median grain surplus per person in good year (kg)	16.5	37.5
Diversification index[b]	5.0	3.7
Livestock		
% farms with livestock	41	49
% farms with cattle	23	46
No. cattle per person	0.52	1.64
Residency		
Median no. of years on site	14	14

[a]Man-equivalent labor units. Rates all laborers according to a system of coefficients developed by Michael Collinson of Ukiriguru Research Station in Sukumaland. Adult males are rated as 1.0 and other age-sex groups according to their estimated abilities for physical labor.

[b]Hectare

[c]The number of crops growing on 10 percent or more of the cultivated area.

Hazard experience

Do farmers in the lower moisture zone live in a "riskier" environment, or at least believe that their environment is more risky compared with their fellow farmers in the higher moisture zone? In particular, do they perceive a greater drought hazard than do the farmers in the moister areas? This question was approached in several ways through interviews.

Farmers' responses to the request to name all the bad droughts they could remember that had occurred at their present locations do not indicate any difference in the number of serious droughts experienced. This may be true, but it may also be true that the question itself and the way it was handled by the interviewers influenced the number of droughts mentioned.

A greater distinction between the areas appeared in the farmers' estimates of the number of "bad" years out of the total number of years lived in the place. Yet the difference between 13 percent for the wetter zone and 20 percent for the drier zone is small and perhaps not significant.

In trying to relate the magnitude of droughts to moisture availability we sought to determine the nature of damage suffered in a "serious" drought. The replies reveal some important characteristics of farming in the two areas. The cash crop on which most of the farmers base their damage estimates is cotton. The different methods of planting cotton result in very different crop densities so that it is reasonable that the higher-density crops, as sown in the more humid areas, are more subject to drought damage than the less densely sown cotton. The density of cotton in the wetter areas averages 62 percent of the recommended spacing of 57,400 plants per hectare (23,000 per acre); in the drier areas it averages 43 percent (Hulls, 1971). Fifty-three percent of the wetter-zone farmers reported substantial damage to cash crops, and only 26 percent of the drier-zone farmers did so.

The nearly identical proportions (60–64 percent) of farmers reporting substantial damage to food crops appear to reflect a similar risk level in the two moisture availability zones. This may be a result of the different food crops grown in the areas as well as other variations in farming practices.

In the case of cattle, the responses are indicators of the different environmental conditions. In the drier zone, where dry seasons are longer and droughts apt to be more severe, 62 percent of the owners reported substantial damage, while the figure was 37 percent for owners in the more humid zone.

The proportion of families (47 percent) reporting it common for them to experience hunger is greater in the drier zone by about one-third. This again suggests that the risk level (from all hazards) is higher in the drier areas, although the size of the difference is not great.

Together these six indicators lend some support to the hypothesis that drought risk and damage are greater in the areas of lower moisture availability, but it is not strong support. The two parts of Sukumaland, with only small differences in rainfall characteristics, exhibit small and indistinct differences in farmers' hazard experiences. Variations in farming practices involving both adaptations and adjustments to the drought hazard could be responsible for this situation.

Adaptive capacity

The term "adaptive capacity" is used here to refer to permanent elements of the farming system, in contrast to the term "adjustment" which is used in reference to those activities practiced only during or in direct response to a specific drought occurrence.

Two sets of measures serve as indicators of adaptive capacity: first, characteristics of farming systems; and second, the well-being of farm families.

The measures of adaptive capacity of the farming systems provide a fairly clear distinction between the two types of areas (Table 12–3). Only the diversification index and the proportion of farms growing "drought-resistant crops" indicate a greater adaptive capacity in the wetter zone. The higher diversification index is the result of larger plantings of cassava and a variety of legumes in the wetter areas. Most of the legumes are interplanted with other crops, a reflection of the higher intensity of the agriculture in these areas.

Cassava is the crop that makes the bulk of the difference in the "drought crop" index, there being little difference in the proportions of farms growing millet/sorghum or sweet potatoes. One factor adding to the higher number of cassava growers in the wetter zone is

Table 12–3. Adaptive capacity[a]

Indicators	Moisture availability zones	
	Moderate High	Moderate Low
Farming system		
Diversification index	5.0	3.7
% farms growing		
Cassava	65	20
Millet/sorghum	36	30
Sweet potatoes	39	31
At least one of above crops	82	61
Median grain surplus per person in a good year (kg)	17	38
Average no. plots per farm	0.28	0.34
Average no. cattle per person	0.52	1.64
Average no. hectares cultivated per person	0.26	0.38
Measures of well-being		
Nutrition index[b]	9.6	7.0
Gross food crop returns per ha (Shs.)	653	596
Gross crop returns per ha (Shs.)	551	624
Net crop income per MEQ (Shs.)	341	525
Net farm income per MEQ (Shs.)	440	725
Net per capita income (Shs.)	251	347
Material wealth index[c]	1.5	1.8
% farms classified below average for site (by interviewer)	13.0	4.7

[a]The values given are the means of the values for the study sites in the respective zones.

[b]Rating of diet quality based on the frequency of consumption of beans, milk, meat, and fish during an average week. Milk, meat, and fish were deemed of equal nutritional value and beans slightly lower. Maximum possible score = 28 (each of the four items consumed every day of the week).

[c]One point given for possession of any of the following: bicycle, radio, modern shoes, watch, wick lamp, clothes iron, metal bed, crop sprayer, wooden table, motorized transport. Two points given for a pressurized lamp. Maximum score = 11. Both this and nutrition index were developed by Robert H. Hulls for SIRP use.

cassavas being grown as a cash crop in much of Geita District (sites 23 and 24). A second factor of importance is the substitution of grain surplus for cassava as drought food in drier areas. There is a high negative correlation ($r = -0.65$) between size of the median per capita grain surplus (in a good year) at the ten sites and the proportions of the farmers growing cassava.

All other indicators of the adaptibility of the farming system suggest greater adaptive capacity in the drier areas. The per capita grain surplus is more than twice as large in the drier areas as in the wetter zone.

The number of plots per farm is an indicator of adaptive capacity for most of Sukumaland because it reflects the extent to which different soil types with differing moisture capacities are used. The drier zone has a median number of plots about 25 percent greater than the wet areas. This difference becomes more significant when considered along with the fact that the flatter soils to the southeast have less marked soil variations than the other parts of Sukumaland.

Similarly, the drier zone shows a greater number of hectares cultivated per person and more cattle per person. Cattle serve the functions mentioned earlier, not the least of which is drought insurance. Even if they die in a drought, cattle can provide meat or cash (from the hide) for their owners and they can always be sold or traded. The three times greater number of cattle per person in the drier zone is clear evidence that this one type of adaptation has a considerably larger role than in the wetter zone.

The significance of hectares cultivated per capita is not so clear. Given that the "moderate/low" moisture availability zone is drier, a larger cultivated area per capita is necessary to keep the risk level even. The most that can be said of this statistic alone is that the difference is in the direction of maintaining or decreasing the degree of risk.

Although the measures of well-being do not explain how the farming systems adapt to the hazard zone, they do provide, as a group, a fairly reliable indication as to how well the farmers have succeeded in adapting to the two types of areas. For only two of the eight measures does the humid zone show a higher value than the drier zone, these being the nutrition index and the average food crop returns per hectare. The relatively large amounts of fish in the diet of persons living near Lake Victoria are probably sufficient to make up the difference between these two figures. In the case of returns per hectare to food crops, the wetter zone would be expected to be higher because the higher rainfall should permit more intensive use of land.

Why are the returns per hectare from all crops not also higher? The answer to this question is not entirely clear. The land is used more intensively as most cotton is planted on ridges and spaced more closely. It is suspected that poorer soils used for cotton account for much of this difference. Lower prices for other cash crops may also be relevant.

The returns per hectare for cash crops and the three income statistics together provide considerable support for the hypothesis that superior economic opportunity is available in the drier zone. It seems reasonable that these higher economic returns would make it easier for the residents to absorb the losses from droughts or other hazards.

The remaining two measures of well-being also indicate that farmers in the drier areas are better off. The material wealth indexes are so similar that little confidence should be placed in that distinction, but the interviewers' rating is clearer. They judge the ratio of below-average farms to be nearly three times as high in the wetter zone as in the drier. Note, too, that this is the only statistic other than the material wealth index that reflects more than a single year's output.

Taken as a whole, the various measures of adaptive capacity of the farming system in Table 12–3 suggest greater adaptive capacity in the drier areas, the larger number of cattle per person being the most significant

of these. The measures of well-being generally reinforce this conclusion and suggest that the higher-hazard areas presently provide greater economic opportunity to farmers.

Drought adjustments

Respondents were asked about the following 11 adjustments on the drought study questionnaire:

Do nothing;
Adjusted planting time;
Drought-resistant crops;
Thorough weeding;
Cultivation of larger areas;
Rainmakers;
Work for wages;
Tie ridging;
Planting on *mbuga* (low-lying, heavy soils with high moisture capacity which may become waterlogged in years of normal rainfall);
Sending cattle to other areas to graze;
Selling livestock to buy food.

Omitted from the list are some adjustments that are obviously part of the Sukuma way of life. Among these are sending children to kinsmen, moving house, storing bumper crops, and planting dry. This is regrettable because it makes this part of the study less useful than it might have been.

In attempting to evaluate the responses to the adjustment inquiries on the questionnaire, three aspects are identified: (1) the number of adjustments mentioned by the respondents, (2) the number of adjustments they said they practiced (had adopted), and (3) those adjustments which have been adopted by 50 percent or more of the farmers in the two moisture availability zones.

There is not much difference in the responses from the two zones. In each case the median number of adjustments mentioned is one, and of adjustments adopted is six. There is also scarcely any difference in the number mentioned and adopted by the tenth and ninetieth percentile farmers.

Adjustments practiced by 50 percent or more of the farmers are listed in Table 12–4. Again there is almost no distinction to be made between the zones. All of the four adjustments listed as common in the drier, but not the wetter zone, are practiced by 10–49 percent of the farmers in the latter area. The smaller number using adjustments related to livestock there is to be expected because fewer farmers are livestock owners. The responses to the tie ridging inquiry are a puzzle because tie ridging is very rarely practiced. One explanation is that the interviewers at times confused tie ridging with normal ridging and the responses thus refer to the latter.

Unlike adaptive capacities, adjustment practices reveal scant differences between the zones. It appears that any variances in responses to drought by farmers in the two zones are not in the kind or number of adjustments practiced.

Table 12–4. Adjustments practiced by at least 50 percent of farmers

Type of adjustment	Available moisture Moderate High	Classification Moderate Low
Accept or self-insure loss		
Work for wages to buy food	+	+
Sell cattle to buy food		+
Plant larger areas	+	+
Eliminate moisture waste		
Weed plots more	+	+
Tie ridge		+
Change moisture requirements		
Plant drought-resistant crops	+	+
Affect source		
Hire rainmaker		+
Change location		
Plant in wet places	+	+
Move cattle		+

Conclusion

The evidence suggests that although the environmental gradient existing in Sukumaland is fairly gentle, drought hazard experience and perceived risk are slightly higher in the areas with lower moisture availability. There is a distinct difference in the adaptive capacities of farms, those in drier areas appearing to be more capable of sustaining drought effects. However, there is no discernable difference in the short-term adjustments to drought practiced by farmers in the two zones except for those related to livestock ownership. Sukuma farmers have responded to the difference in environmental conditions by making permanent alterations in their farming methods, but not by changes in their use of short term adjustments to the drought hazard.

An explanation for this situation may lie with the risk levels in the two areas. A greater variety of adjustments to drought might be expected in a drier zone than in a wetter one, but this should occur only if the risk level is different or if the farmers in the two zones are from different backgrounds. The latter is not true for Sukumaland. The question then is, could the risk level be similar in the two zones? The variations in adaptive capacities provide grounds for arguing for similar risk levels. Certainly the greater adaptive capacities evidenced by drier zone farms would tend to reduce riskiness in the drier zone.

Although the drier zone is generally well adjusted to greater drought risk, not all portions of it are. Some areas in the northern part of the zone have farm characteristics more similar to farms in the wetter zone—fewer cattle and smaller areas of food crops planted per person—so that the system in these areas is less able to bear the burden of a drought. With less land now available for Sukuma expansion, population densities may be expected to increase throughout the drier zone. When this happens, the rest of the zone will be forced, as the northern areas already have been, to change their farming practices in ways that lessen the system's ability to endure drought. This would result in large increases in personal and community losses in the region, and, from the national point of view, it would mean a significant decrease in the amount of cotton planted. To avoid this situation the government must either avoid major increases in population densities in the drier zone or bring about general increases in the productivity per hectare of farms in there.

In Natural Hazards Research Working Paper No. 16 several reasons for human occupance of areas of recurrent hazard were suggested. Two of these hypotheses are supported by the results of this study, which has shown that economic opportunity is higher in the drier zone and that farmers in the drier zone do maintain high ratios of reserves in order to cover potential losses.

To what extent might other factors explain the variations between the two zones discussed here? Two possibilities were thought to possibly be important: length of residence and availability of land. Checks on median years of residence (Table 12–2) and some indexes of land availability (SIRP, 1971), however, indicated no variation between the zones on these factors.

References

Berry, L., Hankins, T., Kates, R. W., Maki, L., and Porter, P. (1971) "Human adjustment to agricultural drought in Tanzania: pilot investigations." Toronto: University of Toronto, Department of Geography, Natural Hazards Research Working Paper No. 19, p. 17.

Brooke, Clarke. (1967) "Types of food shortage in Tanzania." *Geographical Review* 57:333–357.

Cory, Hans. (1951) *The Ntemi: Traditional Rites of a Sukuma Chief in Tanganyika.* London: Macmillan.

Cotter, Fr. George, M. M. (c. 1967) No title (collection of Sukuma proverbs). Shinyanga, Tanzania.

Hulls, R. H. (1971) "An assessment of agricultural extension in Sukumaland, western Tanzania." Reading, England: University of Reading, unpublished report.

Malcolm, D. W. (1953) *Sukumaland.* London: Oxford University Press for the International African Institute.

McLaughlin, Peter F. M. (1971) *An Economic History of Sukumaland, Tanzania, to 1964: Field Notes and Analysis.* Fredericton, New Brunswick: Peter McLaughlin Associates.

Ministry of Economic Affairs and Development Planning. (1968) *District Data,* Dar es Salaam.

Natural Hazards Research. (1970) "Suggestions for comparative field observations on natural hazards." Toronto: Univ. of Toronto, Dept. of Geography, Working Paper No. 16.

Rounce, N. V. (1949) *The Agriculture of the Cultivation Steppe,* Cape Town: Longmans, Green.

Sukumaland Interdisciplinary Research Project, (1971) University of Dar es Salaam, Research Report No. 40 of the Bureau of Resource Assessment and Land Use Planning.

Tanzania, United Republic of. (1971) *The Economic Survey, 1970–71.* Dar es Salaam: Government Printer.

Woodhead, T. (1970) "A classification of East African Rangeland: II. The water balance as a guide to site potential." *Journal of Applied Ecology* 7:647–52.

13. Northeast Tanzania: comparative observations along a moisture gradient

JOOP HEIJNEN
University of Utrecht

R. W. KATES
Clark University

From Mombo to Kulasi, a crow would fly the 40 kilometers (25 miles) in 3 hours, a Land Rover would labor all day, and a bus would not attempt it. Along the way the landscape shifts from dry open woodland to moist woodland, to rain forest, to moist woodland, to dry open woodland again; the elevation more than triples from 400 to 1,400 meters (1,300–4,600 feet); the annual rainfall doubles from 675 to 1,350 millimeters (27–54 inches). Cutting across the Western Usambara mountains of northeast Tanzania with its dense, predominant Shambala-speaking smallholder farming population, this study seeks to trace those variations in farmer activity relative to drought that seem to coincide with differences in the steep gradient of available moisture. By limiting these observations to a very short transect across a homogeneous population, this particular study seeks to relate the variations in environmental opportunity to the well-being of the population, the adaptive capacity of local agricultural systems, the drought-hazard experience, and the adjustments undertaken to minimize drought impacts.

Expectations of hazard, well-being, and available moisture

The road from Mombo to Kulasi follows a steep gradient of moisture available in the soil plants. What expectations may we have as to the drought experience of farmers along this gradient and its relationship to their overall well-being? Can we turn to intuition or theory to suggest these relationships?

Consider first our expectation of how drought hazard might vary along the gradient of environmental potential. A simple expectation based on the dominance of environmental factors would be for drought experience to correlate inversely with available moisture given an absolute reduction in moisture and a general association of greater variability with diminishing precipitation. But hazard is as much a function of the agricultural and social system as of nature. Porter (1965, p. 411) states this well in the east African context:

> Subsistence risk is not given in nature, it is a settlement negotiated between an environment and a technology. Just how much risk an individual or a community can tolerate, how often a failure of crops or decimation of herds can be borne, is a problem that each culture must solve. A community has institutional and technical means of coping with risk. It can tighten its belt, develop surpluses, or raid neighboring territory. Danger to the individual can be decreased by sharing out risks, through dispersal of fields, timing of harvests, cattle deals, and the like.

Irrigation, either by advanced technology or natural oases, is an example of successful "negotiation" and its net effect may be to virtually eliminate the drought hazard for the population involved. Our expectation of the conflicting dynamics of the process involved can be seen in greater detail considering moisture and well-being.

The notion of some overall well-being of population is indeed nebulous. Under the rubric of well-being we can subsume the biological needs of the populace for survival, food, water, shelter, absence of disease, economic desires for material goods and opportunities, and social needs for security, respect, and leisure. The mix of needs, desire, and hope is not constant; some factors are even contradictory, such as leisure and economic opportunity.

Nevertheless, given some vague but shared notion of relative well-being, let us examine its relationship to available moisture in an agricultural system with a strong subsistence component. An environmental theory might suggest the relationship of Fig. 13–1a. Here our expectation is for well-being to increase with available moisture, not linearly and with local variability, but following some sigmoid curve suggesting a minimum threshold of moisture to improve the lot of the area's inhabitants, rapid increase of well-being with moisture, and then asymptotic diminishing benefits until levels of excessive moisture limit or diminish well-being. On the road from Mombo to Kulasi, we would expect relative well-being to rise with elevation and rainfall, perhaps modified but essentially not changed by other factors.

Cultural theories would deny us this tidy relationship, indeed they would prepare us for a null relationship as shown in Fig. 13–1b. Such theory acknowledges differences in environmental opportunity, but assert that these will be compensated for by culture. The settlement pattern, the demographic mix, the agricultural system, indeed the concept of well-being itself—all will adjust to the level of environmental opportunity. On the arid end of the continuum, settlement will be sparse, population numbers small, and livelihood will have a pastoral orientation. A smaller, more mobile society will substitute extensive quantities of sparse resources for intensive use of resources of higher quality. For the limited numbers supported, diets with easy access to milk, meat, and blood may be superior to those of better-watered areas, and wealth—measured in livestock—may be considerable.

Alternatively in the most favored range of the continuum the abundance of fertile soil and water will encour-

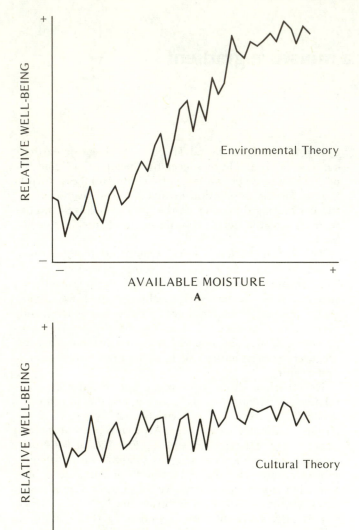

Fig. 13–1. Relative well-being and available moisture theoretical relationships: (a), environmental theory; (b), cultural theory

age permanence of settlement, and with little out-migration, rapid population growth will lead to fragmentation of land, declining fertility, and diminished well-being. Thus, culture theory would lead us to expect a relative constancy of well-being along the road from Mombo to Kulasi, mobility increasing the access to resources on the dry margins, while, as rainfall increases, social and demographic pressures also increase, reducing well-being below the level of environmental opportunity.

Complicating the human ecological patterns that emerge from the interplay of natural environment and human culture are cultural opportunities superposed external to the transect from Mombo to Kulasi. These differential opportunities may include the accessibility of the major north-south road that passes through Mombo, the level of encouragement given to cash crops of coffee, tea, or cardamom by the structure of world markets, the alternative employment opportunities in sisal, or leadership differences in agricultural instruction or political exhortation, even the differential values of Islam and Christendom, both of which claim substantial numbers of Shambala adherents.

The expectation then is for complexity and some unclarity; the intellectual path from Mombo to Kulasi is no less difficult to traverse than the road itself. There is a steep gradient of environmental opportunity, subtle variation in cultural opportunity, and a changing pattern of settlement and agriculture seeking to adapt to these differences.

Data collection

This paper relies heavily on an analysis of responses from 254 farmers to an extended questionnaire containing about 170 questions dealing with six major topics: characteristics of the respondents, their households, sites, and farms, their perception of and experience with drought, their patterns of adjustments to drought. It was administered primarily during May 1971 following a particularly dry crop year by a group of university students from the rural research group of the Bureau of Resource Assessment and Land Use Planning of the University of Dar es Salaam working along with a survey group of the Lushoto Integrated Development Project composed of young men mainly with primary school education. The entire group were Shambala-speaking and administered the questionnaire in that language at ten sites. In the remaining three sites containing some non-Shambala speakers a slightly different Swahili version was used.

The sample is nonrandom, and in clusters. Thirteen sites were chosen to provide a transect of the moisture gradient. In each site a minimum of 20 interviews were sought and 6 interviews less in total were obtained. Instruction for choosing respondents was to secure a point of entry in the community, usually through the local leader (the Tanzanian Government party, TANU, has leadership at the level of the householder) and then interview areally around that household to fill the quota. In most areas, 20 respondents would include a 50–100 percent sample of the location, given the prevailing pattern of settlement.

To those familiar with the hazards of the cross-cultural research we urge a cautious outlook. We have grouped the data in only three moisture classes to give larger samples. We have relied on ordering statistics rather than their cardinal values (e.g., the use of the median as a descriptive statement) to minimize effects of extreme values. A large body of supplementary research material including other surveys has been used for verification; one of us, Heijnen, can draw on observations during several years of intensive work and study

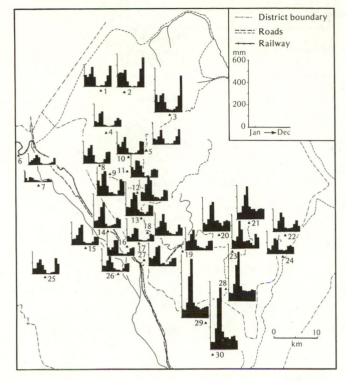

No. of years of rainfall records

1 Tema farm (12)	11 Mlombora (3)	21 Mazumbai (32)
2 Shagayu (10)	12 Magamba (29)	22 Kulasi Pln. (7)
3 Lwandai (20)	13 Lushoto (54)	23 Bumbuli Hospital (8)
4 Manolo (2)	14 Mazinde Hs. (24)	24 Magoma Sisal (16)
5 Malindi (17)	15 Mazinde (31)	25 Toronto Sisal (3)
6 Buiko Rly	16 Kikwajuni (18)	26 Mombo C/Stn. (9)
7 Lake Manka (3)	17 Mabogo (19)	27 Mombo Estate (20)
8 Shume (37)	18 Ubiri (14)	28 Balangai (37)
9 Gologolo (11)	19 St. Michael Soni (9)	29 Sakarre Estate (42)
10 Hambarawei Lr. (6)	20 Herkulu (28)	30 Ambangulu (37)

Fig. 13–2. Rainfall distribution

within the region. We ask our reader then to share in the caution, and to ask from the data, as we will, for trends and explanations rather than isolated bits of statistical significance. And to accept the residual uncertainty, not knowing whether it arises from the complexity of culture and environmental interaction or the faulty nature of our research instruments.

From Mombo to Kulasi: site descriptions

The 5,000 square kilometers (1,900 square miles) of the Western Usambaras and surrounding plains contain a highly variable landscape and moisture gradient. In Fig. 13–2, the exceptional spatial and seasonal variability of the rainfall pattern is shown. For the transect itself, only six rainfall gauges are available nearby. Four of these are part of the Meteorological Department network (shown on Fig. 13–2 as numbers 21, 24, 26, and

27), one has been discontinued (Vuga), and one was recently started (Kongoi).

The thirteen study sites are listed in Table 13–1 and shown in Fig. 13–3 along with the moisture classification. Had rainfall alone been used to classify sites into high, moderate, and low available moisture groups, few would have been classified as moderate. But the presence of irrigation at Mombo (701) and Mailitano (803) or seasonal wetlands at Kulasi (713) supplements the precipitation and increases the moisture availability.

The typical household studied is made up of six persons, with one or two children below the age of 5 and, in many homes, at least one elder. Little difference in the available farm-labor force (calculated as man-equivalents) is found between moisture zones, but a significantly greater number of multiwife households are found in the area of low moisture availability. Some non-Shambala speakers are also found in that zone containing two sites on the plains.

Farmers cultivate typically between 1.5 and 2 hectares (4–5 acres) fragmented into an average of five to six plots with a third to a half planted in maize. The range of yields (5:1 to 10:1) between good and bad years for this staple is considerable even if one tempers the farmer's estimates. The ownership of cattle is exceptional rather than common.

The agricultural system practiced in the Usambaras is complex, the crops are many, the variation in land and season considerable. Altitude, notably in its effect on temperature during the dry season; rainfall, its pattern and amount; location, the availability of irrigable land and floodplain—all complicate the cropping pattern. As the rainfall pattern changes into a more clearly demarcated wet and dry period, options for the farmers are fewer and especially the Vuli crop (early rains) is at risk. In the north, on the other hand, the Mwaka crop (main rains, April-May) may be more likely to fail, as in sites 809–811. Whether or not the farmers can plant in the Kitivo (valley floor, floodplains) is naturally dependent upon the presence of such land. For instance, sites 806–712 have very little if any land in this category. If the rains are early, or late, the pattern of activities is shifted accordingly.

These then are the raw summary qualities of environment, life, and livelihood along the road from Mombo to Kulasi; however, to convey a real feel for place, one must travel the road itself.

Mombo irrigation scheme (701)

Available moisture: moderate. At about 4 kilometers (2½ miles) from Mombo settlement, toward the Pangani River, the Mombo irrigation scheme takes water from Soni River, which has a year-round flow. The scheme is poorly managed and underutilized. During the dry season only a few small plots are cultivated, chiefly to grow tomatoes. As a rule, the land is used only during the wet season, when it is plowed by tractor. Most participants

Table 13-1. Usambara transect study sites

Site No.	Name	Est. elevation (m)	Est. annual precipitation (mm)	Supplemental moisture source	Available moisture classification
701	Mombo	410	700	Irrigation project	Moderate
702	Mombo	415	700		Low
712	Mashewa	410	750		Low
713	Kulasi	380	750	Seasonal wet lands available	Moderate
803	Mailitano	710	850	Traditional irrigation	Moderate
804	Vuga	1,220	1,100		High
805	Mponde	1,400	1,200	Stream nearby	High
806	Bumbuli	1,220	1,200	Soils, slopes favorable	High
807	Handei	1,400	1,300		High
808	Upper Kongoi	1,200	850		Moderate
809	Middle Kongoi	1,070	750		Low
810	Lower Kongoi	915	700		Low
811	Mlingano	505	700	Floodplain with high water table	Low

have 1 hectare (0.4 acre) of rice and 1 hectare of maize on the scheme and have additional fields outside where—in part—the same crops are grown. Irrigation is regarded mainly as an insurance against unexpected droughts. Part of the rice is sold; maize, cassava, and rice are the main food crops.

Mombo off-scheme (702)

Available moisture: low. In the same villages where the participants of the irrigation scheme live—called Kwe-sasu and Jitengemi—other families have settled. They grow the same crops and make use of the surplus water whenever possible for irrigation. Yet they are much more dependent upon the marginal rainfall of about 700 millimeters (28 inches) per annum, as water is available only in limited quantities. Much of their rice is grown in poorly drained areas where the water gathers during the wet season.

Mailitano (803)

Available moisture: moderate. Halfway between Mombo and Soni, along the winding gravel road, a few clusters

of houses are perched on the steeply sloping ground. Before the colonial era the local Shambala, ordered by their chief, constructed long irrigation channels and furrows. In this village the system was later expanded to cover the whole valley floor and the lower slopes. The higher land, mostly above the site of the village, cannot be irrigated and depends solely on a rather unreliable rainfall, averaging some 850 millimeters (31 inches) per annum. Nearly all families, however, have access to the irrigated sections of the valley and make full use of this opportunity. The fields are used intensively, especially during the dry season. Tomatoes and—to a lesser de-gree—onions, green peppers, and some ladyfinger are produced for cash mainly from July to December-January. Maize and cassava are the chief food crops. As elsewhere, however, part of the money earned is used to supplement the diet, mainly on such items as bread, oil, and fish.

Vuga (804)

Available moisture: high. Not much is left of the old splendor, when Vuga was the seat of the powerful Shambala chiefs. The actual site of their residence can

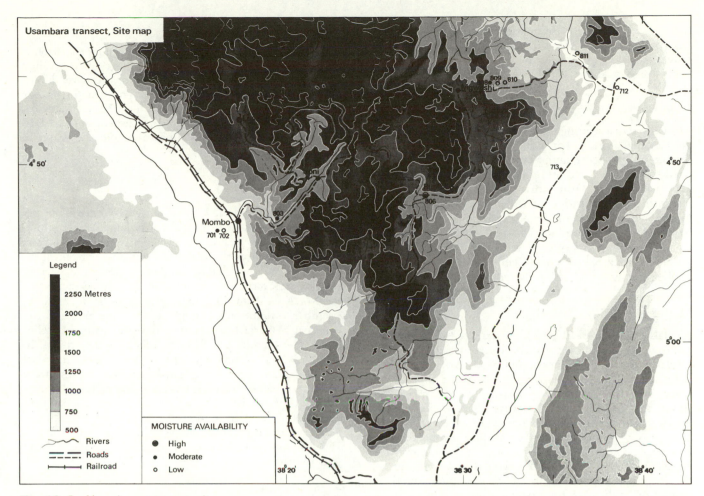

Fig. 13–3. Usambara transect, site map

barely be traced today. Vuga is only a divisional headquarters and a center of Lutheran missionary activities,
surrounded by a few shops and a great number of small
clusters of homesteads. The population density is high,
nearing 200 per square kilometer (520 per square mile).
Rainfall, at over 1,100 millimeters (43 inches), is favorable for arable agriculture, so that the risk of total crop
failure is minimal. The chief disadvantage of the site is
its location, some 5 kilometers (3 miles) off the main
access road into the Usambara Mountains, although
there is a reasonable road connection. Coffee and wattle
are the most important cash crops; maize, beans, bananas, and cassava the chief subsistence crops.

Mponde (805)

Available moisture: high. Isolation is a much more pronounced characteristic of the densely settled Mponde
area. There are only a few poorly stocked shops. The
only reasonable road connection is with Bumbuli, in the
north. Until a few years ago, coffee was the only cash
crop of some importance. Altitude, exposure, and rainfall are, however, favorable for tea, which was first
introduced in 1964. Today tea covers large sections of

the slopes. The result has been that the food growth is
sufficient only in favorable years. Often the farmers
have to use part of their proceeds from the tea to buy
maize and other foods. The valley floor does provide an
opportunity to grow additional food, like beans and
potatoes, if need arises. In 1970, a development agency,
LIDEP, started a vegetable scheme here which was intended to provide additional cash, but especially to
improve the local diet.

Bumbuli (806)

Available moisture: high. Bumbuli in former days was
the court of the heir to the Shambala chief's throne. At
present it is a divisional headquarters and a center for
the Lutheran mission, which during the early sixties
built a large modern hospital here. Unlike Vuga, Bumbuli retained a central place function for the eastern
section of the Usambara Mountains, with a number of
wholesale and well-stocked retail shops, transporters,
"hoteli," etc. With only brief interruptions, the road to
Soni can be used the year round. Soils and climate are
favorable for farming. Like Mponde, Bumbuli receives
approximately 1,200 millimeters (47 inches) per annum.

Coffee and wattle are the main cash crops; maize, bananas (partly in the coffee fields), beans, and cassava are grown for food. As a rule the slopes are steep and level land is very localized.

Handei (807)

Available moisture: high. Near the main watershed, Handei, a small cluster of houses, is situated on gently sloping or even level land. Some of the men work on the nearby Mazumbai tea estate to supplement their moderate cash income from coffee and (some) vegetables. Marketing opportunities for these vegetables are limited to the local markets of the Mgwashi and Bumbuli. The road to Soni via Bumbuli is often impassable for cars during the rainy season. Water for domestic use is a problem, since nearly all available water is in stagnant pools which are dirty and polluted.

Kongoi (808–810)

Available moisture: (Upper), moderate; (Middle, Lower), low. Across the watershed, a mere track leads from Mgwashi down to Mlingano/Masheqa. After a shower, only a Land Rover in low gear would attempt to descend and the driver would still have his misgivings about the decision on the way. The missionaries at Kongoi have a rain gauge from which they tapped an average of 946 millimeters (37 inches) during the past two years. The Mission and the nearby school overlook a deep V-shaped valley. From their station one can see the forest trees grading into shrub and open woodland vegetation, indicative of the rapid drop in rainfall below the mission site, to 700 millimeters (28 inches) or even less in places. After the coolness of the highland forests, the observer would not expect to find permanent settlement in this dry and hot wilderness. But below the Mission, perched on protruding ridges he sees three clusters of houses, at altitudes of 1,220, 1,070, and 915 meters (4,000, 3,500, and 3,000 feet) respectively. The past two years have been hard. Twice the maize failed partly, as in the highest settlement, or completely as lower down. Wild leaves, a few pieces of cassava, and a bittersweet extract from wild roots are the main items of the meager diet. The children look severely undernourished, as is shown by the reddish glow of their hair. The area is remote from the district center and no relief food has been forthcoming. A number of young people have left to look for work or to find a better place for farming. Water for domestic use is far away. A few cattle constitute the only barrier against disaster and death. The elder people are resigned. "Formerly there was a rainmaker, who made it rain. He left, because some people did no longer believe in him. That's why." Except for some coffee in the highest fields, there is no cash crop and even the food supply is extremely insecure. There are no large herds of cattle. People live at a bare subsistence level.

Mlingano (811)

Available moisture: low. A somewhat similar situation is encountered in Mlingano, a small village on the valley floor. But the vegetation is more luxurious here, due to a relatively high groundwater table. The same is true for the cultivated fields in the valley. In 1971, for instance, the maize had dried out, but the cassava appeared to withstand the drought. Groundwater and the water in the nearby rivulet are salty.

Mashewa (712)

Available moisture: low. The village is remarkable because of the presence of a relatively large Arab community, not found anywhere else in the mountains. The village site was on the old trade route which ran north of the mountains. Today the Arab traders and African shopkeepers serve the population of a large section of the Northern Plains and some nearby settlements in the Usambara Mountains. Drought has hit the area and the Korogwe District authorities have distributed relief food. The market has little to offer: some dried fish, meat, cassava, onions and ladyfingers. The nearby sisal plantation has been abandoned and the impression is that the standard of living has gone down, probably as a result of the crisis in the sisal industry. Maize and cassava are the main food crops; there is no cash crop of major importance. The annual rainfall is estimated at 750 millimeters (30 inches) with major variations from year to year.

Kulasi (713)

Available moisture: moderate. Near the sisal estates of Magoma and Kulasi a farming community is populated by ex-workers. Although the rainfall is approximately the same as at Mashewa, or marginally better and more reliable, the nearby floodplains afford insurance against total crop failure. Maize, cassava, beans, and sweet potatoes especially are grown in the floodplains, the latter on ridges or mounds. No specific cash crop is available, money is obtained by selling food surpluses, if the harvest permits. Cash income is low; only a few houses have corrugated iron roofs.

Hazard experiences of Usambara farmers

The Usambara farmers' recall of the occurrence of drought parallels the gradient of available moisture but they report that the severity of an individual drought event provides comparable losses of crop and cattle to all regardless of location. Farmers were asked to estimate the number of "bad years" encountered in the years that they lived at the site (if less than 10), or the years since independence (10) or the years since World War II (25 years). From these estimates, the proportion of "bad years" experienced and the median estimates for each moisture group was calculated (Table 13–2). The median number of droughts experienced, recalled by farmers, showed the same pattern. The proportion of "bad years" and the number of droughts recalled increases by 50 percent as one moves from high sites to medium sites and again from medium to low sites. The 100 percent difference between high and low is only slightly greater than the inverse of the average difference in annual rainfall, seasonality, or the aridity index.

Turning to questions that seek to measure the effects of drought events, a different pattern emerges. Similar percentages of farmers report substantial (greater than 21 percent) damage to major food crops in each moisture zone, although some difference is reported for cattle. These results support the transactional view of drought hazard; i.e., given the prevailing levels of adjustment, events thought of as drought have similar effects across zones. However, a third measure, the frequency of household experience of hunger, strongly differentiates between the high-moisture zone where the experience is rare (15 percent) and the moderate and low zones where it is common (47 percent and 49 percent). For any given drought, crop loss may be similar; but hunger is dependent on the overall resources available, and these are highest in the high-moisture zone. And this in turn can also be modified by accessibility. The isolated settlement at Kongoi (808, 809, 810) and Mligano (811) suffer more than the more accessible dry-plain sites of Mashewa (712) or Mombo (702).

In sum, farmers' perception of the frequency of drought inversely correlates with broad moisture zones, but the prevailing pattern of adjustment leads to similar patterns of severity when a drought occurs. In the high-moisture zones, the lesser drought frequency, combined with other resources (money, cattle, longer plant-ing seasons), reduces considerably the overall stress placed on the household.

Adaptation and adjustment to drought hazard

In the face of drought, Usambara farmers choose from a wide range of purposive actions designed to control or modify the shortfalls of available moisture, to make their crops or cattle less vulnerable to the lack of moisture, and to bear and share the losses of crop and animal production.

When asked what they do when the rains are late or insufficient, farmers suggest a number of preferred actions, the number varying with the available moisture gradient as shown in Table 13–3. The number of adjustments mentioned is greatest in the moderate area and the range measured by the first and ninth decile is considerably greater in the moderate- and low-moisture areas. When asked from a checklist of up to 19 adjustments which they actually employ, the number of adjustments reported by farmers as adopted shows less variation between zones.

While the variation in the number of adjustments adopted is small, a few distinctive differences emerge in examining the favorite adjustments of each moisture zone as shown in Table 13–4. Farmers in all three zones favor seeking alternative cash sources or employing meager savings, employ the labor-saving device of ceasing effort if it appears useless, advocate good practices of early planting with the rains and weeding, undertake increased plantings of cassava and irrigation, and practice the supportive exercise of prayer. In addition to these universal adjustments the high areas, with their more favored environment, even during drought, can employ various alternate options of scattering plots and planting, and seeking low and seasonally wet places for catch crops. The valley floor at Mponde (805) and the adjacent marshy areas at Handei (807) serve this purpose. In the moderate- and low-moisture areas, the

Table 13–2. Usambara transect: measures of hazard experience

Measures	Available moisture		
	High	Moderate	Low
Recurrence			
Median farmers' estimate, proportion of "bad" years	0.20	0.30	0.40
Median farmers' no. of drought years recalled	2	3	4
Loss			
% reporting experience, at least substantial damage:			
To major food crops	96%	81%	95%
To cattle	71%	93%	95%
Stress			
% reporting common family experience of hunger	15%	47%	49%

Table 13–3. Usambara transect: number and range of adjustments mentioned and reported adopted

Indicator of adjustment	Available moisture classification		
	High	Moderate	Low
Total number mentioned			
Lowest decile 10%	2	0	1
Median 50%	4	7	4
Highest decile 90%	5	10	10
10%–90% range	3	10	9
Total number reported adopted			
Lowest decile 10%	8	9	7
Median 50%	13	13	12
Highest decile 90%	15	19	15
10%–90% range	7	10	8

Table 13–4. Usambara transect: type of adjustments reported adopted by at least 50 percent of farmers

Type of adjustment	Available moisture classification		
	High	Moderate	Low
Accept self-insure loss			
Work for wages to buy food	+	+	+
Sell cattle to buy food			
Use savings to buy food	+	+	+
Store more than one season's food when crop is good	+		
Distribute and share loss			
Move to another farm	+	+	
Ask help from friends and relatives		+	+
Ask help from the government		+	+
Eliminate moisture waste			
Weed plots	+	+	+
Stop planting when rains are not enough	+	+	+
Change moisture requirements			
Plant drought-resistant crops	+	+	+
Affect source			
Pay for rainmaker			+
Pray	+	+	+
Change location			
Have plots in different places	+		
Plant in wet places	+		
Move cattle			
Improve moisture storage and distribution			
Irrigate	+	+	+
Schedule for optimal moisture			
Plant without rain			
Plant only when enough rains come	+	+	+
Staggered planting	+		

Table 13–5. Usambara transect: indicators of adaptive capacity

Indicator	Available moisture		
	High	Moderate	Low
Farm size			
Per man-equivalent (ha)	0.50	0.74	0.61
Maize-cassava area per person (ha)	0.10	0.21	0.18
Crop diversity			
No. of crops, 10% land	3.1	2.6	2.5
Locational diversity			
No. of other plots	2.2	2.7	2.8
% household heads, 1 wife	14%	13%	35%
Drought-resistant crops			
% of units drought-resistant Crop	32%	55%	57%
% of land in cassava	11%	15%	20%
Normal surplus			
Median farmer estimate of grain surplus in good year (kg)	120	360	300
Cattle			
% farm units with cattle	34%	24%	16%

greater frequency of drought increases reliance on others for help. At Mashewa (712) food was provided and readily accepted by district authorities, and at Mlingano (811) by the LIDEP through the study team. The recourse to rainmaking in the low area is probably not coincidental.

Overall, with these few differences, the purposive strategies of adjustment differ little. It is rather in the everyday agricultural practice that the significance of the available moisture gradient is reflected. We can only draw on a few of these differences in what we might call the adaptive capacity of the agricultural system to deflect, absorb, or buffer considerable amounts of environmental stress or deprivation with a minimum of harm. From the data we have selected a number of measures of such built-in capacity for adaptation (Table 13–5).

The most significant of these indicators is the concept of the normal surplus developed by Allen (1965) with reference to subsistence cultivators. Allen suggests that in the face of the variability of crop production, often related to precipitation shortfalls, subsistence cultivators produce for what they need in a below-average year. This provides a considerable surplus in an above-average year and a modest surplus in a normal year. Overcapacity in production is the norm.

The surplus is not easily disposed of in economically rewarding ways and considerable human effort is required to create it. Thus, there is a real cost in maintaining the excessive productive capacity, as necessary for survival as it may be. A measure of the surplus was obtained by asking farmers in the three moisture areas what was their surplus of grain remaining just before the new harvest in a good year. Their answers provide a measure of overcapacity.

We would also expect differences in moisture to be compensated for by differences in farm size and the scatter of locations. The former, substituting quantity of area for quality of available moisture, the latter hedging the risk of localized failure. One might also expect a greater variety of crops to be planted in the low zone for the sharing of risk by differential crop moisture requirements, and a larger proportion of such areas to be planted in cassava, the major famine food staple. Cattle, one of the few ways of long-term storage of vegetation (in the form of beef) or storing wealth (in the form of the potential monetary value of the cattle), may serve as an indicator of a farmer's reserve.

The expected differences in all but two, crop diversity and cattle, of the indicators are large between the high-moisture zones on the other. The normal surplus of grain in the moderate area is three times that of the high moisture zone, and is almost as large in the low area. The per capita area devoted to the staple crops of maize and cassava about doubles, the amount of cassava increasing with less moisture. The number of plots located away from the immediate household is greater in the moderate and low area, and multiple households are much more common in the low area. The measure of crop diversity (major crops with over 10 percent of cultivated area) favors the high area, reflecting the greater security of wealth and environment which allows or encourages the planting of cash crops as at Vuga (804) and bumbali (806). Cattle in the Usambaras appear to be primarily a wealth measure rather than a storage alternative.

Thus, along the Mombo-Kulasi road, it is not so much the purposive adjustments in response to a specific drought as the adjustments embedded in the fabric of everyday life that serve to mediate the differentials of environmental opportunity. How successful these mediations are can be examined in measures of well-being.

Indicators of farmer well-being and opportunity for improvement

In the language of science we scarcely identify the quality of well-being or at the very least, relative well-being, let along measure it. Therefore, it is with considerable caution that we present a few measures that are derivable from our data: measures of nutrition, wealth, and subjective interviewer appraisal (Table 13–6).

By international standards, Usambara farmers are poor. The measure of nutrition credits eating three meals, makes distinction between usual and unusual staples, maize and rice, the availability of any protein, or having sugar, oil, or other such simple dietary ingredients. The index of material wealth identifies wealth with the ownership of a bicycle, a radio, or home improvements such as a metal roof or a cement floor. The ownership of cattle is another indicator as is the interviewers' classification of households relative to the living quality of the area.

It is a measure of the overall poverty that, using the simple nutrition index, the median Usambara farmer food consumption is only half of the maximum value of 38. The median farmers owned less than one of the material items on a five-point scale. In a country where the number of cattle equal the number of people only a fourth of the families had any cattle, where theoretically the interviewer measure of wealth should have shown two-thirds of all families average or above average for the area; significant differences were found in one of the zones.

All four of the measures indicate a tendency for well-being to diminish along the available moisture gradient, but they do not discriminate equally. On the measure of nutrition, the moderate and low areas are similar, while on measures of interviewer assessment it is only the low areas that show any substantial difference from the expected percentage in the high and moderate area. Crude as the indicators are, the association is clear. This lends considerable substance to our expectations on the basis of dominant environmental theory, that greater well-being is related to greater moisture.

Comparing measures of well-being as ratios between high and low zones with similar ratios of available moisture, we see in Table 13–7 that the mean difference in available moisture as measured by four indicators exceeds the mean difference in well-being by some 18 percent. In line with the theoretical expectations, we suggest that this is an extremely crude measure of the degree to which the steepness of the environmental grade can be moderated by the attributes of culture; namely, the process of adaptation of the Usambara farm system to the expected shortfalls of available moisture and the excessive pressure placed on the environmentally attractive high moisture area. The resource using, space adjusting, hazard controlling or avoiding practices of the Usambara farmers seem to only slightly moderate the differentials of environmental opportunity while providing minimal sustenance to all.

The environmental opportunity can be dampened in two directions: the resource base is overburdened in time in the more favored areas, and the effort to guarantee survival in the more vulnerable areas requires that much of the energy and capacity be directed to subsistence. What might be done to improve moisture management so as to increase the utilization of the favored resource base and to allow for greater diversification of effort in the less favored areas?

In the high-moisture zone, improvement lies in the

Table 13–6. Usambara transect: measures of well-being

Measure	Available moisture		
	High	Moderate	Low
Median nutrition index[a]	24	17	17
Mean material wealth index[b]	0.97	0.86	0.58
% units with cattle	34%	24%	16%
Interviewer's assessment, % of farms average or above[c]	63%	64%	51%

[a]Index weights variety, number, and nutritional content of meals eaten day previous to interview with maximum value of 38.

[b]Index is a maximum of 5 summing equally possession of bicycle (1), radio (1), galvanized metal roof (1), cement floor (1), and other expensive item (1).

[c]Interviewers were asked to classify farm households as below average, average, or above average for living quality of the area.

Table 13-7. Ratios of high- to low-moisture availability zones on indicators of available moisture and well-being

Moisture indicator[a]	Ratio of high to low	Well-being indicator	Ratio of high to low
Average annual precipitation	1.78	Median nutrition index	1.41
1/coefficient of variation	1.38	Mean material wealth index	1.67
Aridity index	2.14	% units with cattle	2.12
Length of season	2.28	% of farms average, above average	1.23
Mean ratio	1.90	Mean ratio	1.60

[a]Based on the average of two stations adjacent high-moisture areas and two stations adjacent low-moisture areas except for the coefficient of variation, which is based on one station for each area.

more intensive use of the land for production of irrigated vegetables on well-tended terraced plots. A major effort to develop this activity has been underway encouraged by the Lushoto Integrated Development Project (LIDEP). Twelve hundred farmers were involved during 1971 in commercial production of vegetables, working in collective groups including one of our study sites at Mponde (805). The project organized horticultural advice, seed distribution, transportation, and marketing to Dar es Salaam, some 400 kilometers (250 miles) away. Prospects for expansion lie in shipments to Europe by air. A self-sustaining operation by the farmers would be a major achievement and an open door to a new stage of resource utilization (Heijnen and Kreysler, 1971; Heijnen, forthcoming).

Improvement in the moderate-and low moisture zones involves building on the survival wisdom of the traditional culture but combining this with judicious inputs of social organization and technology. Collective production in the form of *Ujamaa* (communal) villages permit increasing the land area source, utilizing more diverse environments, supplementing diets with pond fish or chickens, and developing modest cash cropping with the increased security of the staple production base. An improved short-season maize variety with higher yield potential under dry conditions is needed. Evidence from Kenya suggests that it can be developed (Wisner and Mbithi, chapter 11).

For all areas, the local market for maize needs improvement. As of 1971, the local marketing arrangements seemed inadequate to provide assurance to farmers that would permit further diversification. During a drought, as demand increased and local supplies diminished, prices for locally available maize doubled or trebled. For areas of potential famine, relief grain was given away, but for areas of shortage only, no relief was given and the burden of drought was carried by the householder. However, it is often the case that the shortage of grain is relatively local and adequate supplies exist elsewhere. Yet no provision was made in the government controlled cooperative marketing apparatus (applicable beyond local markets) to funnel available supplies back to local markets to bring the free market price down. Nor is there local provision for interseason storage, all surpluses being sent out of the area. Recent news reports of decisions made by the National Executive Committee of TANU suggest that greater efforts for decentralized storage and distribution will be made.

The most vulnerable period in the life of a developing country is when the reliable mechanisms of the folk society are weakened by change and when the new institutions and practices are still in their formative stage. The agriculture along the road from Mombo to Kulasi is conservative agriculture; it needs change, not only to improve well-being but to prevent rural involution. Successful change will draw from the folk wisdom, institutionalizing, expanding, and improving on the considerable knowledge of the Usambara farmer to both diversify and provide reserves in space and time.

References

Allen, W. (1965) *The African Husbandman.* London: Olive and Boyd.

Heijnen, Joop. (forthcoming) *National Policy and Economic Development: A Case Study of the LIDEP Vegetable Component in Lushoto District, Tanzania.*

———, and Kreysler, J. (1971) "Cooperative vegetable production schemes in Lushoto District." Paper presented at the East African Agricultural Economics Society Conference.

Porter, P. (1965) "Environmental potentials and economic opportunities—a background for cultural adaptation." *American Anthropologist* 67:409-20.

14. Coping with drought in a preindustrial, preliterate farming society

HERB DUPREE
University of Michigan

WOLF RODER
University of Cincinnati

This is a report on field observations on human adjustments to drought events in the vicinity of Yelwa, Yauri Emirate, North-Western State, Nigeria. Traditional crops are grown and the farming pattern appears well adjusted to the environment. Drought is a minor hazard, but edaphic conditions promote differential impacts among farmers in the same village. Policy recommendations focus on regional economic development rather than efforts to relieve drought.

Study area

Questionnaire data were collected in Yauri Emirate, North-Western State, Nigeria from three villages within 3 kilometers (2 miles) of the town of Yelwa. These villages, Tondi, Yabo, and Baha, are on the Niger River, now dammed as Kainji Lake (Fig. 14–1).

Fig. 14–1. Location map

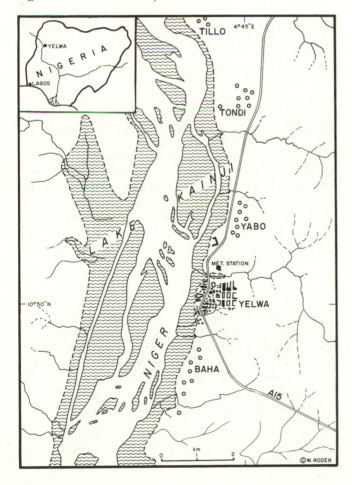

Climate

The study area is located in a region of savanna climate characterized by a summer rainy season with high temperatures and a winter dry season with milder temperatures. Frost does not occur, and climatic interest centers on the availability of moisture for plant growth.

Yelwa is the site of one of the few complete meteorological stations in the entire region. Long-term climatic data, and the soil moisture balance calculated according to the Thornthwaite method, are displayed in Fig. 14–2 for the average and for a low-rainfall year. Annual

Fig. 14–2. Water balance at Yelwa

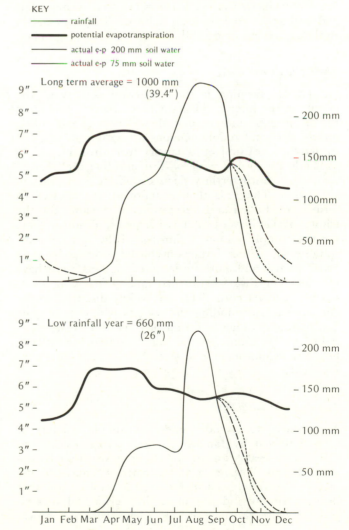

average rainfall amounts to 1,000 millimeters (39 inches) but years with rainfall as little as 450 millimeters (18 inches) have been experienced. Yelwa is a short distance north of the region of a regular two-peak rainfall regime characteristic of equatorial stations. Thus, two-peak rainfall regimes occur in some years, and the long-term curve exhibits a shoulder during May-June. This represents the planting period. A shortfall of precipitation at this time of the year may lead to complete failure of the germinating crop and necessitate replanting.

Soils

The nature of the soils is strongly related to the parent materials, of which the most important are Nupe sandstones and gneisses of the basement complex. The textures of local soils are highly varied; most of them range from sand through sandy to sandy loams, although clays and clayey loams are not absent. Water-holding capacities of soils are of crucial importance inasmuch as the main food crop, sorghum, continues to ripen into the dry season in December. Farmers are well aware of this and will select for this crop fields on level land and in swales, reserving sandy soils for quicker-ripening millet.

The agricultural cycle

Subsistence farming is the livelihood activity of the people in the sample villages. Guinea corn (sorghum) is the most important crop, followed by bulrush millet, maize, rice, and onions. Onions are grown in irrigated gardens and not subject to damage from drought.

Lands are cleared with ax, fire, and hoe, and planted with crops until fertility declines, about 3–8 years in succession. They are then abandoned and allowed to return to the natural vegetation succession. On the whole, availability of land is adequate, although some farmers complain about a shortage of the best and most suitable land. Since farmers prefer lakeside residence in order to fish and cultivate irrigated gardens, most have to walk long distances to their upland fields. In consequence, Guinea corn fields are widely dispersed in the countryside surrounding the villages wherever suitable land is found which is not already claimed by others. A radius of 16 kilometers (10 miles) around each village, however, would include at least 90 percent of the farmers' fields.

The drought situation

The Yelwa area is not subject to complete failure of rains followed by famine. An interesting insight into differential evaluation of the drought situation is provided by the insistence of the emir of Yauri and the divisional officer, head of the local administration, that drought simply did not occur in Yelwa and that no problem with shortage of rains ever exists. In contrast,

149 of the 150 farmers interviewed replied in the affirmative to the question "Do people in this place have any trouble with drought?"

Complete crop failure does not occur because even in the lowest rainfall year enough moisture is available to raise a crop of millet, which is relatively drought-resistant. Harm to farming activities from a shortage of water occurs rather as a result of the distribution of rainfall in time and space, and as a result of decisions on where, what, and when to plant.

Most crops are planted in May and June with the onset of the rainy season. Timing is crucial because a short drought after the planting rains may lead to partial or complete loss of the germinating seed or young plants. Such an event necessitates not only replanting, but generally results in a decrease in the total crop harvested. A second vulnerable period is the end of the rainy season. Low rainfall in October or an early end to the rains will result in lowered crop yields.

Figure 14–2 illustrates the situation for a low rainfall year (1957). Planting rains were delayed until June, followed by inadequate precipitation in July. Moisture available for soil water storage did not exceed 85 millimeters (2 inches) by September, replenishing fully only low-capacity soils. In consequence, a sandy soil with low moisture capacity made more water available for evapotranspiration and plant growth in October than a loamy soil with higher capacity. Millet ripening in October in a sandy soil would provide a good yield, while Guinea corn in better soil would suffer.

The incidence of damage differs. Seventy-three percent of the farmers sampled saw the major damage occurring in the maize crop, a relatively vulnerable plant and a minor crop in this region. In contrast, only 21 percent felt that major damage would occur to Guinea corn, and 12 percent saw damage to millet as important. Rice is planted well after the rainy season is underway and naturally wet or swampy areas are selected. Consequently, only 5 percent of the sample saw major damage from drought as occurring to rice. In this way the cropping pattern is adjusted to the environment. Only a few immigrant families attempt to raise yams, a staple crop further south, whose tubers only attain small size in the vicinity of Yelwa.

The farmers

Hausa is the *lingua franca* of the region spoken by all adults, who also tend to speak a local language at home. In the sample, 27 percent identified themselves as Muslim, speaking mainly Hausa, another 37 percent identified themselves as Muslim, speaking Gunganchi, and 35 percent as Gunganchi speakers following their traditional religion. One head of a family claimed Christianity as his religion.

Literacy is very low. Three farmers indicated that they have some knowledge of reading, and another ten that they had learned a little Koran Arabic, but 91 percent of the sample are illiterate. Consequently, no records are kept of past droughts, acreage farmed, or yields returned. Farmers are hesitant to generalize about their experiences, and claim no knowledge of experience of others, or elsewhere.

The questionnaire study

Perception of hazard

Ninety-nine percent of the farmers interviewed perceive drought to be a problem, through farmers' views on the occurrence of the last drought differ. Thirty-eight percent said there was a drought a few years ago, 43 percent opted for many years ago, and 9 percent indicated as far back as when they were small boys, while 7 percent said they did not know. Probably different concepts of what constitutes a drought underlie these assessments, since 14 percent claimed that there is a drought almost every year, 43 percent see drought as occurring often, and another 43 percent consider it a rare event.

Perhaps the best insight into farmers' assessment of the recurrence of drought is afforded by the story which forces them to choose between indicating that droughts occur (1) in series, (2) at regular intervals, (3) at random, or (4) not again at all. In response to this query 25 percent chose the most realistic evaluation, that droughts occur at random, but 75 percent opted for not at all. Discussion with farmers brought out the fact that most perceive the accuracy of the former choice, but hope that the latter will be true. Moreover, they feel that it would be evil to speculate on future droughts, that fate is in the hands of Allah, and that a good man would wish for no more droughts in the future. The whole question of hazards and what might happen in the future is one on which they would prefer not to speculate at all since they feel at the mercy of God and unable to do much about it.

Drought is not the most important hazard of this area. On being asked to volunteer the disadvantages under which they labor, farmers mentioned drought as frequently (36 percent) as damage from locusts and other pests (32 percent) and shortage of the best farmland for clearing (39 percent). Other hazards mentioned were sickness (11 percent), invasions of weeds (9 percent), and shortage of drinking water from wells (13 percent). Further problems arise from the fact that these hazards tend to interact; weeds and pests are seen as being especially plentiful during drought, and well water is seen by some as causing illnesses.

Perhaps a more helpful understanding of drought perception can be had from their general view of environmental hazards. Preoccupation with problems of the social environment can affect a farmer's perception of the drought situation. As an example, farmers in Yabo were more concerned about social problems (30 percent) than drought (22 percent).

Evaluation of damage

Only 17 percent indicated that they could expect all their crops to fail in a drought, most thought that half their crops would fail (56 percent), while 26 percent indicated that less than half of their crops might fail. These replies give as much insight into what farmers consider constitutes a drought, as to what actual damage they might suffer. A larger proportion of Muslims expect all or half their crops to fail in a drought than among those following the traditional religions. It can only be speculated that this is related to the treatment the different groups expect to receive from the hands of their gods.

When a drought occurs 80 percent expect their households to suffer, 46 percent expect food shortages accompanying crop failures and a similar percentage expect to experience anxiety and fear in such times.

Opportunities for adjustment

Farmers were asked what they would do to mitigate the damages of a drought event (Table 14–1). Virtually all of them indicated that their chief response would be simple to bear the losses, i.e., to suffer and starve until better times came along. An equal proportion would turn to God and pray to Allah or the traditional gods for help at such time. Beyond bearing the losses, 59 percent believe they could get help from relatives, 55 percent might look for work in the town of Yelwa, but not further afield, 84 percent would try to enhance their income by selling handicraft items, cut firewood, or grass for sale in the market, and 75 percent would aim to increase their food supply by fishing. Other activities in which farmers might engage are planting late cassava (19 percent), selling some of their few belongings, pots and clothes (11 percent), or consulting with a local medicine man (3 percent). A very few, notably the village heads, have regular wage income (7 percent).

Crafts practiced by farmers in their spare time and in the dry season include blacksmithing, boat ferrying, music, weaving, mat and rope making, fish-trap making, and fence making. Altogether 13 percent of the sample practice a craft. Onions are grown by 97 percent, and 52 percent fish the river regularly.

None saw his irrigated onion crop as a means of increasing income, yet 96 percent indicate that they would plant more onions in case of shortfall in food crops. When it becomes clear that the grain crops will be small, most of the onions have been set out, though a late crop could be planted at the risk of low market prices. Although 25 percent of those interviewed own cattle, none anticipated selling a cow or ox to buy food, though this occurs under pressures of hardship.

One possibility of warding off hardship is the storage of food crops from one year to the next. This presents difficulties in this climate and may not be possible because of obligations of sharing food with relatives.

Table 14–1. Response to drought losses in Yelwa area

| | Percentage indicating they would take this action | | | |
	Volunteering adjustment	Indicating yes, when asked	Indicating no, when asked	Don't know
Would suffer and starve	37	62	1	0
Would pray to Allah or traditional gods	4	57	1	1
Would turn to relatives for help	8	51	41	0
Would attempt to seek laborer job	24	31	43	2
Would sell firewood, crafts, or grass	41	43	15	1
Would do more fishing	36	39	24	1

Only 31 percent indicated that they might store grain from one year to the next; 69 percent indicated they would not. The impression conveyed is that drought events are in the hand of God, so there is little point in anticipating trouble, and the subsistence level of living rarely allows for the storage of crops from one year to the next.

Farmers were asked to whom they would turn to discuss a community problem. The village headman was mentioned by 65 percent, 7 percent would look to the elders, and 11 percent to God. In contrast, when asked to whom they would turn for help to recover from drought losses, 43 percent indicated they would turn to God, and 28 percent to family and friends. Twenty-one percent thought it worthwhile addressing their problems to the government or village head, only 12 percent thought of the possibility of government giving relief to drought sufferers, while 6 percent thought the government would help with prayers. Seventy-eight percent did not know how the government could help.

The unawareness of government relief is further borne out by the fact that only 16 percent identified the distribution of grain with drought relief. Such help is viewed as charity, which is part of the traditional role of the leaders, who therefore must have more resources. It is also seen as part of the broader web of mutual aid and support during times of need.

Conclusion

Farmers in the Yelwa area see themselves at the mercy of the elements and in the hands of God. They know that drought can come again in any year and that its occurrence cannot be predicted. Although they would like to deny that it will ever come again, some farmers did mention that when locusts are unusually heavy one can expect drought. When faced with drought or other natural disasters, such as a plague of locusts or other insect pests, their chief response is to pray to God and to the ancestors according to which religion they follow.

Aside from prayers, they know that their families will suffer. They anticipate doing everything in their power to avert and mitigate the worst consequences of disaster, but they do not have many opportunities to increase their income or to increase the amount of food available to their families. Aside from a small group of farmers in the villages of Baha and Yabo, migration to farm elsewhere or to seek a job for cash income is limited. Those who own cattle may sell a cow or ox. Those who have craft skills may work at them more diligently in the hope of selling more items for cash. Everyone fishes to put more food into the cooking pot. Others may attempt to sell firewood in town or cut grass to feed the livestock in the market, or they may look for jobs as laborers in Yelwa.

Farmers expect family and friends to help them bear losses and to get over hard times. A few are unaware that the government might help out by distributing famine relief grain. Underlying the farmers' thought is an assessment of possible drought disaster having a differential impact in the area. Few if any will experience total loss of their crops. Even in a bad year some farmers will bring in good crops and some households will have plenty to eat, while other suffer from drought. Thus, there will be opportunities for mutual aid among friends and not every able-bodied man will be scrambling for supplementary income in the small town of Yelwa.

Recommendations

The traditional farming pattern appears well adjusted to the environment. Crops grown elsewhere in Nigeria which would be more vulnerable to drought, e.g., maize and yams, have not entered the cropping pattern to any great extent. Different crops and soils, use of wet lands for rice, and irrigated gardens, fishing, and crafts provide some but insufficient alternative incomes in years of low rainfall. Disturbance of this adjustment by modification of the farming system or the introduction of alternative crops should be considered only with great care.

Remoteness of the region and the low hazard potential argue against a policy of organized public relief. Low population densities, poor transport facilities, and limited government structures would make this costly per head of the affected population.

General economic development and tying this region more closely into the Nigerian economy would provide the best protection against future need. This recommendation aims at increasing employment and income alternatives for the farmer, and is thus in keeping with what the people regard as possible and traditional courses of action. More and better roads and transport services and a greater volume of trade would enable farmers to obtain a higher return for their products during good years and meliorate experiences during bad. Increased employment opportunities in years of low rainfall, e.g., in trade or road maintenance, would present alternatives for families suffering shortages.

Acknowledgments

Acknowledgment for help and moral support is due Alhadji Mohammadu Tukor, emir of Yauri; the divisional officer, Mr. Dalhatu Zurmi; and Father Ces Prazen, O.P., of the Roman Catholic mission, Yelwa. Financial support was provided by the University of Cincinnati Internship Program and Faculty Summer Fellowship, and by the Commission on Geography and Afro-America of the A.A.G. We owe a great debt to the conscientious interviewing by our field assistants, A. Dung Bingel, Istefanus Thomas, and David Fodeke. Special thanks are due Professor Imevbore, University of Ife, for his crucial support at the initiation of the study.

References

Dupree, Herb. (1972) "Responses to stressful stimuli in traditional society: Yelwa, Nigeria." Cincinnati: University of Cincinnati.

Roder, Wolf. (1970) *The Irrigation Farmers of the Kainji Lake Area.* Rome: United Nations Food and Agricultural Organization.

15. Individual and community responses to rainfall variability in Oaxaca, Mexico

ANNE V. KIRKBY
University of Bristol

Parts of the valley of Oaxaca are so marginal for agriculture that the number of years in which harvests fail outnumber those in which they succeed. In an average year, it is an area of *permanent drought,* defined by Thornthwaite (1963) as an area where precipitation throughout the year is less than potential evapotranspiration. Despite this, agriculture is today the principal means of livelihood, and in the past the valley formed one of the earliest centers of agriculture in Mesoamerica, supporting one of the greatest pre-Conquest civilizations in Mexico.

This paper investigates the relationships between rainfall variability as recorded at meteorological stations, peasants' perceptions of rainfall patterns, and their annual decisions regarding where and when dry-farmed subsistence maize is planted. It suggests that one of the main adjustments to a hazardous environment, of which the variability of rainfall is one important factor, is found in the social structure of peasant village society. The village community acts as a reservoir of pooled resources to even out the effects of rainfall variability and production losses in the long term, and provides feedback to influence the annual agricultural decisions in the short term.

Study area

The valley of Oaxaca is a landlocked region of some 700 square kilometers (270 square miles) of flatland set in the southern highlands of Mexico (Fig. 15–1). It is defined by forested slopes rising to 3,000 meters (10,000 feet) altitude, beyond which are steeply dissected and heavily forested mountain ranges which discourage anything but small, isolated settlements and shifting agriculture. The valley floor lies at 1,500 meters (5,000 feet) elevation and is drained largely by ephemeral streams. Where they have not been obliterated by long-continued agriculture, remanants of xerophytic cactus and thorn-forest vegetation remain. Although it lies within the tropics, its high altitude means that temperatures are temperate rather than tropical, and

Fig. 15–1. Oaxaca Valley, Mexico

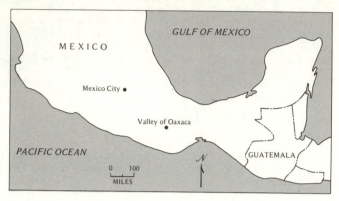

Fig. 15–2. Mean monthly temperature, evaporation, and precipitation for Oaxaca de Juarez

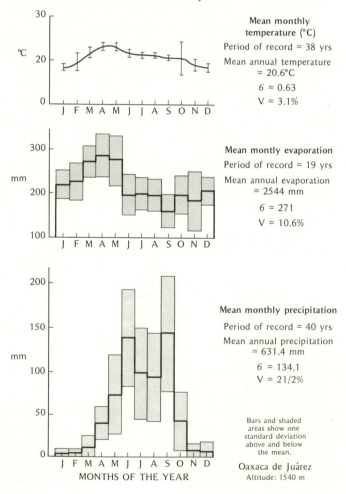

Mean monthly temperature (°C)

Period of record = 38 yrs

Mean annual temperature = 20.6°C

σ = 0.63

V = 3.1%

Mean montly evaporation

Period of record = 19 yrs

Mean annual evaporation = 2544 mm

σ = 271

V = 10.6%

Mean monthly precipitation

Period of record = 40 yrs

Mean annual precipitation = 631.4 mm

σ = 134.1

V = 21/2%

Bars and shaded areas show one standard deviation above and below the mean.

Oaxaca de Juárez
Altitude: 1540 m

vary about 5° from a mean of 20° C. (68° F.) for every month of the year (Fig. 15–2). However, the agricultural potential of this area is offset by its low rainfall (630 millimeters or 25 inches per annum) and scarcity of irrigation water.

Agricultural production is mainly dependent on the output of small and scattered peasant holdings which are farmed by traditional methods using ox plow, hoes, and machetes (long-bladed knives or cutlasses), and are hampered by low capital resources. Except for the capital, Oaxaca de Juárez, which is an important regional and tourist center, the valley is a poor rural area. It is comprised of villages with usually less than 1,000 inhabitants which are linked by dirt roads that frequently become impassable to wheeled transport except oxcarts in the rainy season. Most homes are single-roomed adobe houses without water or sanitation: health is poor; illiteracy is still 50 percent, and Zapotec (the local Indian language) is still spoken in many homes in preference to Spanish (1960 Oaxaca state census).

Landholding is both communal and private with sharecropping on a 50–50 basis a common practice. Private holdings are fragmented by bilateral inheritance so that a peasant typically holds or sharecrops about 2 hectares (5 acres) of land in five scattered plots, from which his net income (if the land is under maize and if the production were sold) is about 3,700 pesos (U.S. $300). This is the equivalent of a year's wage labor at local rates. Although most peasant households grow some of their own food needs, they also derive additional income from craft specialization such as pottery or weaving, from the production of cash crops in villages where some irrigation is possible, or from the provision of services within and between communities.

Rainfall: its characteristics and importance for maize production

The two most important characteristics of rainfall for agriculture in the Valley of Oaxaca are (1) that the total rainfall is only marginally enough for dry-farmed maize in an average year; and (2) that this marginality is aggravated by high variability of rainfall totals *between* years and of rainfall distribution *within* years.

The rainfall pattern is one of dry winter months from November to March (mean monthly rainfall less than 10 millimeters or 0.4 inch) and a summer rainy season which characteristically begins in late April to May but does not become well established until June. The rainy season has two peaks: in June (mean rainfall of 137 millimeters or 5.4 inches) and September (mean rainfall of 144 millimeters or 5.7 inches) (Fig. 15–2). There is a mean water deficit in every month of the year ranging from about 100 millimeters (4 inches) in the growing season (May to November) to almost 2,000 millimeters (79 inches) at the end of the dry season in May (Fig. 15–3). Dry-farming techniques to concentrate water in the soil and reduce the density of crops are therefore

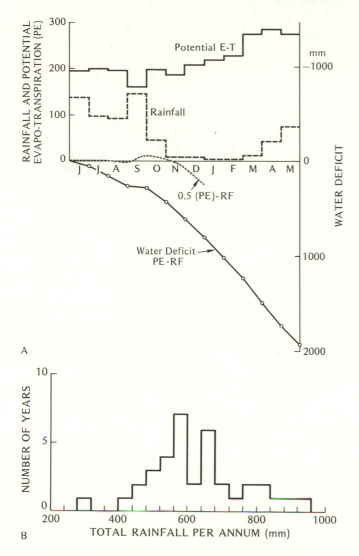

Fig. 15—3. A. Mean annual pattern of water deficit and B. Variability of annual rainfall for Oaxaca de Juárez

The latter can vary from an almost symmetrical double-peaked mean pattern to extreme single peaks as in 1933 and 1948.

The amount and distribution of rainfall in any year are crucial to peasant agriculture because (1) dry farming is the basis of the economy and unirrigated maize is the staple crop, and (2) water availability is the most important single factor affecting maize yield.

Any form of irrigation is available to less than half the valley floor area; within this sector, water is actually applied to less than 20 percent of the cultivated land. Even where irrigation water is available, it is preferentially applied to either cash truck-farming crops or to alfalfa (and to a much lesser extent to tobacco and sugarcane). However, maize is by far the most important and widespread crop, occupying over 90 percent of many unirrigated areas and comprising, on the average, between 30 and 50 percent of the cultivated land even where irrigation water is available.

A series of two-way analyses of variance were run for maize yields measured in 164 milpas from all parts of the valley and including a wide range of soil type, water use, physiographic zone (mountains, piedmont, high alluvium, and present floodplain), slope angle, and degree of surface erosion. The results showed that differences in water availability were the most important single factor considered in its effect on maize yield. Furthermore, given sufficient water, similar yields (approximately 2—3 metric tons per hectare or 0.9—1.3 tons per acre) can be obtained from any area of the valley, in any physiographic zone, on any slope (0°—16°), and on soil of any texture.

Thus water is the key to increased agricultural production. For subsistence maize production water must come directly from rainfall or from floodwater. Rainfall variability affects the time of planting, the location of crops, and the ultimate success of the harvest. Without irrigation, accurate judgment of the rainfall pattern is one prerequisite for the successful cultivator.

Field methods

Data were obtained for (1) peasant cultivators' perceptions of rainfall variability which included their memory of past wet and dry years, their belief in rainfall patterns, and their basis for predicting future rainfall in order to decide when and where to plant their maize; and (2) the amount and distribution of maize planted in 1966, 1967, and 1968.

The three components of peasants' perceptions of rainfall variability were obtained from informal interviews with a random sample of 45 cultivators who were met as they worked in their fields, without any prior arrangement. No interpreter or formal questionnaire was used and no notes taken in the presence of the informant. Questions were asked in any order as they occurred naturally in the conversation, and the informant's name was not taken unless he offered it. This

necessary, and their use makes the curve for 0.5 (potential evapo-transpiration) minus rainfall more relevant to the present farming situation. Based on this second curve there is on the average just enough water to support agriculture between June and October (Fig. 15—3).

The degree of variability in monthly rainfall between years is illustrated by the standard deviations shown in Fig. 15—2. For example, in June, which is the main planting month, mean monthly rainfall at Oaxaca over the last 40 years of record is 137.4 millimeters (5.4 inches), with a standard deviation of 55.1 millimeters or 40.2 percent. Extreme recorded values for June range from 12 millimeters (0.5 inch) in 1922 to 245 millimeters (10 inches) in 1957. The variability of total annual rainfall (Fig. 15—3) and the distribution of rain within the growing season are also of vital importance.

method seemed to minimize the informants' two main concerns—that his information would find its way either back to other families in his village, or to government officials—and elicited the most open and friendliest responses. All informants were male, working milpas which they either owned or sharecropped on a long-term arrangement, and grew at least some maize for their own family's consumption. Their ages ranged from the early twenties to the early sixties, although no record was made of the ages of individuals. All referred to themselves as campesinos (peasants).

The amount and distribution of maize planted was obtained by field survey of six sample areas which were selected to represent between them the total variation of physical geography and resource use found within the valley of Oaxaca. Each sample area covered 2–5 square kilometers (0.8–1.9 square miles), within which physiographic variables (geology, soil, hydrology) were mapped in detail and the maize yields of several fields were measured just before harvesting. The crops planted in each landholding unit (totaling some 3,000 units) were also mapped at a scale of 1:7,000 and these crop surveys were made in each of three years 1966–68.

One difficulty in comparing the two sets of data—perception of rainfall and area of maize planted—is that, because holdings are very scattered, the land-use data do not necessarily include all the holdings of the peasants interviewed. Another source of error is that the points for which rainfall is measured are very few and the distribution of rainfall recorded at a meteorological station may be different from that falling on a particular field elsewhere in the valley. The fragmentation of holdings across acreas with different moisture characteristics means that in each year a peasant cultivator is likely to experience a range of rainfall for his holding or allows him always to remember the most extreme conditions that occurred on just one of his fields is uncertain.

Perception of annual rainfall

Recall of past annual rainfalls

Informants were asked to name which years were the wettest and driest they could remember. Six individuals (30 percent) could not name an extremely wet year compared with 66 percent who could not name an extremely dry year. Of those who did give at least one year, 55 percent correctly gave 1951 as the wettest they had experienced, but only one individual correctly remembered 1954 as the driest year. Other years remembered as being exceptionally wet were 1943, 1958, and 1966; and 1967 was frequently recalled as being unusually dry. Meteorological records show that 1958 and 1966 were unusually wet years, in agreement with the cultivators' recall, but 1943 appears to have been a dry year (Fig. 15–4 and Table 15–1). One year which was not mentioned as a wet year was 1954, although it had a higher rainfall than that of 1966.

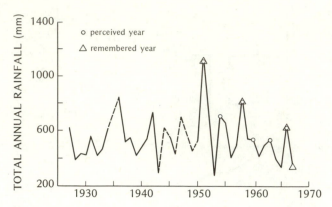

Fig. 15–4. Patterns of rainfall memory and perception superimposed on meteorological records for Tlacolula, Oaxaca

Table 15–1. Recall of past rainfall years and perception of pattern in rainfall variability between and within years by a small sample of peasant cultivators in the valley of Oaxaca (N = 45)

	%
Wettest year recalled	
1951	55
Other	15
None	30
Total[a]	100
Driest year recalled	
1954	3
Other	31
None	66
Total	100
Perception of pattern	
Cyclic events between years	25
Autocorrelation with last year	33
Random events between years	42
Total	100
Belief in spring rainfall as predictor of summer growing season rainfall	
Yes	95
No	0
Don't know	5
Total	100

[a]This total includes those informants questioned in 1968 only because 1969 was an exceptionally wet year and all those interviewed in 1970 could recall only 1969.

It appears that wet years are recalled more easily than dry ones. A dry year is one in which widespread crop failure would be expected except in those areas on the floodplain where a water table exists within 2 meters (6½ feet) of the surface, and where irrigation water is available to divert onto maize to save the harvest. A wet year is generally a good year for agriculture, although high rainfalls over a short period of time can produce widespread and uncontrollable flooding. These floods may destroy one crop but they provide sufficient soil moisture to support a good crop later in the year or early in the following year. Like gamblers, therefore, it seems that peasant cultivators remember their successes rather than their failures; or perhaps it is only the drama of the floods that they recall.

As would be expected, memory of past rainfalls is a function of recency and magnitude of the event. Furthermore, the most common pattern of rainfall memory is a series of years which begin in 1951 and are most frequently recalled in descending order of magnitude through time, in addition to which the rainfall of the last year is probably recalled whatever its magnitude (Fig. 15–4). In more general terms these data could be interpreted to indicate (and nothing more) that memory of some salient natural events may begin with an extreme event which effectively blots out recall of earlier events, and acts as a fixed point against which to calibrate later ones. Thus it could be that memory of rainfall for the Oaxacan peasants sampled is stepped; that is, the last wet year is remembered, then the year before that which was even wetter, and so on. Some corroboration of this possibility was found in 1970 when the occurrence of a very wet year in 1969 appeared to blot out the memory of previous wet years including 1951.

Perception of pattern in past annual rainfall

Informants were asked if they thought that wet years occurred regularly every few years; if wet years came in pairs; or if they thought annual rainfall was random and therefore unpredictable.

Fifty-eight percent thought they could detect regularity in the rainfall pattern. Of these 25 percent thought that wet years were cyclic and 33 percent said they believed that wet years came in pairs; particularly that a really wet year was followed by another less wet, but still higher than average rainfall. However, 42 percent believed that annual rainfall between years was completely random and several in this group mentioned cyclones as being important factors.

The group who believed that wet years come in pairs may be influenced by the higher yields that often occur in the year *following* a wet year as a result of recharging of the soil moisture. Of the 25 percent who thought that wet years occurred at regular intervals, most suggested a 3-year cycle; two suggested a 4-year cycle; and two individuals volunteered longer cycles of 5 and 7

years duration. The most commonly perceived cyclic pattern of 3 to 4 years is not dissimilar from the actual intervals between wet years (greater than 540 millimeters or 21 inches per annum) since 1951; these are 3, 4, 5, 3, and 3 (shown in Fig. 15–4 as perceived years). It is possible therefore that this perception of a cyclic pattern is partly a function of idealizing the rainfalls experienced within the time period of recall. In general, as might be expected, more informants were able to express a belief in the regularity or randomness of rainfall patterns than could recall specific past events (Table 15–1).

Almost all informants said they believed the rainfall in the early part of the year (January to June) gave some indication of what the rainfall would be like during the growing season (June to November) although they varied in how much reliance they admitted to placing on this relationship. Informants who believed they could see patterns in rainfall between years expressed faith in this early-late rainfall relationship in addition to the other patterns.

Validity of pattern perceptions in relation to meteorological records

When the two types of pattern perception (autocorrelation between years and cycles of 3, 4, 5, and 7 years duration) are tested against the measured rainfall totals recorded at Tlacolula and Oaxaca de Juárez, neither belief is found to be statistically valid. With the provisos mentioned above (that higher than average *crop yields* may come in pairs and that a 3-to-4 year cycle is an approximation of actual experience within the period of recall from 1951 to 1970), annual rainfall, as far as it has been measured for the past 40 years, follows no discernible pattern.

The same is not true for patterns of rainfall *within* any single year. Plotting the rainfall of January to April against the rainfall of May to December produces a significant correlation (0.01 level). If the early rainfall is greater than 80 millimeters (3 inches) the growing season rainfall has an 80 percent chance of being greater than 600 millimeters (24 inches), which would be generally recognized as a wet year. If the spring rainfall is low, between 20 and 40 millimeters (0.8–1.6 inches) in total, the later rainfall has a 50 percent probability of being less than 420 millimeters (17 inches). This would mean a dry year and considerable crop failure. By June reliability has increased so that if June rainfall is more than 150 millimeters (6 inches) then rainfall throughout the growing season has an 85 percent chance of being above average. Even with a knowledge of only the first four months rainfall (January to April), 50 percent of all wet years and 40 percent of all dry years can be predicted. The probability of correctly predicting wet growing seasons (greater than 600 millimeters or 24 inches) or dry ones (less than 420 millimeters or 17 inches) by chance is 25 percent each so that knowledge of early rainfall patterns represents a considerable advantage to cultivators.

Annual variations in the pattern of dry-farmed maize production

In order to determine the amount and direction of land use change, the areas changing from one crop to another, which had been obtained from field surveys, are put in matrix form. The transitional matrices are treated as first-order Markov chains and equilibrium distributions of crops are calculated. The results showed that

1. The transitional matrices are significantly different from each other;
2. The amount of change that occurs in any one year is generally large enough to significantly alter the land-use pattern within a very few years (each annual change involves 10–20 percent of the total area);
3. The land-use changes which occur are greater than those expected from fallowing cycles and crop rotation, and from changes in cash-crop production related to known economic trends (Kirkby, 1973).

The variations in planted maize area for the three years 1966–68 were tested to see if they could be related to peasants' perceptions of pattern in annual rainfall. Assuming a belief in a pattern of autocorrelation between pairs of years, or in any of the cyclic patterns suggested, the meteorological records were used to calculate the expected rainfalls of 1966, 1967, and 1968 (that is, wet, dry, or average rainfall during the growing season and during the whole year). These expectations were compared with the directions of land use changes observed for the three years, but no correlations could be found.

The observed variations in maize area in each sample of the valley were also plotted against different proportions of early rainfall that fell in each of the three years (January to April, January to May, January to June, and June by itself). The best fit relationship is with the rainfall for the month of June rather than for any other period considered (Fig. 15–5). Early rainfall in the year serves two functions.

1. It helps predict rainfall in the growing season;
2. It provides definite information on soil moisture conditions up to the moment of planting.

Since both functions are best served by information on rainfall later and later in the year, and maize must be planted by the end of June or beginning of July in order to avoid drought at harvest time, it follows that June rainfall is the preferred basis for prediction.

There are also factors operating in the opposite direction and tending toward an early planting date:

1. Field preparation, particularly of long-fallowed areas, requires time and labor so that the decision to plow for planting must be taken before June;

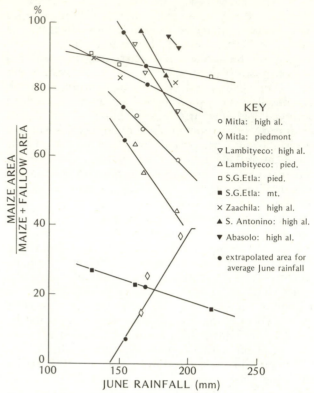

Fig. 15–5. Relation between observed planted maize area and June rainfall for nine sampled areas

2. To obtain maximum infiltration the first plowing should ideally anticipate the first main rains;
3. The cost of hiring ox teams to pull the plow rises steeply between April and June and their time is more heavily booked. The result is that poorer cultivators tend to be forced to plant earlier in the year with less rainfall information and greater risk, and thus tend to stay poorer.

What is more unexpected is the way in which June rainfall was found to be related to maize planting; that is, when June rainfall and therefore the expectation of a wetter year and higher yields is increased, the *area under maize is reduced rather than expanded* (Fig. 15–5).

This inverse relationship is found in all unit areas sampled except one. It therefore holds for a wide variety of physical environments from poor dry farming on very marginal land to moist high water table and canal irrigated fields. This variety of physical setting and slope gradients reduces the effectiveness of one possible control—that high June rainfall leads to sufficient waterlogging of the fields to prevent planting. The only area sampled for which the inverse relationship does not hold is the Mitla piedmont (Fig. 15–5).

The Mitla sample is the driest and most marginal area for agriculture within the valley for Oaxaca, with an average annual rainfall of less than 550 millimeters (21.7

inches) per annum. It was also selected because the valley is so narrow at this point that three physiographic zones with very different moisture characteristics can be included within a small area (dry piedmont slopes; less dry, flat high alluvium; and narrow floodplain liable to flooding), thus providing Mitla peasants with a real choice of which type of land to plant each year. This area was therefore used to test the effect of local variability in the physical environment on the relationship between rainfall perception and patterns of maize planting.

In the three parts of the Mitla sample area (piedmont, high alluvium, and low alluvium) agricultural behavior in deciding which fields to plant with maize can be best related to June rainfall, as was found for all other areas sampled. In the Mitla piedmont, however, this behavior is to increase maize planting in expected wetter years in contradiction to all other areas including those of the alluvium immediately adjacent to it. Two explanations appear likely:

1. Substitution is occurring between the long-cultivated, underfallowed alluvium and the long-fallowed piedmont in those years wet enough to obtain some yield from the piedmont;
2. Much of the piedmont is privately owned and constitutes a spiritual extension of the peasant in the same way that all his landholding does (Redfield, 1956) so that whenever moisture conditions in the piedmont offer a chance of some return, he wishes to plant there.

Deliberate substitution of maize planting between physiographic zones is likely strategy for the Mitla area, particularly, because the area as a whole is so marginal that at some point, given sufficient rainfall in both areas, yields on long-fallowed piedmont will be as great or greater in any one year than yields on under-fallowed alluvium, and such inversions of yield have been measured. However, information from other parts of the valley suggests that planting maize in the more marginal areas in wet years does not generally stem from a desire to substitute maize area between physiographic zones but reflects a desire to farm the land when possible. In the words of one Mitla man:

> If a man owns a milpa he shouldn't let it stand idle all the time because it was not given to him for that purpose, but he should plant it when God gives him the rain.

The Mitla piedmont represents a category of land which is so marginal for agriculture that maize cannot be grown there successfully in most years and not continuously year after year. On this type of land, Oaxacan peasants try to maximize returns by literally gambling their maize seed whenever the chances of any returns seem possible on the basis of rainfall in June. Thus a more complex, but still oversimplified pattern emerges of two types of agricultural strategies based on perceived relationships between early and growing season rainfall:

1. In the continuously cultivated main agricultural zone, maize area is *decreased* as early rainfall and expectation of higher yields increase;
2. In the area of sporadic cultivation in the marginal agricultural zone, maize area is *increased* as early rainfall and expectation of higher yields increase.

Perceived range of alternative adjustments

In addition to the area of maize planted, many of the agricultural decisions taken each year are made in response to expectation of drought or flood. In such conditions of uncertainty, the ultimate success of the harvest is largely the result of correctly matching planting conditions with rainfall conditions. A simplified scheme of the decisions involved is given in Fig. 15–6. Apart from judging the timing of the rains, which will determine exactly when to plant, cultivators adjust their annual strategies in several other ways. They have two main types of local Indian maize to choose between (*violento* and *tardon*; the first has a shorter growing season and survives drought better; in a wetter year, *tardon* gives higher yields); they can vary the ratio of maize to beans within a single field (beans tolerate drought conditions better); and they can vary the density of maize plants in the field (section B of Fig. 15–6). These adjustments were suggested by cultivators during interviews. They have also been confirmed by field observation and measurement. The proportions of maize to beans are significantly different between physiographic zones, and measured densities of maize plants range from 30,000 to 80,000 plants per hectare (12,000–32,000 per acre).

In the longer term, cultivators can try to adjust their landholding distribution by buying land, sharecropping it, or obtaining new fields through marriage alliances between families. Any of these means may be employed to extend or improve the range of land held in relation to available water (section A of Fig. 15–6).

Another common adjustment is to control and distribute floodwater from the summer flash floods which usually rise and fall in the space of a few hours. Floodwater schemes within the valley show great variety. Some involve only the construction of one or two walls, 1–2 meters (3.3–6.6 feet) high and a few meters wide, across a small valley in the hillside behind which soil and soil moisture are concentrated by overland flow. Other schemes involve the construction of dams over 10 meters (33 feet) high across the beds of the mainstreams and are linked to a complex system of earth canals. Groups of peasants cooperate to build and maintain these systems. Floodwater flowing along tracks and roads is also diverted onto the fields. As far as it seems possible with their present techniques, capital resources, and social structure, Oaxacan peasants use all available floodwaters to mitigate the effect of drought on their crops.

In addition to practicing these adjustments to reduce

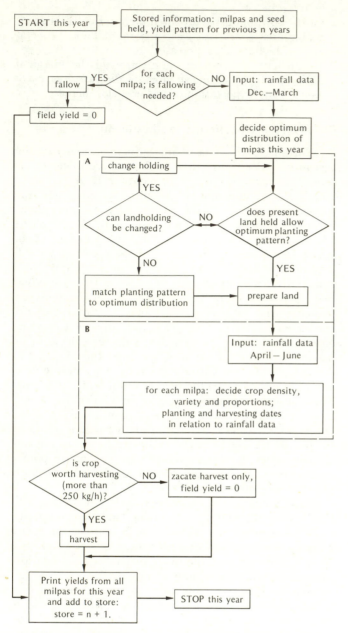

Fig. 15–6. Simplified flow diagram for maize planting decisions in dry farming

probably come by them both at the expense of others and by evil means, and are therefore not to be approved of.

Beyond these adjustments, peasant cultivators see few further possibilities for themselves. Additional adjustments are perceived as coming primarily from central government sources. The construction of dams for floodwater and canal irrigation schemes in a few villages recently has been partly financed and designed by the state. This has encouraged an attitude of looking at such schemes as panacea for all the ills that presently beset peasant life. Many cultivators seemed to regard a government-built dam as the key to increased agricultural prosperity, whether or not the streamflow would justify it. Other suggested solutions are improved roads, the provision of cooperative facilities for marketing produce, deep wells and pumps, improved strains of crops, a move toward mixed farming with dairy cattle, and, of course, the ever-present desire to have more land redistributed among the peasants.

Saarinen (1966) found for the Great Plains that the perceived range of choice of alternative operations is relatively narrow and that the range was increasingly restricted as aridity increased. Not only are Oaxacan peasants situated at the marginal end of the environmental spectrum but their conservative attitude to change further restricts their perceived alternatives. Their reverence for maize as the staple of life and their traditional view that a man should produce some of his family's own food provides considerable constraints on alternative modes of production. Thus maize is grown on poor land where other crops would tolerate the conditions better; it is cultivated on good land capable of irrigation which could be more profitably devoted to intensive vegetable production; and craftsmen who could be fully employed pursuing their craft for profit devote some of their time to growing maize for their families to eat.

Social adaptation to uncertainty in the environment

Village society in the valley of Oaxaca shares to some degree the characteristics of "closed corporate peasant communities" described by Wolf (1957). One of these characteristics is that membership of the community entails acceptance of its values even where these are in conflict with the ideal of personal advancement. The aims of the group transcend those of the individual. Community values include participation in religious rituals, the distribution of wealth or surplus within the community, and giving aid to other members of the group.

In Oaxaca, the variability of rainfall and the marginality of the area for dry farming has provided an extremely uncertain environment for peasants who have few reserves of capital or food. Without aid from the community, harvest losses from drought or flood would reduce many families to starvation each year. Village

the effects of rainfall variability, cultivators seek to modify the causal system through the power of prayer. They also leave offerings in caves in the hillsides which were traditionally the abodes of Indian deities and now are occupied by images of Christian saints. Some peasants place crosses in their fields, and the fields may be blessed. Attitudes to luck and bountiful harvests are somewhat tempered, however, by the peasant's conservative tendency to believe that the supply of bounty and good luck is limited. Thus those individuals who seem to have a disproportionately large amount have society in Oaxaca has therefore evolved a more formal system of providing mutual insurance than might be expected in a situation of less uncertainty or greater reserves to meet losses.

Two mainstays of Oaxacan peasant society are the *guelaguetza* and *cargo* systems. The *cargo* system is a hierarchy of civil and religious official positions within the village community which require expenditure of time and wealth on the part of the holder and provide the community with regular, communal celebrations. Men in the village proceed up this hierarchy, a process which can take an average of 30 years since it may take some 3 to 6 years to repay the debts incurred by holding the last position on the ladder. The tenure of these offices, except perhaps for the highest echelons, is not restricted to a small elite but is rotated, in at least some villages, among 90 percent of the men (Webster, 1968). Refusal to serve is considered not only antisocial but a punishable offense. Communal celebrations are provided in the form of fiestas involving several days of eating and drinking given in honor of a saint by religious officeholders or to mark family celebrations. The means to pay for this is the *guelaguetza* system, which mobilizes the stockpiling of surplus from the network of friends and relatives into one central household—that giving the fiesta. This surplus is then redistributed and immediately consumed by members of the *guelaguetza* network during the celebrations. It must be repaid and careful written accounts are kept. In this way a peasant household is assured of instant wealth which can be repaid over the next few years, and the donors are depositing their surplus in a credit account.

The *guelaguetza* network operates in time of scarcity as well as plenty so that any household which has suffered severe harvest losses or whose expenses become too great for its income to bear can obtain food and money on credit or can ask for an earlier loan to be repaid. Such a system depends on its reliability and comprehensiveness. It is threatened by the accumulation of private wealth which is discouraged and regarded as unnatural unless mitigated by lavish contributions to the community coffers and generosity to less fortunate individuals. Thus, despite growing pressures from the outside world, there is still considerable coercion within many villages to conform and enter into the *cargo* and *guelaguetza* systems.

While such group insurance schemes assure the survival of hard-hit individuals in the short term and the continuance of peasant village society in the longer term, they also appear to decrease personal initiative. In particular, they are seen as influencing the observed response to expected higher rainfall; that is, to decrease the area under maize when higher yields per unit area are likely. Partly because surpluses over and above those obtained by neighbors will have to be at least partly shared, and partly because they prefer to pursue a successful *cargo* career rather than an agricultural one, Oaxacan peasants on the average choose to devote their

energies to fulfilling social obligations rather than obtaining the highest economic returns possible. The same attitude is revealed in their methods of cultivation and in their irrigated agriculture (Kirkby, 1973).

The situation is not static. Already a few villages are reducing or abolishing altogether the traditional systems of *cargo* and *guelaguetza*. The younger men particularly try to avoid the obligations of the *cargo* offices and prefer to pursue individual economic advancement at the expense of social disapproval. As this economically orientated group expands, the values of traditional village communities are eroded and this will lead to the collapse of the present structure of peasant society. In the future, adjustment to hazards in the environment such as rainfall variability will not be only to increase economic production (e.g., to increase effort and area cultivated when higher rainfalls are expected). Adjustments will also rest more with the responsibility of the individual at one end of the spectrum and with the state at the other. The present intervening scale of adjustment, *at community level,* is likely to disappear, only to reemerge when extreme and infrequent hazards strike.

Conclusion

The main results of this study of perception and adjustment to hazard in the natural environment by some Mexican peasant communities can be summarized as follows:

1. Memory of specific rainfall events is relatively short and begins with the event of greatest magnitude, against which other events appear to be calibrated.

2. Peasant cultivators hold theories about the regularity of natural events similar to that found in more advanced societies (Burton, Kates, and Snead, 1969). These theories are not supported by strict comparison with available meteorological records but have some validity in relation to the sequence of recent rainfall years. What is more important, these theories appear to be nonoperational for most peasants; that is, they do not act on them.

3. There is a general consensus, which is supported by meteorological data, that early rainfall in the year helps to predict rainfall in the coming growing season. A similar belief is reported for peasants in Yucatan (Burton, Kates, and White, 1970).

4. Predicted rainfall can be related to measured agricultural response (in amount of maize area planted), but for most of the cultivated area, this response is in an unexpected direction; that is, the area of maize planted is decreased when expectation of higher yields is increased. In areas of sporadic cultivation beyond the main agricultural zone, opportunism reverses this relationship, and in expected wet years maize area is expanded.

5. The range of perceived alternative adjustments is narrow. Beyond their present agricultural strategies to

use and conserve all available moisture and floodwater, and to match the distribution and types of crops with expected moisture patterns, Oaxacan peasants tend to see further adjustments as being initiated and provided by the state.

6. In Oaxaca a formal social system has evolved to provide insurance for its members against losses due to flood, drought, or any other economic or social disaster. Participation in the community and receipt of its benefits requires the individual to subsume his own advancement to that of the group. This appears to be effective in influencing cultivators to use the opportunity afforded by the chance of higher yields to spend more time and effort fulfilling obligations within the group rather than individually obtaining higher economic returns.

In this way, the community structure is not only a major and effective form of long adaptation to the nature of the physical and social environment; it also provides feedback on the relationship between perception of natural hazards in that environment and the types of shorter-term adjustments made in direct response to it.

Acknowledgments

This paper is part of a larger study of the ways in which Oaxacan peasants make use of their land and water resources now and in the past (Kirkby, 1973). It is an integral part of the Valley of Oaxaca Archeological Project (Flannery et al., 1967), supported by the Smithsonian Institution in 1966 and the National Science Foundation under Grant GS-1616 to the University of Maryland in 1967, and Grant GS-2121 to the University of Michigan in 1968–70, without whose generous support this work could not have been done.

Climatic data used in this study were provided by Secretaría de Agricultura y Ganadería and Secretaría de Recursos Hidraulicos, (records on file in Oaxaca de Juárez).

References

Burton, Ian, Kates, R. W., and Snead, R. E. (1969) *The Human Ecology of Coastal Flood Hazard in Megalopolis.* Chicago: University of Chicago, Department of Geography, Research Paper No. 115.

——— Kates, R. W., and White, G. F. (1970) *Suggestions for Comparative Field Observations on Natural Hazards.* Toronto: Natural Hazards Research Working Paper No. 16.

Flannery, K. V., Kirkby, A. V., Kirkby, M. J., and Williams, A. W. (1967) "Farming systems and political growth in Ancient Oaxaca." *Science* 58:445–54.

Kirkby, A. V. T. (1973) "The use of land and water resources in the past and present Valley of Oaxaca, Mexico." *Memoirs of the Museum of Anthropology* No. 5. Ann Arbor: University of Michigan.

Redfield, R. (1956) *Peasant Society and Culture.* Chicago: University of Chicago Press, 4th imp., 1965.

Saarinen, T. F. (1966) *Perception of the Drought Hazard on the Great Plains.* Chicago: University of Chicago, Department of Geography, Research Paper No. 106.

Servicio Meteorologico Mexicano.

Thornthwaite, C. W. (1963) "Drought." *Encyclopaedia Britannica* 7:699–701.

Webster, S. S. (1968) "The religious cargo system and socio=economic differentiation in Santa Maria Guelace." Unpublished manuscript.

Wolf, E. C. (1957) "Closed corporate peasant communities in Mesoamerica and central Java." *Southwestern Journal of Anthropology* 13:1–8.

16. Drought in South Australia

R. L. HEATHCOTE
Flinders University of South Australia

The role of drought in Australian history has been documented only relatively recently, and the details of human adjustments to drought stresses have been only locally considered (J. Foley, 1957; Gibbs and Maher, 1967; Bureau of Agricultural Economics, 1969; Reserve Bank of Australia, 1968; Campbell, 1968). Some of the problems of such a general study have been documented elsewhere (Heathcote, 1969), and it is the aim of this chapter to provide some preliminary results of a more detailed study of drought in southern Australia.

This study was undertaken as part of a comparison between the adjustments to agricultural drought of crop and livestock farmers from a developed nation (Australia) and an underdeveloped nation (Tanzania) over the period 1944–70.

"Agricultural drought" is defined as a shortage of

water harmful to man's agricultural activities. It occurs as an interaction between agricultural activity (i.e., the demand) and natural events (i.e., the supply) which results in a water volume or quality inadequate for plant and/or animal needs.

For the Australian crop-livestock economy sample, the Mallee area of eastern South Australia and north-western Victoria was chosen (Fig. 16–1) and four sites astride the state border were outlined, two of high drought risk and two of medium drought risk (Table 16–1).

The aims of the study were to test the basic hypotheses of the Natural Hazard Research Group and to provide information for a comparative study with the Tanzanian data. The report here only concerns the first of these two aims.

Sources

Two basic sources have been used. Archival search of official and private documents, to assess the historical sequence of drought occurrences and reactions to them over the period 1944–70, was the first source. Here, private diaries, newspapers, official production statistics, and legislation provided chronologies of drought occurrences, indexes of intensities of concern, and some clues to drought impacts locally and statewide.

The second source represented field evidence, obtained by both informal unstructured interviews with officials, newspaper editors, senior citizens, and "barflies," and application of a questionnaire (based upon that outlined in Natural Hazards Research Working Paper No. 16, 1970) to the four study sites after modifica-

Fig. 16–1. Physical characteristics of Mallee study area. Source: *Atlas of Australian Soils* (1960), sheet 1.

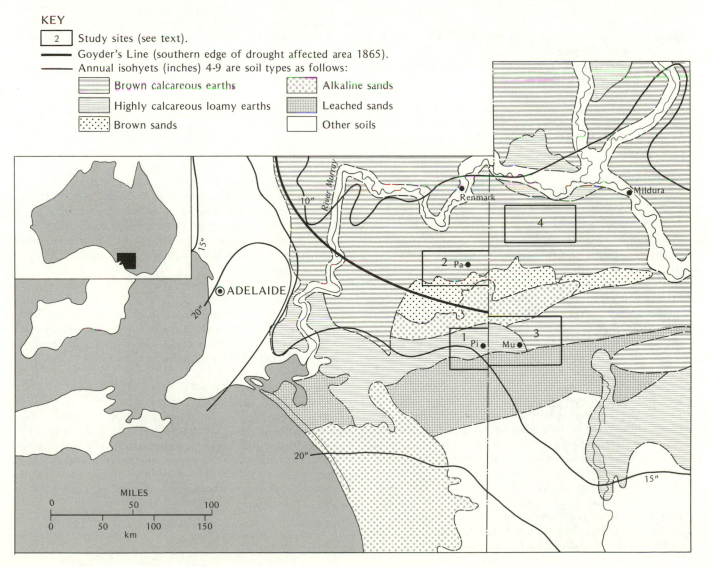

KEY

| 2 | Study sites (see text).

Goyder's Line (southern edge of drought affected area 1865).

Annual isohyets (inches) 4-9 are soil types as follows:

Brown calcareous earths   Alkaline sands

Highly calcareous loamy earths   Leached sands

Brown sands   Other soils

Table 16–1. Mallee survey site characteristics

Site	Mean annual rainfall[a]	Drought frequency[b]	Drought risk[c]
Pinnaroo, South Australia	341 (13.4)	43	Medium
Paruna, South Australia	275 (10.8)	65	High
Murrayville, Victoria	321 (12.6)	c. 43	Medium
Millewa, Victoria	267 (10.5)	c. 70	High

[a]Precipitation in millimeters (inches).

[b]"The number of years in a hundred in which the season of continuously effective rainfall is less than five months." Trumble (1948).

[c]Researcher's estimate.

tions from a pilot survey. Initially about 120 interviews per site were hoped for, but the high cost of rural on-farm interviewing (about $10 per successful interview) plus the limited time available reduced the final numbers to half that figure. Field application of the basic interview was by a team of 6–10 student interviewers under my supervision. All farms were visited and a sample of at least one-third of all farmers in each site successfully interviewed.

The Mallee study area

The Mallee ecosystem

The Mallee area derives its name from the dominant vegetation form, the mallee, being species of eucalypts with many stems branching at or near ground level from single underground lignotubers. The heights vary up to 9 meters (30 feet) and crowns may form a complete cover in the denser "scrubs" or bread into isolated clumps in the more open stands. A shrub and grass understory is usually present, but varies considerably in density.

Within the original area of this vegetation, local topography, soil, and moisture appear to have supported variations in species. On unconsolidated mainly eastwest trending sand dunes which may reach 60 meters (200 feet) but are more usually about 10–20 meters (30–65 feet) above the surrounding country, a mallee heath occurred. Here *Eucalyptus dumosa* and *E. incrassata* formed stunted mallees with a porcupine grass (*Troidia irritans*) and occasional flowers. On the undulating to flat plains sloping to the Murray River on the north, a deep red-brown sand carried pine (*Callitris spp.*) on ridges and belar (*Casuarine spp.*) on lower country originally. On the lower plains between the dunes more open stands of mallees with grasslands and herbs occu-

pied red-brown clay loam soils often with limestone at or near the surface. Where surface water lay after rains, a claypan or salt pan fringed by succulents or small saltbushes varied the patterns.

Apart from isolated holes in the few exposed areas of limestone and on the claypans, surface water was nonexistent and no water courses crossed the area. Average rainfalls decreased from about 380 millimeters (15 inches) in the south to about 250 millimeters (10 inches) in the north. A winter maximum is standard but annual totals have been recorded as low as 25 millimeters (1 inch) and as high as 480 millimeters (19 inches). Below the surface, however, at depths as shallow as 45–60 meters (150–200 feet) potable water could usually be found.

The history of Mallee land use[1]

For convenience, the sequence of human use of the Mallee may be divided into three basic periods. Prior to about 1840 the area was exclusively occupied by aboriginal tribes with a hunting-gathering economy; from about 1840 to the end of the century aboriginal land use gave way to Anglo-Australian pastoral land use; and after about 1900 pastoral land use in turn gave way to agriculture.

Aboriginal land use before 1840. Despite the paucity of records there is sufficient evidence to suggest that aboriginal groups moved through the Mallee, particularly after rains, when isolated rock holes and some of the claypans held a little water, hunting gray and red kangaroos as well as smaller rodents, emus, and Mallee fowl. Moving in from the relatively densely populated Murray River frontages, the aborigines appear to have fired parts of the Mallee as part of their hunting technology, and indeed some of the open grasslands or grassed open mallee stands may have resulted from repeated aboriginal burning.

Apart from the possible effects of their use of fire, and the few campsites (little more than old hearths, fired bones, and worked stone chips) there was little apparent evidence of their impact on the landscape and the first white explorers may be forgiven in thinking they were exploring an uninhabited country.

Anglo-Australian pastoral land use about 1840–1900. Initial Anglo-Australian contact with the Mallee came from exploration down the Murray Valley from New South Wales in the 1830s and by exploration in the 1840s down the creeks flowing into the Mallee from the southeast. Faced by an apparently endless and virtually impenetrable scrub devoid of any surface water, the initial reactions were negative and the area was considered to be generally useless for settlement.

1. Detailed documentation of the history of Mallee land use has been omitted here, but a valuable lead-in is Harris (1970.)

In fact, pastoral settlement began to occupy the fringes of the area from the 1840s onward, moving livestock first onto the Murray River frontages where permanent water and useful natural grasslands offered a base for periodic use of the Mallee scrubs south from the river after seasonal rains, especially in the 1850s. A similar pattern was established by pastoralists approaching from the south, east, and west, using first creeks and then the temporary lakes or claypans as bases for head stations. Both sheep and cattle were introduced in this way.

Apart from the grazing of domestic livestock, which tended to concentrate upon the pockets of natural grassland and frontages to the Murray River, the impact of human occupation was initially scanty. Wells were dug and horsepowered "whims" raised the water into adjacent troughs; rough homesteads and yards for the livestock were constructed of local timber; and in later years woolsheds were added. These islands of settlement, however, were isolated in a sea of scrublands, linked only by rough sandy tracks barely passable for much of each year, the grazing land unfenced and generally undemarcated. Of greater significance to the ecosystem was the arrival of rabbits in the 1870s probably from the south and west. By the late 1870s, stations were being abandoned as fodder was no longer available in the face of competition from the rabbits.

Attempts to revitalize the pastoral land use in Victoria by legislation in 1883 provided some stimulus to renewed occupation, but innovations in agricultural technology elsewhere together with the pastoralists' experience of the variable character of the Mallee were to provide a new appraisal of the agricultural possibilities of the Mallee in the 1890s, which led eventually to the opening up of the Mallee for agricultural settlement from about 1900 onward.

Anglo-Australian agricultural land use 1900–70. Prior to the 1880s agricultural land use in both South Australia and Victoria had generally avoided the areas of Mallee scrub. Apart from reservations about the fertility of the soils and low rainfalls, a major problem was the difficulty of land clearance, since the mallee was expensive to cut down, and tended to sucker almost indefinitely afterward unless the roots were grubbed out.

The 1880s, however, saw several innovations in agricultural technology which enabled a revision of traditional attitudes to the Mallee:

1. The stump jump plow—developed in 1876—enabled plowing to go on simultaneously with land clearance and enabled crops to be sown in the minimum time.
2. Land clearance costs were significantly reduced by the development of mallee rollers; now one man, a roller, and a horse team could knock down scrubs faster and more efficiently than a team of axmen.
3. Harvesting was made easier, cheaper, and faster by the wheat headers and threshers which McKay com-

bined in his machines from the late 1880s onward. With these machines the wheat ears could be harvested (headed) from among the regenerating mallee shoots prior to burning the stubble to kill the mallee regrowth.
4. Drought-resistant wheat varieties began to appear, more suitable to the lower rainfall country. Early-maturing varieties such as Purple Straw had been developed in the 1860s in South Australia, and American hybrids provided further genes for local innovations in the 1880s.

As a result of these innovations and their successful application to other mallee areas in South Australia and Victoria, pressure to reconsider the agricultural potential of the study area was building up in the 1890s. As a result, legislation in both states opened up the area to agricultural settlement from about 1900 onward.

Land was surveyed, roads opened, railways built. The farmers moved in, began to clear the scrub, build farmsteads, reap their first crops, and cart the bags of grain to the rail sidings for export. This pattern, modified in detail but basically similar in components, was repeated for the Mallee as a whole over the period 1900–30. In detail, the Pinnaroo site (site 1, Fig. 16–2) was mainly surveyed in 1904 with later additions to the south in the 1920s; the railway was completed by 1906; and agricultural settlement began about 1904 and developed rapidly up to the First World War. The Paruna site (site 2, Fig. 16–2) was occupied by settlers coming south into the scrubs from the Murray River after 1906, and most settlement dates from the arrival of the railway in 1914. The Murrayville site (site 3, Fig. 16–2) was surveyed and occupied soon after the Pinnaroo railway was opened, and apparently several South Australian farmers were pioneers here. The railway from Ouyen was opened in 1908 and most of the agricultural settlement took place between 1908 and the First World War. The Millewa site (site 4, Fig. 16–2) was the last to be occupied, the railway from Redcliffs reaching Werrimull by 1923 and Morkalla by 1925.

The sequence of agricultural settlement which resulted can be summarized in three periods. The years 1900–30 saw the rapid increase of immigrant farmers and their families, the spread and increase of rural population, and the appearance of small towns, usually at strategic railway sidings and usually laid out by government surveyors. The mallee scrub was rapidly cleared and replaced by grainfields, where wheat was the main grain, with oats for the horses the second crop. Dry-farming practices were being introduced and bare fallowing without fertilizers became the standard cultivation technique. Droughts in 1914, 1918, and the late 1920s brought severe hardship and government drought relief measures.

The period from 1930 to about 1947 saw increasing economic hardship on the farms with the depressed world wheat prices in addition to local yields depressed

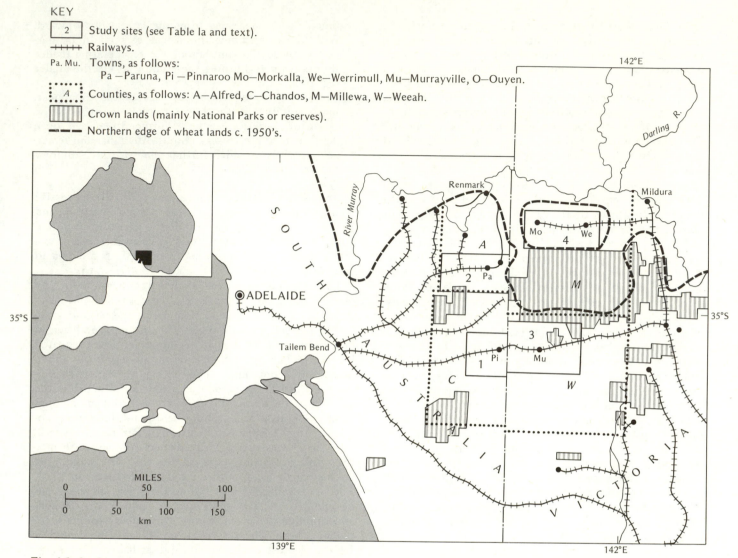

Fig. 16–2. Settlement patterns in Mallee study area. Sources: South Australian and Victorian government records; *Atlas of Australian Resources* (1957), map, "Dominant Land Use"

by further droughts in the 1930s and 1944–45 and an increasing intensity of soil erosion from old dunes reactivated by land clearance and left in bare fallow all the dry summers. Repeated requests for drought relief from this and other Australian wheat-growing areas led to a national reconsideration of the plight of the wheat farmers on these economically and environmentally "marginal lands." With Commonwealth funds, both Victoria and South Australia applied a Marginal Lands reconstruction scheme to the northern half of the study area (including sites 2 and 4) whereby bankrupt farms were bought up and reallocated to increase the size of neighboring surviving farms to allow a mixed wheat-livestock system to be introduced.

After about 1947 soil conservation measures and increasing prices for wheat and wool stabilized the outflow of population and reduced the soil erosion hazard. Mixed farming benefited from increased use of fertil-

izers and pasture improvement from nitrogen fixing legumes. Land clearance was renewed but the rural depression of the late 1960s brought further amalgamations of properties and quotas on wheat purchases encouraged a further diversification of production. By the 1960s, moreover, several of the smaller urban centers appeared to be below the threshold of successful service provision and empty shops were to be found in all the central places within the study area in 1971.

An innovation in land use was the creation of wildlife reserves and national parks in portions of the uncleared Mallee covering the higher sand dunes, from the 1960s onward.

Land use systems in the 1970s

Before considering the detailed contrasts between the four study sites, a general description of the contemporary farming in the Mallee area as a whole will provide a relevant background. A recent survey provides information for a small sample of farms in both the South Australian and Victorian portions of the Mallee.

The following section is based upon a recent survey of wheat growing in Australia over the 3-year period 1964/5 to 1966/7. The South Australian area "Region 3, Murray Mallee" includes basically the whole area east and south of the Murray River and is therefore a less representative area than the Victorian "Region I, Mallee" which is more directly restricted to the Mallee proper. Twenty-six farms were sampled in South Australia, 25 in Victoria (Anonymous, 1969).

In terms of terrain the differences across the state border are insignificant, over 80 percent of the land being flat or undulating and suitable for cropping and the bulk of the remainder suitable at least for grazing. Only 1–2 percent was classified as rugged and waste, i.e., unsuitable for either grazing or cropping.

Similarly, in the condition of the land, differences across the border were minimal. Both state areas showed over 80 percent of the land had been cleared of the original mallee vegetation, while the remaining area retained the original vegetation cover. Comparing the estimates of terrains with land condition, there would seem to be scope for a further clearance of some 12–14 percent of the average farm land if required for cultivation, but little new land clearance is being currently undertaken.

The average size of farms is higher in the South Australian portion, about 1,300 hectares (3,172 acres) compared with about 1,100 hectares (2,776 acres), although the most common size in both states is 400–800 hectares (1,000–1,999 acres), and 70 percent or more of the total farms in both states fall into the size group 400–1,200 hectares (1,000–2,999 acres).

This contrast is a reflection of the contrasting significance of cropping and grazing across the border. The South Australian farms have a higher proportion of land in improved grazing (lucerne, sown grasses, and clover pastures) and a resultant smaller proportion in grain crops (45 percent in grazing as opposed to 9 percent in Victoria, and 13 percent in grain crops as opposed to 30 percent in Victoria). The higher proportion of land in grain crops in Victoria is further reflected in the higher proportion of land fallowed there (21 percent as opposed to 3 percent in South Australia).

The number of livestock, particularly sheep, carried on the South Australian farms is four times (over 1,700) the Victorian figure (400) and further reflects the greater reliance upon crops for income in Victoria than in South Australia where livestock and livestock products are more important. Thus crops provided 82 percent of gross income in Victoria, but only 39.6 percent in South Australia, where wool provided 42.0 percent compared to only 6.8 percent in Victoria.

Most farms are family-run production units, with family partnerships forming 67–79 percent of the ownership and 79–91 percent of the management in Victoria and South Australia respectively. Owner-operators made up most of the remaining management.

In terms of economic returns, the Victorian farms appeared to have a slight advantage, providing a greater gross return per acre, a greater gross return on capital, and greater return per labor input. Net income was higher in Victoria and the level of indebtedness appeared to be lower. In part this may have reflected a marked contrast in productivity—yields of both wheat and barley being markedly higher per acre in Victoria.

Testing the hypotheses in the Mallee[2]

Hypothesis 1. "Occupance of hazardous areas is rational according to the inhabitants."

If this were true, we might expect that all farmers would show an awareness of the drought hazard, that they would consider the hazard either not dangerous or dangerous but tolerable, and that they would be aware of other areas of lower- or higher-hazard risk.

Almost all farmers interviewed (89 percent of the 181) claimed that their locality experienced droughts and when asked to estimate the effects on a five-point scale from "very serious" to "of no consequence," 42 percent considered the effects "serious" while another 42 percent considered them "tolerable." Significantly, when analyzed by sites, the farmers on the high-risk site 2 stressed the serious nature of drought effects (46 percent) whereas fewer of medium-risk site 1 thought the effects serious (only 30 percent), and more thought the effects were tolerable (48 percent) compared with 38 percent in the high-risk site.

Almost all farmers (90 percent) expected drought to come again, while 81 percent thought it would come again but did not know when. Only 11 percent thought there was a regular cycle of droughts, while 7 percent thought drought would come soon.

Knowledge of alternative areas with less drought hazard and providing equal economic opportunities was claimed by 64 percent of farmers, and 89 percent of these correctly identified areas of lesser drought hazard when probed.

Questioned whether they would work on the farm many more years, 80 percent said they would, and only 16 percent claimed that they would not remain.

From the above, therefore, it would seem that the great majority of farmers are well aware that they live and work in a drought-risk zone and that they appear to have accepted this situation even though they are aware of less hazardous locations for their activities.

Hypothesis 2. "Response to hazard in Australia would be of the technological (industrial) type—attempting more human control over nature rather than attempting to work more in harmony with nature."

---

2. Preliminary analysis has been completed only for sites 1, 2, and 3, and several other hypotheses remain to be tested for these and the fourth site.

If true, human adjustments in the Mallee should show the dominance of technological innovations in response to drought stresses; if the hazard stress remained constant through time the impact ought to be reduced as a result of technological solutions and adjustments over time; and finally, areas of highest risk ought to show maximum use of technology.

Both farmers and officials of the South Australian Department of Agriculture claimed that as a result of increased mechanization of farm operations and implementation of soil conservation measures, the impact of droughts has been considerably reduced. By comparison with the 1940s, when horse teams provided most of the tractive power, the widespread use of tractors or self-propelled machinery provided greater flexibility to meet drought stresses.

To illustrate the technological changeover, sites 1 and 2 had a total of 1,327 and 910 horses respectively in 1944/45. By 1966/67, the last year that horses were counted, the figures were 13 and 49 respectively. For tractors the figures in 1944/45 were 80 and 36 respectively, and by 1966/67, 285 and 174 respectively, more than one for each farm in the area. The main period of mechanization appears to have been the late 1940s and early 1950s.

Once horses were replaced there was no longer any need to grow oats or buy their feed. With tractors, all the field operations could be carried on faster and for longer periods. It was claimed, for example, that with an eight-horse team a man could sow 8 hectares (20 acres) per day, with a large tractor and multiple drills 101 hectares (250 acres) per day could be sown, and one farmer claimed to have sown 1,212 hectares (3,000 acres) in 10 days with the help of his sons. Reaping time could be more than halved when rates for a self-propelled header with a 4.3-meter (14-foot) blade (20 hectares or 50 acres per day) were compared with horse-drawn strippers using 3-meter (10-foot) blades (8 hectares or 20 acres per day). This increased mobility and speed of operations helped the farmer take advantage of useful rains at seeding time and enabled him to save more of the crop if storms threatened his standing grain. Also important was the ability of tractors with large low-pressure tires to sow the loose, eroding sand ridges—places which were often impassable to a horse team.

The widespread use of Wimmera and associated ryegrasses (*Lolium spp.*) was claimed to have successfully stabilized all but a few of the sand ridges which in the 1944–45 drought were sufficiently mobile to cut most north-south roads in the study area. Field observation noted some ridges still active in 1971, but no roads were affected and occurrences were said to be rare. Use of stubble mulching rather than bare fallow had reduced most areas of soil erosion also.

Interviewing farmers in the Mallee, it soon became apparent that we were interviewing many individuals rather than sampling a group community. When asked whether friends and neighbors could do anything to prevent drought damage, 67 percent of farmers replied no; only 23 percent thought they could be of assistance. When asked whether their neighbors made any different adjustments to drought stresses, most replied no or not to their knowledge and a question attempting to probe details of neighbors' adjustments failed to achieve any useful responses, apparently because of lack of knowledge or interest in neighbors' affairs. Help in time of personal emergency, illness or accident, was noted, but apart from that Mallee farmers appeared to keep to themselves even in drought.

In drought, although farmers claimed to talk over their problems with family (30 percent), friends (25 percent), or the local farmers' union (United Farmers and Graziers) (21 percent); they went mainly to their stock firms (45 percent) or banks (26 percent) for specifically financial help to carry on after a drought. Only 9 percent looked to government for help. The majority (69 percent) claimed that the other two sources of finance (stock firms or banks) were either partly or completely successful in financing their further operations.

In terms of expertise in cereal cultivation, the majority of farmers showed that they followed the official recommended wheat varieties. Only 16 percent in 1970–71 were growing wheats recommended earlier than 1966. These few growing "out-of-date" varieties contrasted with the 43 percent growing only varieties recommended after 1966 (which included 8 percent growing only the recommended 1971 varieties), and the other 30 percent growing some "out-of-date" alongside the post-1966 varieties. While accepting the Department of Agriculture recommendations, most farmers were at the same time hedging their bets by growing several rather than only one of the most recent varieties.

While field interviewing achieved some indication of the role of technology, a further complementary source was found in the archival sources. Content analysis of the local newspaper at site 1 provided a chronological index of drought reporting which was compared with the sequence of wheat yields, wool clips, and sheep numbers alongside values of stored soil moisture obtained from a CSIRO water balance computer program.

The program provides weekly values of stored soil moisture using weekly rainfalls and these data for site 1 were summed to week 26 and week 43 of each year (beginning and end of the season respectively) in an attempt to obtain a finer measure of drought conditions than were annual precipitation figures (Keig and McAlpine, 1969).

The result (Fig. 16–3) shows the overall relationships. Despite the lowest rainfall and soil moisture figures for the period in 1967, the newspaper reports were only one-third the intensity of 1944–45, wheat yields were higher than 1944, and sheep numbers did not decline markedly. Apparently more extreme drought stresses in 1967 could not result in the same loss of production or

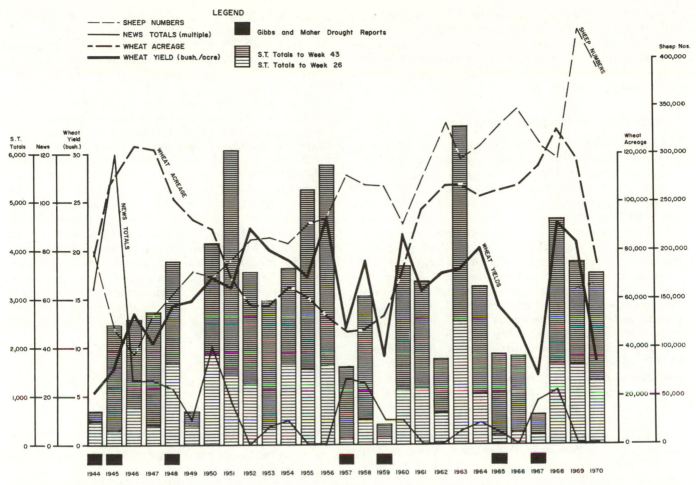

Fig. 16–3. Trends in drought impact indicators—Pinnaroo (site 1)—1944–70. Key: Sheep numbers for County Chandos. News totals (multiple)-index of reports of drought occurrences in local newspaper. Wheat acreage and wheat yield for County Chandos. Drought reports: occurrence of droughts, Gibbs and Maher (1967). Totals—soil moisture accumulated figures to end of weeks 26 and 43 of each year Keig and McAlpine (1969), for Pinnaroo

create as much local interest as the 1944–45 drought. While this preliminary analysis will need to be checked, particularly to see if other factors were affecting for example the newspaper coverage (editorially there was no change 1944–70), it suggests some support for the claim that droughts are no longer the threat they once were, and technology may have provided the reasons.

Measuring drought impact

Prior questionnaire surveys of popular attitudes to and knowledge of drought in South Australia (unpublished surveys by geography students of Flinders University in 1970–72) had suggested that drought was seen to have many varied associated impacts, ranging from environmental (e.g., heat waves, water salinity, and dust storms) to economic (shortfalls in retail sales and business activity), social (community concern and appeals for help), and demographic (population migration). Farmers in the

Mallee similarly saw drought's impact as wide-ranging. For most however, the impacts were on crop production (54 percent classed this as the major drought damage), with damage to land by erosion noted by 11 percent, losses of animals noted by 9 percent, and the effects on people of economic hardship noted by a further 9 percent. For their crops, 20 percent said they had lost the whole crop on occasions, while 55 percent claimed their losses had been substantial. For their livestock, only 5 percent claimed total losses, while 40 percent claimed substantial losses.

Comparing "good"-year production figures with drought-year figures, modal figures for wheat showed a drop of 80 percent from high to low yields, a 67 percent drop for barley yields, and increase in sheep deaths from modal 21 head in good years (blowfly strike, poisoning, lambing deaths) to 51 head in drought years. For the few farmers keeping cattle, the drought-year losses were less than the good years (1 as opposed to 1.5); the explanation for both the reverse here and the insignificant sheep losses (actually just over 2 percent of the herd in both good and drought years) was that in drought conditions, livestock would be sold off before they could die, whereas in the good seasons "they could, and did, die anytime."

Recommendations for drought policies

It became obvious during interviewing in the field that we were talking to the survivors of a silent battle which had been going on in the Mallee ever since the first Anglo-Australian settlers moved onto their farms some 60–70 years ago. Official statistics showed that between 1944 and 1970 the number of farms in the area declined by 14–40 percent. Only on the northern fringe where irrigation lands were opened up post-1945 were any increases in rural holdings noted. Originally, the size of farm blocks was about 130–400 hectares (320 or 620 up to 1,000 acres) but by the time of our surveys only 18 percent were less than 400 hectares (1,000 acres), 29 percent were 400–800 hectares (1,000–2,000 acres), 32 percent were 800–1,600 hectares (2,000–4,000 acres), and the remaining 20 percent were over 1,600 hectares (4,000 acres). Amalgamations had been officially encouraged in the late 1940s and early 1950s in order to achieve economies of scale and to provide sufficient profits to create an economic unit. Since then private amalgamations by sale and bequest appear to have continued the process. Thus in theory the farmers remaining have had the benefit of this process of weeding out the inefficient; as survivors they have at least the aura of successful resource managers.

Most in fact claim to have weathered several droughts. Forty-three percent claimed to have experienced four or more droughts and this tallies with the 42 percent of the farmers who had been in the area throughout the period 1944–70. At least 92 percent had experienced the last major drought of 1967 and only 4 percent claimed no drought experience. These are experienced men (85 percent had known no other job and 63 percent were over 40 years of age), well aware of the impacts of drought and not overly concerned by it. That is, if it does not last more than one season.

Virtually all farmers (98 percent) agreed that the effects of a drought depend upon its length. The majority (64 percent) claimed that the worst length of time for a drought to last was between 1 and 3 years or effectively over two seasons. One season's drought losses could be borne but further failure in the second season posed greater problems because finances had been absorbed by the first year's losses, and their own supply of seed would have become exhausted so that seed would have to be bought for the third season's sowing.

The initial impression from this survey, and one which will have to be checked by further analysis of the data both from the field results and from some as yet untapped archival sources, is that farmers in the Mallee appear to be coping successfully now with drought stresses, as a result particularly of innovations made in the 1940s and 1950s. While further adjustments in sizes of holdings and economies of scale of operations may come about as the aging population sells out to neighbors, there seems to be no strong case for government relief policies to be instituted except when drought happens to coincide with low prices for primary produce. At that time some financial support—e.g., low-interest loans, to enable farmers to keep operating for the second or third year of stress—may be necessary, although there will be need to beware of merely prolonging the activity of inefficient managers. What may prove to be of more concern to governments in the Mallee may be the problem of the cost of provision of minimum social services and the question of the future of the small rural retailing centers with population densities declining to perhaps less than 0.4 per square kilometer (1 per square mile) as a result of further farm amalgamations. There has been already some suggestion of "town farming" (residence of farmers in town) becoming a feature of this area (Williams, 1970) and this might be a partial answer..

References

Anonymous. (1969) *The Australian Wheatgrowing Industry: An Economic Survey 1964–65 to 1966–67*. Canberra.

Atlas of Australian Resources. (1957) Canberra.

Atlas of Australian Soils. (1960) Canberra.

Bureau of Agricultural Economics. (1969) "An economic survey of drought affected pastoral properties New South Wales and Queensland 1964–65 to 1965–66." Canberra: Wool Economic Research Report No. 15.

Campbell, D. (1968) *Drought: Causes, Effects, Solutions*. Melbourne: Cheshire.

Foley, J. C. (1957) "Droughts in Australia: review of records from the earliest years of settlement to 1955." Melbourne: Commonwealth Bureau of Meteorology, Bulletin No. 43.

Gibbs, W. J., and Maher. (1967) "Rainfall deciles as drought indicators." Melbourne: Commonwealth Bureau of Meteorology, Bulletin No. 48.

Harris, C. R. (1970) "Mantung—a man-land study in the Murray Mallee of South Australia." *Proceedings of the Royal Geographical Society of Australasia* 71:1–26.

Heathcote, R. L. (1969) "Drought in Australia: a problem of perception," *Geographical Review* 59:175–94.

Keig, G., and McAlpine, J. R. (1969) "WATBAL, a computer system for the estimation and analysis of soil moisture regimes from simple climatic data." Canberra: CISRO Division of Land Research, Technical Memorandum 69/9.

Natural Hazards Research. (1970) "Suggestions for comparative field observations on natural hazards". Boulder: Univ. of Colorado, Inst. of Behavioral Science, Working Paper No. 16.

Reserve Bank of Australia. (1968) Rural Liaison Service, *Physical and Financial Effects of Drought: A Survey in Northern New South Wales*. Sydney.

Trumble, H. C. (1948) "Rainfall, evaporation and drought frequency in South Australia." *South Australian Journal of Agriculture* 52:55–64; Supp. 1–15.

Williams, M. (1970) "Town-farming in the Mallee lands of South Australia and Victoria." *Australian Geographical Studies* 8 (2):173–91.

17. Decisions by Florida citrus growers and adjustments to freeze hazards

ROBERT M. WARD
Eastern Michigan University

Agricultural managers throughout the world are often confronted by problems originating in the atmosphere. It is easy to cite specific cases when nature has provided too much of one weather element, not enough of another, or does not act in an expected manner during the plant's growth cycle. Since orange trees were planted in the vicinity of St. Augustine by Spanish explorers, Floridians engaged in citrus production have contended with the threat of cold-air damage to their crop. During the twentieth century, serious freezes occurred in 1917, 1934, 1940, 1957–58, and 1962. At these times many growers' responses to low temperatures were inadequate to allow salvaging their fruit or saving the lives of their trees. In an effort to understand more thoroughly how growers have coped with freeze hazards, and to account for spatial differences, one hundred were interviewed to determine their adjustments to potential freezing conditions.

Fig. 17–1. Spatial distribution of sample counties, Florida

The sample was selected from five counties possessing a wide range of freezing occurrences in central and south Florida (Fig. 17–1). Marion, Hardee, and the inland portion of Hillsborough counties are most prone to this hazard. Groves in Hendry and Indian River counties have experienced fewer freezing temperatures due to both proximity to the ocean and southerly location.

Caution should be exercised when interpreting the interview data because complete adjustment, partial adjustment, or lack of adjustment must be realized as part of the grower's entire agricultural situation. The writer agrees with Burton (1962) when he states:

> A dominant characteristic of agricultural occupance . . . is that the occupance responds to a variety of conditions among which the flood hazard is not often of primary significance . . . In making an appraisal of the resource potential of his farm, the operator places weight not only on external factors such as the markets and transportation, but also on the whole resource complex over which he exercises control, and not simply on the individual parts of it.

Although it is possible for growers to defend successfully against each instance of freezing conditions by utilizing grove heaters or annual comprehensive insurance policies, long-term cost/benefit analyses often prove to be more important than preventing the total loss of a single crop. Personal pride, lack of working capital, and failure to accept or be aware of innovations are among other factors that help influence growers' attitudes toward preventing crop damage.

During an extended period of time, agricultural land-use patterns have evolved, and will continue to do so, as man adapts to changes in his assessment of the weather/plant/market syndrome. Before examining specific actions to counteract possible weather damage to crops, it is helpful to review several reasons for man to engage in an activity that contributes to the promotion of a hazardous situation. An understanding of this decision by growers necessitates an awareness of their psychological, economical, and philosophical attitudes.

Growers' commitments to a risky site and occupation

Some risk taking may be attributed to an attitude of superiority in which man scoffs at the idea of being subservient to nature. In combat vis-à-vis nature, his ingenuity rises to the occasion and eventually he expects to be victorious. A second suggestion to explain hazardous behavior patterns may be found in the joy of

accepting a challenge. This characterizes the frontiers-
men and explorers who head into the unknown in
search of freedom, fame, and fortune as they pit themselves against the elements of their Creator. McNeil
(1968) suggests that certain individuals engage in stressful situations as a form of pleasure seeking or ego
building. What may appear superficially to be mystify-
ing behavior by growers may have a logical explanation
when a more thorough search is made into their person-
alities.

In man's drive for economic gains, often he tries to
assume the role of the ecological dominant (Tuan,
1971). If this view is correct, farmers confront certain
events of nature as thwarting their efforts to earn a
livelihood. Growers are aware that nature not only influences their yields, and consequently their returns from
the market, but indirectly their physical and mental
being. Florida horticulturalists often believe that nature
is to be manipulated through the application of technological advancement. Many of them reason that as the
artificiality of their environment increases, their economic returns should also increase.

The commitment to faith and fate may take the form
of a predestined attitude in which growers believe that
the Creator has decided whether man can expect a good
or bad outcome. In such circumstances, the environmental extremes will not greatly influence or interfere
with his future. This fatalistic attitude is also evident
among growers who do not subscribe to the concept of
God's plan, but who are willing to accept whims of

nature as being beyond their control to alter or sup-
press. Without precluding attempts to modify the hazardousness of the situation they are resigned to the
outcome.

Cognition and perception

Citrus growers are often long-term residents of Florida
and have established a mental image of a normal cropping season. Their cognizance of weather elements includes deviations from a norm. The eccentricities or
extreme conditions of nature are measured in the context of frequency, size, duration, time (day and season),
degree of uncertainty, and other characteristics. The
mental construct of frequency of hazardous weather
could explain the grower's behavioral pattern, his perception of current environmental conditions, and his
prediction of future events.

The initial analysis shows that Florida orange growers' estimates of the frequency of fruit and wood damage follow a predictable pattern (Table 17–1). Counties
with relatively high frequencies of cold weather are also
the counties in which growers estimate high frequencies.
A comparison of the subjective frequency of the two
measures of cold damage by growers to the frequency
described by scientists reveals that growers both exclude
and supplement portions of their environment. Their
underestimation of the frequency of damaging tem-
peratures(- 3.3° C. or 26° F.) in colder counties may be
related to the relative lack of significance that they

Table 17–1. Frequencies of orange freeze damage to various parts of the plant by January 9 (numbers in percent of years out of 100)

| County | More prone to freezes | | | Less freezes | |
	Marion	Hardee	Hillsborough	Hendry	Indian River
Growers' mean estimate of fruit damage	20	20	17	9	8
Actual frequency of −3.3° C. (26.1° F.) by January 9	30	20	26	7	8
Mean grower error using algebraic sign	−10	0	−9	+2	0
Growers' mean estimate of wood damage	11	11	8	4	4
Actual frequency of −6.7° C. (20.1° F.) by January 9	5	3	3	0	0
Mean grower error using algebraic sign	+6	+8	+5	+4	+4

attach to this event. Many oranges can be sold to juice concentrate plants if they are processed within a few days after the freeze. This allows owners to avert the complete loss of a 1-year income; therefore, the event lacks an extreme negative re-inforcement.

In relation to tree damaging temperatures (−6.7° C. or 20° F.) growers appear to lose their optimism, since they have overestimated the frequency of the hazard. It is suggested that growers' responses exemplify the concept of conservatism (Phillips, Jays, and Edwards, 1966). It is found that often individuals tend to avoid guessing the extreme, but modify their answers toward the middle of a continuum.

Agricultural commitment and constraints

The problem of decision making is approached by Kolars (1965–68), who recognizes two distinct levels of decisions which can be applied to agricultural studies: those made during periods of evaluation and those made at times of commitment. During conditions of evaluation, the agriculturalists have a wide range of choice. Following their commitment to a specific type of agricultural pattern they are forced to follow through on a rather well-defined sequence of events. The most elementary example occurs during the evaluation period when growers are in a position to select any crop or combination of crops which they believe will be most profitable, or to choose not to plant a crop. Until they plant the seed, they have not committed themselves to the whims of nature, labor, and/or markets.

In the case of vegetable and citrus growers each of the two levels of decision making is well illustrated. Citrus growers, by the nature of their long-term investments, are initially placed in the category of committed individuals; whereas vegetable growers who represent short-term investors have more flexibility in their choice of crops and planting dates, and even enjoy the option of quickly abandoning farming during a particular season. As soon as the decision is made to begin planting, vegetable growers are as committed during the cropping season as citrus growers, since both groups will depend upon their cropland for a source of income. At each stage in a crop year, growers are confronted with a decision. Once they act, they are committed to another chain of events during which they evaluate an additional set of circumstances in order to be committed again.

Time becomes an influential constraint in the decision-making process. Even though there is a proper time and a set of optimum conditions for most agricultural activities, some actions can be postponed for a limited period until a threshold is reached and action must either be taken or ignored. Under such conditions growers can foresee the need for a decision and are not pressured for time. This tends to reduce stress and permits them to choose the most promising course of action. In some cases, as time decreases and pressure increases, individuals respond to the challenge by exhibiting great ingenuity vis-à-vis the impending dilemma.

Contrary to this series of events is the confrontation of a virtually unexpected hazard or a point at which time has ceased to act as a positive stimulus for creative thinking. This situation usually requires immediate action, anxiety often becomes exceedingly intense, and hasty decisions may reflect less sensible behavior and a feeling of desperation.

During some occasions, growers may have a large amount of new information thrust upon them at a critical time in the decision-making process that has the effect of quickly changing their set of alternatives. Under such circumstances a new disease-resistant plant may be announced just prior to the time they plant their seed but its ability to withstand cold air and extreme moisture conditions is unknown. During the midst of the cropping season, government policy changes may be revealed that drastically alter the market price of crops. The time constraint is again evident and the information overload may lead to confusion and irrational action.

Before elaborating upon the economic constraints of capital and marketing conditions, it is helpful to examine a model of growers' decisions when they are forced to cope with hazardous weather (Fig. 17–2). Two levels

Fig. 17–2. Model of growers' decisions in coping with hazardous weather

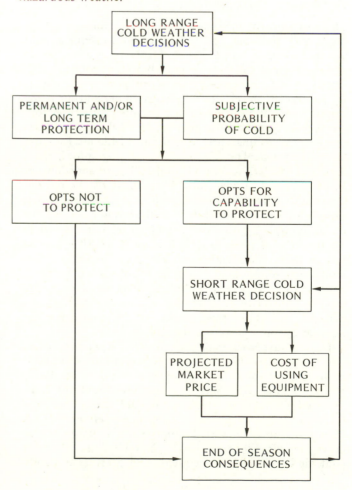

of decisions are related to low-temperature problems in an attempt to illustrate the decision process. First, as a part of the long-range planning for cold weather, growers must contemplate their methods of initiating permanent of long-term protection, such as selecting favorable sites, purchase of protective equipment, or determining their degree of dependency upon successfully obtaining a good harvest. Second, they must consider the probability of cold-weather damage to their fruit and plants, and subsequently, methods of reducing or eliminating possible damage during the time of impending danger.

Long-term decisions and adjustments

While administering a questionnaire schedule to orange growers, it became apparent that when interviewees were queried about possible means of preventing damage, most neglected to mention long-term action. The truth is that many growers have made some type of long-term response to future freeze conditions. Several of these adjustments include the use of slopes, location near large bodies of water, proper selection of scion and rootstock variety, migration to warmer sites in south Florida, and diversification of investments.

Topography as an adjustment

The best adjustment to potential freeze damage is selecting a site that has had a history of infrequent cold weather. This is considered by growers as a multiscale decision since variation occurs between counties, between groves, and among sites within groves. The majority of freezes in Florida are a result of cold air being trapped near the surface during a temperature inversion. As a consequence most growers in cool counties have utilized the rolling surface and the highlands for their citrus plantings. The importance of topography is most apparent in Marion, Hardee, and Hillsborough counties. In Marion and Hardee only 10 percent of the respondents stated that the majority of their citrus was grown on low or level land, while in Hillsborough, 35 percent of the sampled growers who located on flatland did so on the coastal plain around Tampa Bay. All groves in Indian River and Hendry counties were located on flatland since rolling landscapes are totally absent at these sites. It can safely be assumed that if these two counties were as subject to cold weather as those in the northern part of the state, they could not be considered for large-scale commercial production.

During advective colds such as the one experienced in 1962, plants on the upper parts of slopes often received more damage than those at lower elevations. The latter group of sites are somewhat sheltered, and the trees incurred less direct attack from cold winds than was received by plants on hill crests. Lower-elevation sites also have the quality of holding warm air given off by grove heaters (Phillips, 1969). Nevertheless, advective cold air is relatively infrequent, and growers in cooler

counties recognize the necessity of planting on sloped land to avoid the ill effects of temperature inversion.

Influence of saltwater and inland lakes

The presence of surface water is one of the prime elements that contributes to the location of certain crops in Florida. During the interviewing, large percentages of orange growers in the two coastal counties and also in Marion County exhibited an awareness of the influence of water as a moderating factor on low temperatures at their groves. Indian River growers unanimously acknowledged this influence. This is attributed to the proximity of their groves to the Atlantic Ocean. Eighty percent of the sampled growers in Hillsborough County indicated the same awareness. All operators in the latter county who replied that nearby bodies of water did not affect their plant temperatures were located more than 13 kilometers (8 miles) from Tampa Bay in an area of rolling lands.

Marion County provides one of the most interesting geographic relationships of the study. The majority of orange plantings are on the south and southeast sides of inland lakes (Fig. 17–3). Those 10 percent of the growers who did not plant on the slopes in Marion County were situated on favorable downwind sites near bodies of water. Without this moderating influence many growers stated that they would be forced to change their farming enterprises to livestock management.

Thirty-five percent of the interviewees in Hendry County stated that their grove temperatures were influenced by water. Most of those men were located on the south bank of the Caloosahatchee River. They asserted that it was 2°–5° C. (3.6°–9° F.) warmer on their side of the river during times of cold weather. Several believed that some moderated temperatures in the county were a direct result of air passing over the Gulf of Mexico.

Selection of scion variety and rootstock

Once a location has been chosen, growers must select a rootstock and scion variety. It is logical to suspect that the two questions of site and plant characteristics are concurrently resolved. A situation may occur where growers are confronted by a choice between a good producing plant which is less tolerant of low temperatures and a plant which produces less fruit but is more tolerant of cold air. The men who have selected the site with the greater risk are further forced to decide whether they should plant all low-yielding/high-tolerance plants or mix their plantings in order to increase the yield.

One method available to orange growers is to produce fruit that matures before it is subjected to freezing weather. In Marion, Hillsbrorough, and Hardee counties, 95 percent, 70 percent, and 65 percent respectively of the sampled individuals indicated that the majority of

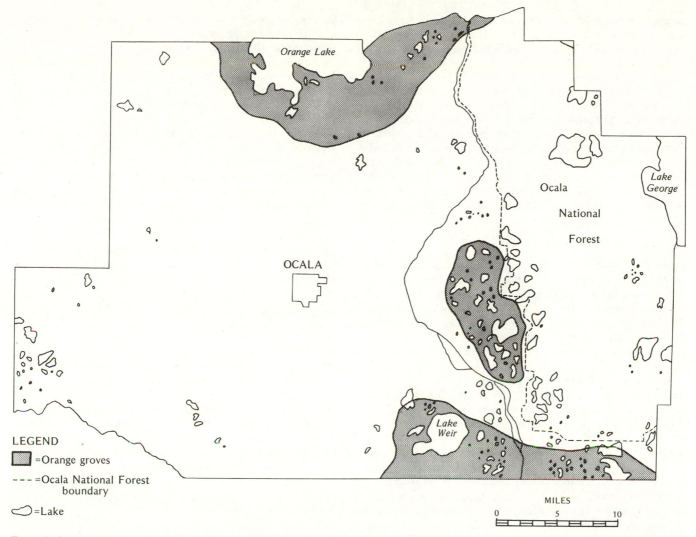

Fig. 17-3. Distribution of orange groves in Marion County

their crop was early and midseason oranges (Hamlin and Parson Brown), while 45 percent of the Indian River growers and 35 percent from Hendry County stated that more than one-half of their oranges were early and midseason varieties. Since almost all growers produce some of each seasonal orange, the extremely large percentage of respondents who rely upon the early varieties in Marion County seems to reveal their awareness of the high probability of cold temperatures. This is interpreted as a significant spatial difference.

The immediate reaction to a freeze year is to initiate preparations that will guard against future low temperatures. This fact was supported by increased purchases of heating units following major freezes, and is also reflected by abnormally large orders for early-bearing fruit trees during the same years. Late-maturing Valencia oranges have long been the most popular variety in Florida, but following the 1962 freeze, the sale by

nurseries of the early-bearing Hamlin variety in 1963–64 and 1964–65 exceeded the sale of Valencias for the first time since the late 1930s (Savage, 1967). A similar phenomenon occurred after the 1957–58 freeze when in 1958–59 there was a larger proportional increase in the sale of Hamlins than in the sale of Valencias. This is indicative of two facts: (1) many trees were killed in the hazard-prone counties, and (2) when growers were confronted with the option of a variety to replant, they often selected early-bearing trees as a means of maintaining their annual fruit production.

The variety of rootstock also provides a clue to the adjustment to several natural hazards. Sour orange root is more tolerant of cold air and high water tables than the other major variety, rough lemon. The growers' choice of rootstock is not simple. Rough lemon, which has a better developed root system, is usually considered to be a larger producer, but it does not always produce as good quality as sour orange. Many men attempt to mix the two varieties if possible. This practice is apparent in Hendry, Hardee, and Hillsborough counties.

Marion County is the best example of a location which has soil conditions that would be best adapted to rough lemon rootstock, but 95 percent of the interviewees have selected sour orange for its added cold protection. Surprisingly few growers in Hardee relied upon sour orange. In the northern portion of the county the slopes permitted the use of rough lemon, but following the 1962 freeze many of these growers replanted parts of their groves with sour orange root for the sole purpose of adopting a more cold-tolerant plant. This was also true in parts of Hillsborough County.

Citrus migration to southern Florida

From the time of the first citrus plantings in Florida along the northeast coast, the trend has been to abandon groves in the north and initiate new plantings in the south. Currently, the majority of Florida's citrus production is located south of an east-west line that bisects the state. The prospect exists that there will be a continued southerly movement of the center of citrus production.

Heathcote (1969) has suggested that droughts have had important and positive roles in shifting Australian settlement. Identical findings have been reported by Brooks (1971), who described population migrations that have resulted from droughts in northeast Brazil. A similar statement regarding freezes can be made about the exodus of Florida agriculturalists from the central interior croplands into south Florida. Here other circumstances were influential in guiding this movement, e.g., inexpensive land and lower taxes; but the attraction of warmer temperatures is one of the principal reasons for citrus plantings to encroach upon extensively used grazing lands of the southern Florida interior.

The writer assumes that the freezes of 1957–58 and 1962 provided an impetus to men who had doubts about the wisdom of concentrating their production in the citrus heartland of central Florida. This receives support from a former Hendry County agricultural agent who reported that three large citrus growers from central Florida purchased over 6,000 hectares (14,800 acres) of land in Hendry County shortly after 1962–63 (Gallo, 1967). He also asserted that soon after the freeze, four other large tracts of land were acquired for citrus production. Furthermore in 1964, it was stated that 24,000–32,000 hectares (60,000–80,000 acres) of Hendry County were being prepared for citrus (Florida Grower, 1964).

Diversification of investments

Some growers have chosen diversification of investments as a means of protecting themselves from a large loss. Following a random selection of growers to be interviewed, many were excluded because they had purchased interests in groves that were widely scattered throughout the state. Although this diversification of investment in orange groves may be attributed to many factors, it is viewed mainly as another deliberate attempt to minimize losses from localized natural hazards.

When the sampled growers were asked to describe their additional agricultural investments, it was discovered that at least 50 and 60 percent of the interviewees in Hendry and Hardee counties respectively were currently engaged in other forms of agriculture. The Hardee County example suggests a sincere effort to reduce their dependence on citrus income. The trend among some growers is to develop pastureland for beef cattle. The fact that one-half of the orange growers in Hendry County have diversified into other forms of agriculture appears to be indicative of corporate attitudes and of individuals with other large financial interests.

Short-term decisions

When a site has favorable climatic conditions the need for other forms of protection is reduced. Growers whose sites are in a location of greater vulnerability must investigate the desirability of other forms of providing crop protection. Often these growers regard heating equipment, wind machines, and other types of technological safeguards as providing economic security (Table 17–2). Mechanical protection devices also fulfill a psychological need by some growers to be actively engaged in combating cold weather. Less committed individuals, who rely more upon luck than wise management, may envision insurance protection as an easy means of averting financial disaster.

Heating equipment

During the interviewing no other adjustment to cold air was mentioned as frequently as the use of heating equipment. Twenty-two of the 24 growers who rely upon artifically produced heat as their primary method of combating the influence of cold air are located in Marion, Hardee, and Hillsborough counties; 13 of these 22 are Marion County residents. Although the sample is small, an application of the Spearman Rank-Order Correlation testing the relationship between frequency of cold weather and use of heating equipment yielded a coefficient of 1.0. This accentuates the fact that growers in cooler counties are aware of possible cold damage to their groves. Conversely, men who farm in warmer counties appear to have assumed a carefree attitude toward low temperature hazards.

The perceptiveness and sound judgment of growers are supported by Norman and Wallis (1971), who have indicated percentages of years that *no* heating would be required at selected stations when using $-3.3°$ C. as the critical temperature. Two of the stations used by the

Table 17–2. Primary method of adjustment by growers to cold air

Adjustment	Marion		Hardee		Hillsborough		Hendry		Indian River	
	Nos.	%	Nos.	%	Nos.	%	Nos.	%	Nos.	%
Use of heating equipment	13	65	5	25	4	20	1	5	1	5
Use of water	0	0	2	10	4	20	3	15	9	45
Healthy trees (cultivation practices)	2	10	6	30	5	25	5	25	0	0
Other adjustments	0	0	1	5	3	15	0	0	3	15
Does not prepare	5	25	6	30	4	20	11	55	7	35

present writer are in Marion County (Citra, 33 percent; Lake Bryant, 35 percent); one in Hillsborough County (Riverview, 38 percent); and one in Hardee County (Wauchula, 71 percent). No stations in Hendry or Indian River counties needed heating equipment.

The fact that individuals possess heating equipment does not mean that they distribute it throughout the entire grove or that they always use it. The decision to ignite fires is contingent upon capital expenditures versus the time and perceived extent of the cold weather. Growers consider the availability and cost of labor, projected value of fruit, time of day or night, type of meteorologic freeze, wind direction, rate of temperature decline, advice from the Frost Warning Station, recordings of their own grove thermometers, variety of scion and rootstock, general health of their trees, and their knowledge of "cold pockets" in their groves. Managers weigh the importance of each factor, calculate the risk and consequences of each alternative, and make decisions. These inputs which influence decision making are constantly being reevaluated as the constraints become more binding. When the crisis has passed the effect of the grower's action or inaction becomes a new cognitive input that will influence him during future experiences.

Use of water

Pumping warm groundwater onto grove land will not raise the temperature as much as using heating equipment, but it is a popular form of protection in the warmer, flatland counties and is less expensive. The use of water is especially important in Indian River County where 9 of 13 growers who adjust to cold air consider such water use as their primary method of protection. Even though an application of the Spearman Rank-Order Correlation test produced a significant coefficient, the inference of cause and effect between frequency of cold air and use of water may be invalid.

Growers who occupy level land are equipped to regulate easily the water in their groves. Consequently, surface irrigation systems can serve the purpose of providing a means of cold-weather protection. Six respondents used water as their primary adjustment in cooler counties and most of these are located near Tampa Bay in Hillsborough County where cold air is infrequent.

Sprinkler irrigation systems are also used to protect plants, although if water is not applied constantly to all parts of the plant during cold periods, more damage may occur than if water is not used at all. The underlying principle is based on the fact that heat is released when water changes from a liquid to a solid. This heat keeps the temperature above the danger point of the plant. Even though use of water for freeze protection is advocated by some agricultural scientists, this resulted in much freeze damage to orange groves in Hardee County during the 1962 freeze. So much ice accumulated on trees at that time that the branches broke from their own weight.

Healthy trees and proper cultivation

Eighteen percent of all growers stated that keeping their trees healthy and practicing sound grove management were adequate to counteract low temperatures. No spatial pattern is evident among the growers who use this general technique, although variation occurs among specific types of management. It is commonly known among growers that trees are very susceptible to cold-weather damage during active growth. Respondents in Marion, Hardee, and Hillsborough counties hope to see trees become dormant by early December and maintain that state until late February. A period of warm weather in early January could induce cambial activity and make trees more vulnerable to rapidly lowering temperatures.

In Indian River and Hendry counties, growers often encourage continual growth throughout the year. One manifestation of different styles of cultivation occurs when warm-county growers apply fertilizer in November to keep trees growing. The advantage of uninterrupted growth is viewed as an important locational factor by growers who are expanding into new lands of south Florida. This cultivation practice is never considered by north and central Florida growers unless they are at coastal locations.

A popular means of preventing low-temperature damage by growers is to cover the base of the young tree with insulating material to protect the plant from lethal temperatures. The practice referred to as "banking" or placing soil above the bud union around the base of the trunk is widespread. Although this demands large labor outputs in the fall and spring, differences of several degrees centigrade have been measured when comparing "banked" and "unbanked" trees.

Purchasing insurance

Included in the miscellaneous category of other adjustments is the occasional use of Federal Crop Insurance. This was available to all citrus growers with the exception of those in Hendry County in the sampled counties during 1969. Fourteen of the interviewees acquired policies, although these were seldom relied upon as the primary safeguard against freeze damage. In several cases only part of the crop was insured. Acquisition of policies appears to be more seriously considered by growers who are not totally committed to the returns from their crops and who have other business interests. The remaining growers often voiced complaints that premiums were too high, and since they had already taken other precautions, insurance was a poor investment. No spatial pattern was evident; participation was uniformly distributed among the counties where coverage was available.

No preparation

The data in Table 17–2 reveal that 18 growers in the two warmer counties and 15 growers in the three cooler counties responded that they did not prepare their groves against a possible cold hazard. This suggests that many of the hundred growers correctly perceive the danger, or lack of danger, of cold weather. Although a Spearman Rank-Order Correlation coefficient of 0.6 appears large, it is not significant, but the number of observations is small. It may be asked why so many growers in the cooler counties did not make adjustments. Their answers can often be generically classified under one of four headings: (1) they do not realize the degree of risk, (2) they do not consider it to be economically justifiable to invest in protection, (3) they may not believe that the methods of protection will be adequate, and (4) they may have religious or philosophic reasons. If these serve as explanations for not making adjustments, the growers appear to have established an adequate level of expected financial return and are working in a rational manner to achieve that goal.

Game theory in citrus protection

In Fig. 17–2, short-range decisions are analyzed among growers who have decided to purchase protection equipment. A choice is made during the period of cold-weather risk when the operators must decide whether to use some form of short-term protectant. The choice is not always easy, since the growers must calculate the point at which the projected market price is worth the cost of using or purchasing equipment. In addition, crisis points cause peak labor demands, and the cost of preparing against cold air is also weighed against the reliability of the weather forecast.

A game theory approach helps to explain the growers' options in the climate/plant/market syndrome (Table 17–3). These men are combatants in a game with nature which is viewed on two levels: probable high-intensity damage and probable low-intensity damage. During times of generally favorable growing weather, or when widespread but minor crop damage has occurred throughout the state of Florida, a very high yield should be achieved. This produces a strain on the migrant labor

Table 17–3. Payoff matrix to individual growers under widespread and/or local weather damage (payoff measure in profit or loss)

Individual grower's action. Crop cycle:	Weather hazard. Intensity of damage to most growers:			
	Great. Intensity of damage to individual grower:		Minor. Intensity of damage to individual grower:	
	Great	Minor	Great	Minor
Completed	Large loss	Very large gain	Very large loss	Small gain
Terminated	Small loss	Very large loss	Small loss	Large loss

market, which is largely responsible for both citrus and vegetable harvesting. At the same time the large quantity of fruit results in lower market prices, which may produce lower net returns for the grower. The high-priority decisions that are made during the season with only minor hazardous conditions are focused on labor problems. This poses a frequent question: Will lower market prices for the crop and higher labor costs allow the grower to harvest his crop at an acceptable profit? If the grower decides in favor of terminating his agricultural activity, he loses the money he has already invested. He has guessed that rising labor costs would not warrant the continuation of the normal cropping season, yet there is an incentive to complete the season, and every reasonable effort is made to harvest the crop. If he completes the production cycle, he may be rewarded with a modest gain. When other growers have experienced minor problems but an individual has had locally severe weather, he will probably lose money during the agricultural year, but the amount will be dependent upon his decision to terminate or continue growing the crop.

Conversely, widespread hazardous weather forces another set of decisions. During the warning stages for cold weather, the grower is confronted with the choice of investing in protection, or gambling that his crop loss will be within acceptable limits. Again the decision is strongly influenced by the prospects of a demanding market. Adverse weather has the effect of reducing production for the national market. This has special meaning for Florida citrus and winter vegetable growers who harvest a large portion of United States production. The cost input in protecting the crop from the effects of the weather may be more than offset by higher market prices for those who reach the harvest stage. In addition, the grower is blessed with a more amenable labor situation, since the migrants' choice of employment is severely restricted by reduced production. This indicates that the total effects of a single hazardous event may prove to have built-in advantages (Goldberg, 1968). If the grower opts to ignore the impending hazard, he saves the money which would be used to protect himself, and takes his chances on losing the crop. Should he lose the harvest, he avoids harvest labor costs, but he also has missed the opportunity to make large capital gains.

End-of-season consequences

In an effort to facilitate a better understanding of the perceptual and decision-making processes, the consequences of decisions that are made during times of hazardous conditions can be assigned a positive or negative value, i.e., growers either succeed or fail. If they can properly manage their crops during the hazardous period, their perceptions of the event are reinforced by a correct assessment of the stimulus, or environment, as it is incorporated into growers' memories. Should they misinterpret the stimulus as it passes through the per-

ceptual mechanism, and should the results of their decisions have a negative effect, they lose money and the meaning given to the percept will undergo a degree of alteration. At the end of the growers' crop season, the final outcome is described as forming a loop in the decision-making process, and the results will become cognitive inputs to be evaluated before they commit themselves to another agricultural season.

Summary and conclusion

Most interviewed orange growers in central and south Florida have made some agricultural adjustment to the threat of low winter temperatures. The extent of their actions appears to be commensurate with the frequency of the hazardous event. Their responses indicate the high degree of dependency upon immediate technological adaptations, e.g., the use of heating devices in cooler counties and warm irrigation water in the flatland counties less susceptible to freezes. Further examination reveals a variety of long-term protective actions not indicated during the formal interviews. These include activities such as careful evaluation of site factors, proper choice of scion and rootstock varieties, and diversification of investments. Approximately 25 percent of the growers in cold-hazard-prone counties do not take short-term action to protect their fruit and trees. This is partly attributed to an inaccurate evaluation of the environmental threat by some respondents, while other men expressed economic, religious, or philosophic reasons for not acting prior to freezing weather.

The perception and decision making by Florida growers during times of weather hazards are manifestations of their unique psychological, economical, and philosophical traits, and of their commitment to, and dependency on, their agricultural occupation. The decision to adjust is also a reflection of the grower's constraints as he tries to achieve his expectations by imposing his will upon nature. It is reasoned that most decisions in life are the result of more or less crude estimates of anticipated outcomes, and therefore it is assumed that the growers, consciously or subconsciously, calculate the probability of weather damage. Once they have assessed their alternatives and made their decisions, they review the consequences of their actions and this knowledge becomes an input that will influence their subsequent decisions.

References

Brooks, R. H. (1971) "Human resources to recurrent drought in northeastern Brazil." *Professional Geographer* 23:42–43.

Burton, Ian. (1962) *Types of Agricultural Occupance of Flood Plains in the United States.* Chicago: University of Chicago, Department of Geography, p. 144.

Gallo, T. (1967) "Advantages and problems in developing citrus in S. W. Florida." *Citrus Industry* 48:19.

Goldberg, Ray A. (1968) *Agribusiness Coordination: A Systems Approach to the Wheat, Soybean, and Florida Orange*

Economics. Boston: Harvard University, Graduate School of Business Administration, p. 213.

Heathcote, R. L. (1969) "Drought in Australia: a problem of perception." *Geographical Review* 59:189.

Kolars, John F. (1965–68) "Decision and commitment in Turkish agriculture." *Review of the Geographical Institute of the University of Istanbul.* International ed., no. 11, pp. 37–42.

McNeil, Elton B. (1968) "The ego and stress-seeking in man." In Samuel A. Klausner, ed., *Why Man Takes Chances.* New York: Doubleday, pp. 171–91.

Norman, O. N., and Wallis, W. R. (1971) "Note on frequency of nights when protective heating for citrus may be re-quired." *Weather Forecasting Mimeo*, WEA 71–1, p. 2.

Phillips, L. D., Jays, W. L., and Edwards, W. (1966) "Conservatism in complex probabilistic inference." *IEEE Transactions on Human Factors in Electronics*, HFE-7, pp. 7–18.

Phillips, William. (1969) Personal communication.

Savage, Z. (1967) *Movement of Citrus Trees from Florida Nurseries, July 1, 1928 to June 30, 1967.* Gainesville: University of Florida, p. 2.

Tuan Yi-fu, (1971) *Man and Nature.* Association of American Geographers, Resource Paper No. 10, p. 1.

——, (1964) "Citrus looks south." *Florida Grower and Rancher* 72:13.

18. Frost hazard to tree crops in the Wasatch Front: perception and adjustments

RICHARD H. JACKSON
Brigham Young University

Utah's fruit crop loss from killing frost the past three nights was estimated at $9 million today. (Provo, Utah, *Daily Herald*, 1972)

The Wasatch Front (or Wasatch Oasis) area of Utah is the core of what is commonly recognized as the Mormon cultural region (Meinig, 1965), and is physically and culturally distinct from adjacent areas east or west. To the traveler crossing the United States, it is an oasis in the midst of a dry landscape.

The habitat represents the interface between the Wasatch Mountains of the Rockies to the east, and the basin and range province to the west. These mountains are critical elements in the environmental setting because of their climatic modifications. The region as a whole suffers from moisture deficiency; through their surplus moisture the mountain watersheds make possible intensive irrigation agriculture. Without the mountains land use would be limited to extensive grazing and/or dry-land farming.

On a macro scale landforms and settlement sites are similar throughout the Wasatch Front. Settlements are located in the valleys at the broken edge of the mountains which rise abruptly a further 1,200–2,100 meters (4,000–7,000 feet) above the 1,200–1,500 meter (4,000–5,000 foot) elevations of the valley floor. Micro aspects of the landforms—exposure, orientation, and slope—cause important thermal differences within the region, and farmers' adjustments to these variables have created an intricate mosaic of land-use types.

Soils and climate also can be categorized briefly. Soils are primarily alluvial. If water is available, and proper care is maintained, these soils are quite fertile. The climate is best characterized as being continentally controlled and of low moisture. Average precipitation in the valleys of the Wasatch Oasis is from 300 to 400 millimeters (12–16 inches) per year, but the mountains receive as much as 900 millimeters (35 inches). These averages mask the extreme seasonal variability of precipitation, for in the valleys precipitation has ranged from 200 to 560 millimeters (8–22 inches) in differing years, and the average is rarely realized. The growing season averages about 135 days, but it is also highly variable with elevation and exposure (U.S. Department of Commerce, 1971). Temperatures average about $-3°$ C. ($27°$ F.) in January and $22°$ C. ($72°$ F.) in July. Absolute maxima for the Wasatch Front are over $43°$ C. ($109°$ F.) and the minima are under $-34°$ C. ($-29°$ F.).

The sample area and population

Fruit production in the Wasatch Front is not uniformly distributed among the five counties which make up the region. In 1964, 74 percent of total production was located in one of them, Utah County. This concentration is partly the result of the larger size of Utah County, but increasing suburbanization in the other counties is also forcing relocation of orchards to Utah County. (Between 1960 and 1964, Utah County's pro-

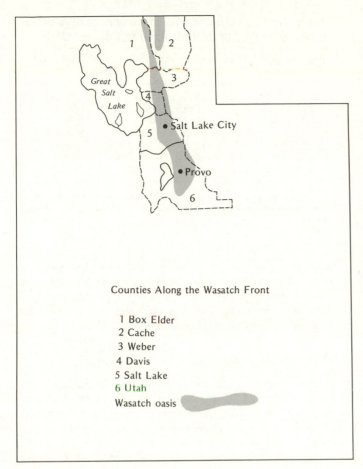

Fig. 18–1. The Wasatch Front in Utah

The nature of the hazard

"Frost" as understood by the orchardists of the Wasatch Front is not the hoarfrost most people associate with the term. Frost refers to freezing air temperatures in orchard levels which occur after the fruit trees have begun to bud or blossom, and is rarely accompanied by hoarfrost. The mechanism which causes frost damage to fruit in the study area is different from that which normally causes frost damage to fruit in California or Florida: primarily radiation frosts which create an inversion, characterized by clear, still air. Approximately 90 percent of the frosts in Utah, however, are advection frosts accompanied by perceptible air movement (Richardson, 1972). The frosts in either case destroy the bud or blossom, and since the tree will not produce new buds, destroys part or all of the fruit crop for that year.

Several factors affect the extent of damage caused by frost. The most important is the state of the buds at the time of frost. The most critical time is the early blossom period, with late bud period just prior to blooming being only slightly less vulnerable. Temperatures of −1.66° C. (29° F.) or lower during the blossom state will destroy the blossoms, and temperatures of −2.22° C. (28° F.) or lower after budding has begun will destroy the entire fruit crop (Barlow, 1972). Since the extent of damage relates to stage of development, the tree species also affects the extent of damage: some bud later than others. The order of budding from earliest to latest is as follows: apricots, sweet cherries, sour cherries, peaches, pears, and apples (Fig. 18–2 indicates the percentage of each tree type in Utah County).

The predominant species are apples and cherries, though cherries are highly susceptible to frost. Moreover, both cherries and apples have been increasing in

portion of fruit production increased from 51 percent to 74 percent.) Urban sprawl also affects Utah County, but the present availability of land allows orchards to be relocated there (Fig. 18–1).

The census lists 653 farms in Utah County having 20 or more fruit trees, comprising a total of 2,558 hectares (6,321 acres) of orchard (*U.S. Census of Agriculture*, 1964). The figure of 653 farms is misleading because of the large number of farms which have only 20 or 30 fruit trees, and do not produce commercially. It is estimated that no more than 100 of these farms are commercial (Barlow, 1972), and the sample was restricted to these commercial farms. The rationale for this was that it was felt that the threshold of viewing frost as a hazard would vary in a direct relationship with the extent of reliance on fruit for income. The individual who is only mildly inconvenienced because he doesn't have fresh fruit to eat from his own trees will view and react to frost much differently than the man whose livelihood is seriously threatened by frost. Approximately three-fourths of the 100 commercial fruit growers were interviewed.

Fig. 18–2. Fruit trees in Utah County. Source: *U.S. Census of Agriculture* (1964)

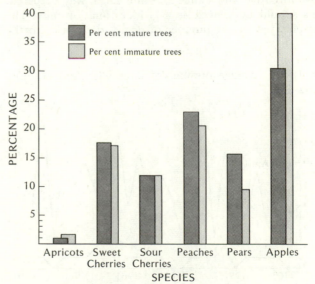

number, primarily because these are the most profitable tree crops.

Another important factor affecting the extent of frost damage is the location of the orchard with respect to physiographic features. Orchards planted on slopes with a southern exposure tend to bud earlier in the spring and are more susceptible to frosts. Orchards planted at the mouth of a canyon receive air drainage which prevents frost damage during radiation frosts. Orchards are rarely located in depressions in the valleys or in river bottoms because these locations are more susceptible to frost as cold, heavy air drains into them at night.

Probability of frost

The critical months for frost in the Wasatch Front are normally April and May. Rarely is it warm enough for trees to bud in March, and it is usually the latter part of April before development has proceeded far enough to cause fear of frost. In assessing the probability of frost, the six stations which report temperatures in Utah County were analyzed. Frost was defined as screen temperatures of $-1.66°$ C. ($29°$ F.) and below since at this temperature damage to orchards begins.

The probability of frost in April is 1.0 and in May it is 0.72. These figures, however, do not indicate the severity of frost. To the fruit grower a frost in which temperatures do not drop below $-2.22°$ C. ($28°$ F.) or persist for an extended period is actually beneficial. Such a light frost will only destroy enough of the fruit to ensure that the balance will not be crowded and can fully develop. The probability of frosts of all magnitudes is high in the first weeks of April. The probability of light or "beneficial" frosts remains high through the first half of May, but severe frosts decrease rapidly through April and May (Fig. 18–3).

To present an indication of the correlation between frost intensity and damage, Table 18–1 was prepared. This should be viewed as a gross estimation only. The complexities of the number of degree-days prior to the

Fig. 18–3. Average number of days with temperatures below $29°$ F.

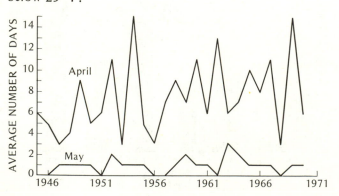

Table 18–1. Estimated loss in percent and value for frosts of varying magnitude in Utah County, Utah

Temperature (°C. and F.)	Crop Destruction (%)	Value of Loss[a] (Millions of Dollars)
-2.22 (28)	10	No loss (beneficial)
-2.77 (27)	10–19	0–1
-3.33 (26)	20–34	2–3.5
-3.88 (25)	35–50	3–3.5
-4.44 (24)	51–64	5–6.5
-5.0 (23)	65–79	6.5–8
-5.55 (22)	80–99	8–10

[a]Expressed in 1972 dollars. Based on values of Utah County fruit crop indicated by Bureau of Agriculture personnel (*Daily Herald*, 1972).

frost occurrence and concomitant state of bud development, slope and exposure of the orchard, and relationship to canyons or other physiographic features preclude a completely accurate statement of relationship between frost intensity and damage.

Perception of probability of frost by farmers

Intensity represented by temperatures of $-3.33°$ C. ($26°$ F.) was selected as the damage threshold for fruit producers. Frosts of lesser intensity normally are accepted by fruit growers because although the probability of their occurrence each year is very high, they do not cause severe damage to individual farmers. Frosts with temperatures of $-3.33°$ C. or lower, however, result in major damage to farmers and are viewed as a definite hazard. Frosts of this magnitude have a probability of occurrence of 0.56 after April 15. April 15 was selected as the critical date since there have normally been too few degree-days for bud development to become critical before then. Thus the probability of 0.88 for frosts of $-3.33°$ C. in the first two weeks of April simply reflects normal climatic conditions for the period; the frosts rarely damage the fruit crop. The period from April 15 to May 15 is the critical period for frost damage.

To determine how well grower's perception of frost probability correlated with actual probability they were asked to indicate how often frosts of varying magnitude occurred. Since trial surveys indicated that frost intensity was viewed in terms of damage rather than as occurrence of specific temperatures, they were asked to indicate probability in these terms (Table 18–2).

Comparison of Tables 18–1 and 18–2 reveals that there is a high correspondence between the probability of mild frosts occurring between April 15 and May 15 and the growers' perception of this probability. Using the loss of one-third of the crop (temperatures of $-3.8° - -3.33°$ C. or 25.2–26° F.) as the initial intensity, it appears that the perception of the growers approaches the actual probability. Growers tend to seriously underestimate the probability of occurrence of more severe

Table 18-2. Perception of probability of frost among fruit growers in Utah County, Utah (percent occurrence of frost of indicated magnitude)

Intensity of damage	Never would occur	Once Every							
		2 yrs.	3 yrs.	4 yrs.	5 yrs.	6 yrs.	7 yrs.	8 yrs.	10 yrs.
Destruction of 1/3 or less of main fruit crop		36	44	11	2	4			2
Destruction of 1/2 of main fruit crop	7	14	36	36	2	4	2		
Total destruction of main fruit crop	7		4	8	15	8	14	33	8

frosts, however. The probability of total destruction (below -5.55° C. or 22° F.) after April 15 is 0.32, but only 4 percent of the growers indicate they expect total destruction once every 3 years. This possibly reflects an attempt to cope with severe hazards by denying or denigrating their existence (Burton and Kates, 1963). In analyzing the response of growers it appears that they are loath to admit total destruction. If some fruit remains, even if there is not enough to be worth harvesting, the destruction is not viewed as total.

Forecasting the occurrence of frost hazard

Although the farmers show a reasonable ability to predict the probability of frost occurrence, they also face the dilemma of forecasting the days on which it will occur. The amount of warning fruit growers receive of impending frost, and the accuracy of these forecasts regarding magnitude and areal extent of potential damage, are of critical importance in their response to the hazard (Hewitt and Burton, 1971). Occurrence and intensity of frost in the Wasatch Oasis is highly predictable, but warning time is short. The U.S. Weather Service in Salt Lake City maintains an agricultural section which deals with the effect of weather on agriculture in Utah. Growers can gain information on temperatures through the weather service or county agent, and local newspapers, radio, and television stations. Information is available on the temperatures expected that night for each fruit growing region in the state. Rarely, however, does this provide more than 8-16 hours of notice of impending frosts.

Because of differences in the site aspects of individual farms, temperatures can differ by several degrees in a distance of several hundred meters. Since such a difference in magnitude is of critical importance, many orchardists monitor the temperature in their own orchards when a frost warning has been given.

Adjustments to frost hazard in the Wasatch Front

The following have been suggested as possible forms of adjustment which can be taken to minimize damage from hazards: affect the cause of natural agent directly, modify the hazard, or modify the loss potential. If damage has already occurred, it is suggested that losses can be dealt with by spreading the loss, planning for losses, or bearing the loss individually. In terms of adjustments available to fruit growers in the Wasatch Front, the first seems impossible. At present there is no known way of directly affecting the cold air masses which trigger the frost hazard.

The frost hazard can be modified, however, in one of two ways. If the frost hazard is the result of advection of a cold air mass, it is possible to heat the air in the orchard, and thus minimize damage. If the frost results from radiation, heating can also be used, or mechanical mixing of the air to destroy the surface inversion can raise orchard temperatures above the critical level.

Several alternatives are available to orchardists to modify the loss potential from frost hazard. These include changing varieties of fruit crops, changing to non-fruit crops, placing a "hot cap" over individual trees to protect them, spraying trees with water, developing a budding inhibitor, or planting on slopes for suitable exposure and air drainage. All fruit growers were aware of all existing strategies for minimizing loss from frost damage. Most (91 percent) had only experimented with part of them, however.

Adoption of strategies

The three strategies which have been most widely used to cope with losses are to modify the hazard by heating, to change varieties of crops, and to consider the physiographic features when planting. In 1967, 42 percent of the orchard area in Utah County was heated. This represented over 50 percent of the commercial orchards in the county (Utah County, 1967). Surveys in the spring of 1972 revealed that only 25 percent of the commercial orchards were heated, however. The reason for this decrease in a short period can be categorized as an institutional application of changing social attitudes. Prior to 1968 heating was carried on in orchards through the use of a variety of mechanisms. These included piles of old tires saturated with petroleum products or open buckets filled with crude oil and set ablaze, small furnaces spaced throughout the orchard and fired with fuel oil, and buried gas lines using liquid petroleum products or natural gas. The state legislature in 1968 passed legislation prohibiting open burning or use of crude petroleum products after 1969. The only techniques now legal are natural gas and liquid petroleum products (e.g., propane, butane), or electric heaters. As a result of this formalization of social attitudes, most fruit growers are abandoning their traditional method of coping with frosts. Of growers interviewed, 70 percent indicate that the costs of legal methods of heating are greater than the returns. Costs of heating with butane average about $250 per hectare per night ($100/acre/night), and costs of installing the system are $740 per hectare ($300/acre) and up depending on the number of burners utilized per hectare (acre). Since heating only increases the air temperature by an average of 4°–6° C. (7°–11° F.) over the ambient air temperature, there is still a fairly high probability that a grower can invest heavily in successfully heating his orchard for several nights only to have an even colder night destroy his fruit. Most growers now view heating as an uneconomic adjustment to frost hazard. However, the state meteorologist maintains that it is still profitable, and that farmers are not economically rational in calculating benefit cost. He maintains that abandonment of heating is an overreaction to the law prohibiting the more primitive methods (Richardson, 1972).

Changing varieties, as an adjustment to prevent damages, is still utilized. A change from highly vulnerable apricots and sweet cherries to more frost-resistant apples and sour cherries has been calculated. Only one-third as many apricot trees existed in 1964 as in 1950, apples increased by 20 percent, and sour cherries increased by 300 percent in the same period. This does not reflect frost hazard only. Cherries are quite vulnerable to frost, but the return per hectare (acre) in good years is the highest for any tree product in Utah. Hence most orchardists grow both cherries and apples, but assume that if the cherry crop is lost they will still have some income from their apples.

The other adjustments indicated above have never been widely used. Spraying with water proved unsuccessful for one grower who tried the method during the severe frost of 1972. Placing a hot cap of plastic or other fabric over individual trees is technically but not economically feasible. Experiments with budding inhibitors are under way, but at present nothing is available for the individual farmer to use in his orchard. Few growers change to other types of crops. Most fruit farmers maintain that fruit production yields a higher return per hectare (acre) than alternative crops even after frost damages are deducted. This is probably true on the hillside locations where trees are the only viable crop, but on more level land it is doubtful. As long as growers accept the idea that fruit production is the most profitable, however, none seems willing to change to other crops.

There exist several methods for dealing with losses after damage has occurred, as mentioned earlier. These include spreading the loss through government or other public and private assistance, planning for losses with insurance or savings, and bearing the loss personally. The type of loss behavior which comes into play is a direct function of the magnitude of the frost damage. Most farmers bear the loss themselves unless damage exceeds 50 percent of the crop. Some (only 20 percent of the sample) maintain a reserve of funds specifically for this purpose, but the balance simply curtail purchases and survive on the remainder of the harvest. If losses are severe, with the harvest being less than 25 percent of normal, loss-sharing strategies are undertaken. Groups involved in this include federal and state agencies, financial institutions, churches, family and friends.

Since all of the respondents to the surveys were members of the Mormon Church, it is important to note its role in spreading the loss. The Mormon Church operates a relief program which provides the basic necessities (food, clothing, shelter, medical care) for all who so need it. When severe frosts occur, assistance is requested from government agencies. This usually takes the form of low-interest or interest-free loans (*Deseret News,* 1972). Financial institutions assist by extending new loans or granting a moratorium on payment of existing loans. Family and friends assist according to their ability. These are all stopgap measures designed to enable the farmer to exist until the following year, and none reimburses him for the income lost by frost. The loss of income is merely viewed as part of the "game" of farming in the variable climatic region.

Summary and conclusion

Fruit production is, and will continue to be, a fundamental element of the agricultural economy of the Wasatch Front. Fruit growers are aware of the high probability of frost hazard in the region, but several factors prevent a change to crops less vulnerable to frost. Of primary importance is assessment of fruit production as the most remunerative agricultural land use. Coupled with this is the inertia common to most farmers, which is reinforced by relative permanency of the crops. Once the decision is made to plant an orchard, the time and capital investment preclude a change for at least a decade. Thus the growers "gamble" that next year will be free of frost and they will gain an adequate return on their investment, rather than change to other crops. The exception to this is selling land for homesites. The hillside locations of orchards are valued as building sites because they are "view lots." Growers fortunate enough to be located near the urban areas often solve the problem caused by frost by selling one or two lots. View lots of 0.13–0.21 hectare (0.3–0.5 acre) sell for $3,000–$15,000 depending on location. A farmer can thus sell 0.4 hectare (1 acre) and manage to exist until the following year's crop.

The growers are rational in their assessment of the probability of occurrence of minor frost damage, but they tend to seriously underestimate severe frosts. This probably reflects an attempt to minimize the uncertainty caused by lost income during severe frosts. The reoccurrence of mild frosts is assessed accurately because the resultant damage does not create as much uncertainty about the life-styles of the growers.

In conclusion, it is important to note the role of social and institutional variables as they affect coping strategies utilized in dealing with hazards. The institution of legislation barring the traditional methods of minimizing frost damage by orchard heating was a direct outgrowth of the ecological movement among urban citizens of the state. The concern of urban dwellers over black smoke produced by any type of open burning led to the enactment of the law. (Had there not occurred a concomitant reapportionment of the legislature giving urban areas greater representation, this could not have happened.)

Growers maintain that it is economically infeasible to adopt alternative methods of heating, but it is questionable whether this has been rationally calculated. The reaction, rather, seems to be an emotional reaction to the imposition of laws by urbanites. It is impossible to say whether this is an attempt to repeal the law by showing the increased losses to frost, or merely frustration at the change from a rural- to an urban-dominated state. This does point, however, to the problems presented to the agricultural sector of our society as it becomes more urban. It also suggests that a possible approach to integrated studies of all hazards in a region should consider them according to occupance types. Frost presents minimal hazard to those who reside in urban places in the Wasatch Front, hence there is no need for coping strategies and little understanding of the needs of the fruit grower. The urbanites' perception of visual pollutants as a potential hazard led to enactment of legislation detrimental to the fruit growers. Hopefully, future studies will analyze hazards in terms of conflicting perceptions of differing occupance types and present greater understanding of the conflicts which result.

References

Barlow, Joel. (1972) Provo, Utah: personal communication with county agent.
Burton, Ian, and Kates, Robert. (1963) "The perception of natural hazards in resource management." *Natural Resources Journal* 3:412–41.
Daily Herald. (1972) Provo, Utah, March 27.
Deseret News. (1972) Salt Lake City, March 28.
Hewitt, Kenneth, and Burton, Ian. (1971) *The Hazardousness of a Place.* Toronto: University of Toronto Press, pp. 135–37.
Meinig, Donald. (1965) "The Mormon culture region: strategies and patterns in the geography of the American West, 1847–1964." *Annals of the Association of American Geographers* 55:191–220.
Richardson, Arlo. (1972) Logan, Utah: interview with state meteorologist.
U.S. Census of Agriculture. (1964).
U.S. Department of Commerce. (1947–71) *Climatological Data; Utah; Annual Summaries.*
Utah County. (1967) *Annual Report of Utah County Agent.*

19. Human adjustment to volcanic hazard in Puna District, Hawaii

BRIAN J. MURTON AND SHINZO SHIMABUKURO
University of Hawaii

In 1955 a flank eruption of considerable magnitude occurred in the inhabited area along the East Rift Zone of Kilauea Volcano in Puna District, island of Hawaii (Fig. 19–1). Since 1955 Kilauea Volcano has erupted 20 times, and only 7 such eruptions have not occurred along the East Rift Zone; but only in 1960 did lava break out again in an inhabited area. In this inhabited area, the eastern part of Puna District, potential losses of property are a feature of life. Although volcanic hazard is ever-present, the area has a population of

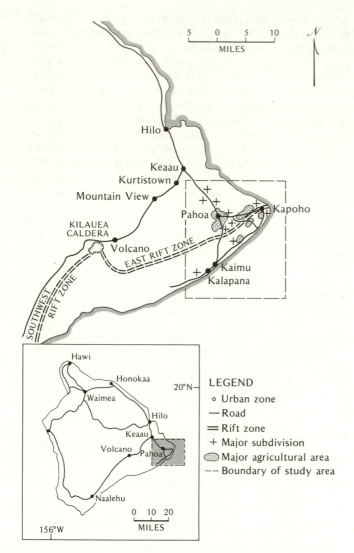

Fig. 19—1. Puna, Hawaii

relationship to differences in experience and personality.

2. What factors are related to awareness of adjustments to volcanic hazard? It is suggested that awareness is a function of individual experience and the magnitude and frequency of volcanic activity.

3. What is the nature of adjustments to volcanic activity? It is suggested that given the character of the hazard as an uncontrollable geophysical event, the range of adjustments will be small, and that both individual and collective action will be taken.

4. What factors are related to evaluation and adoption of adjustments? It is suggested that evaluation and adoption of adjustments are related to experience, and not to socioeconomic variables, such as land tenure and income, or to residence in zones in different hazards.

5. What factors have caused land-use change? It is suggested that although lava flows have destroyed property and crops, recent change reflects economic factors.

Objectives

The study has three major objectives: (1) to assess the importance of volcanic activity to the economy and social organization of eastern Puna; (2) to judge whether volcanic activity should be taken account of in land-use planning; and (3) to understand the circumstances in which people in eastern Puna make decisions to cope with hazards.

Field methodology

The main part of the field research was based on a questionnaire survey containing 39 questions plus a sentence completion test. A version of the questionnaire was field-tested in the study area. Eight households of different ethnic backgrounds were visited. Changes in question order and wording were made as a result of the field test.

Five hundred and eighty-seven households are located in Eastern Puna. These households were divided into ten clusters on the basis of distance from axis of the Rift Zone. A random sample of 15 households was drawn from each cluster, giving a sample of 150 households. This procedure was followed rather than that suggested in Working Paper No. 16 (Natural Hazards, 1970) because of five basic constraints: the majority of the population have schooling above the elementary level; the diversity of ethnic backgrounds in the area; the localized nature of the hazard; logistical reasons, including time, transportation, and funds; and inability to draw a clear distinction between agricultural and non-agricultural households.

Interviews were conducted with heads of households. In total, 101 interviews were completed. Interviews were not obtained in the remaining households because of language difficulties—primarily Filipino's Ilocano (30 percent); no one answering at home—the interviewer

about 1,500—most of whom are engaged in the cultivation of sugar-cane, papayas, orchids, anthuriums, and macadamia. In addition, there are three urban zones, plus more than 20 large subdivisions totaling over 30,000 (largely uninhabited) house lots.

This chapter is concerned with the way in which the residents of this area perceive, evaluate, and adjust to the ever-present volcanic activity. More specifically, it investigates how human use and natural events interact to create problems of management and development.

Questions and hypotheses

Several questions are posed in relationships to perception, evaluation, and adjustment to volcanic hazard:

1. What factors are related to the perception and evaluation of volcanic hazard? It is suggested that differences in perception and evaluation can be interpreted in

visited such places at least three times (50 percent); and refusals (20 percent).

Characteristics of Puna District

Population and people

Eastern Puna had a population of 1,450 in 1970, a slight increase over the 1,326 residents in 1960. Nearly two-thirds (924) live in the Pahoa Urban Zone in the central part of the study area, and about 150 in the Kapoho Urban Zone and 100 at Kaimu-Kalapana.

An important and complicating factor in any study in Hawaii is the diverse ethnic background of the population. On the basis of the interviews, people of Japanese ancestry were the largest ethnic group (33 percent), followed by Hawaiians (28 percent), haoles (Caucasians) (22 percent), and Filipinos (13 percent). Although English (including Pidgin English) is spoken at home by 80 percent of the respondents, Japanese, Hawaiian, and Ilocano also were in use. With respect to religion, Buddhists (31 percent) were the single largest group, followed by Protestant (24 percent) and Catholic (22 percent) Christians.

Occupation and land use

Retired people (24 percent) are common, as are people involved in service activities (35 percent). Only 17 percent of the respondents were engaged in full-time agriculture, although most households interviewed grew flowers and fruit for sale.

Most of eastern Puna consists of waste and forest land. About 1,600 hectares (4,000 acres) of sugarcane are cultivated, a large part of it on, or close to, the East Rift Zone. However, diversified agriculture—papaya, anthuriums, vanda orchids—is of increasing importance. The area under papaya increased from approximately 160 hectares (400 acres) in 1963 to nearly 400 hectares (1,000 acres) in 1971, most of which is flattened a'a lava (rough, spiny, rubbly, clinkery lava) from the 1955 and 1960 eruptions. Vanda orchids and anthuriums, which are not grown in soil, also use flattened, recent lava flows.

A further major feature of land use is 23 large, speculative subdivisions containing 30,781 lots. Several of these lie on or close to the East Rift Zone. The three urban zones of Pahoa, Kapoho, and Kaimu-Kalapana contain most of the population. Pahoa (population 924) is the service center of eastern Puna. Kapoho is essentially a beach-home subdivision and Kaimu-Kalapana, a Hawaiian village. The area is, however, one of the two visitor destination areas on the island and has considerable land zoned for resort development. In future, hotel complexes will be constructed in the area, which is liable to a complex of seismic and volcanic hazards.

Characteristics of Hawaiian eruptions

The outstanding characteristic of Hawaiian-type eruptions is their gentleness. Almost no explosive eruptions occur, although jets of liquid lava do shoot into the air, forming lava fountains that may continue uninterrupted for many days and commonly reach heights of several hundred feet above the vent opening. Lava sloshes, fountains, and spills out of a vent, crater, or cone, and flows downslope toward the sea. Such eruptions are localized and do not affect more than a small part of the area at one time.

Historical occurrence

Since 1750 Kilauea Volcano has erupted more than 50 times (Macdonald and Abbott, 1970; Swanson et al., 1971). Twenty of these eruptions occurred along the East Rift Zone, the latest of which was continuing at the time of writing. Since 1955 only 7 eruptions have not taken place in Puna, although before then flank eruptions were much less frequent (Fig. 19–2).

Fig. 19–2. Eruptions on the East Rift Zone

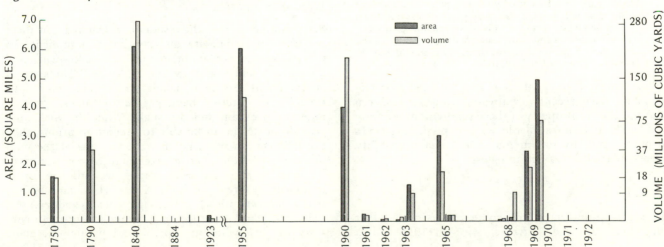

Extent of area subject to volcano hazard

No part of the study area is free from volcanic activity and lava flows. The eruptions along the East Rift Zone since 1750 have covered approximately 170 square kilometers (65 square miles). Activity since 1961 has covered a large area just to the west of the study area but the destructive 1960 and 1955 eruptions covered over 26 square kilometers (10 square miles) of land in eastern Puna with lava. The 1960 Kapoho eruption occurred in one place and covered 10 square kilometers (4 square miles) of the eastern tip of the study area. In contrast, the 1955 eruption broke out in nine separate locations and several flows reached the south coast. These flows covered slightly over 16 square kilometers (6 square miles).

The question of probabilities

In terms of establishing probabilities or predicting eruptions the volcanologists say that they cannot predict where and when an eruption will occur and that while it is possible for them to make statements of statistical probability, those kinds of statements give misleading information to the general public (Peterson, 1971). One such estimate for Puna points out that since 1750 lava has covered 16 percent of the area. On this basis, the likelihood of lava covering any given plot of land on any single year is less than 0.1 percent (Honolulu *Star Bulletin,* 1960). Furthermore, taking all the available evidence over the last 10,000 years the lava flow incidence is once in about 300 years on any given square mile (2.6 square kilometers). Almost all authorities agree that the closer land is to the 3.2–4.8-kilometer (2–3-mile) wide rift zone the greater the danger, but to say that there is risk is quite different from defining the seriousness of the risk. This type of statement of probability has been used by land subdividers and developers to allay public concern about eruptions covering subdivisions. For example, in 1960 the subdivider of Leilani Estates, which lies right on the East Rift Zone, said:

> We are happy we have used your newspaper to inform the public that the volcanic risk is only 0.1 percent; and that leading geologists in the islands, in commenting on security insofar as volcanoes are concerned say it is such in this area, that if people like the land they should buy it. [Honolulu *Star Bulletin,* 1960]

Nonetheless, county officials and Volcano Observatory staff have recently expressed cautious concern about dangers to property if these subdivisions are fully settled. There is no available way of predicting times and places, and any statements of statistical probability for particular places are meaningless.

Losses and damages

Volcanic eruptions have only come into serious conflict with human use and activities twice—in 1955 and in 1960. Lava flows from both of these eruptions destroyed agricultural land, crops, houses, and roads. Considerable information is available on the extent of losses and the damage caused by the 1955 eruption. About 450 hectares (1,100 acres) of cropland were covered and 520 hectares (1,275 acres) of crops destroyed by fire or sulfur fumes. Over 60 homes were destroyed or rendered uninhabitable, and 10 kilometers (6 miles) of road covered by lava (Macdonald and Eaton, 1964). The monetary cost of this destruction was estimated at $2.5 million, of which insurance covered $716,000 (Puna Volcanic Fact Finding Committee, 1955). Of the total, $600,000 was in loss of land, and $1,240,000 in sugar. These figures do not take into account costs to the state and county during the eruption, the loss of wages to people in the affected area, and the social costs of displacement of people from their homes.

No detailed information on damages and losses is available for the 1960 eruption at Kapoho which destroyed a village of 70 structures. This eruption forced the inhabitants of Kapoho Village to move permanently, either to nearby Pahoa or to Hilo.

The residents of eastern Puna are subject to a number of natural hazards other than those directly associated with volcanic eruptions, namely earthquakes, landslides, tsunamis, and droughts.

Perception of volcanic activity

In this section we discuss some simple frequencies relating to perception of characteristics of eruptions, the nature of personal encounters with eruptions, and certain personality factors.

Perception of characteristics of eruptions

Information on three aspects of the characteristics of eruptions was obtained: signs that immediately precede an eruption; the nature of eruptions themselves; and information on the future occurrence of eruptions.

Signs of an eruption

Sixty-one percent of the respondents claimed that there are signs that indicate an eruption is imminent, 15 percent answered that there is no way of knowing, and 25 percent said that they "didn't know." The signs of an imminent eruption include strong and continuous earthquakes, unusual heat, gas sulfur fumes, "funny" rains, calm ocean, and dry spells. While the volcanologists do not claim to be able to predict eruptions, they say that certain signs, notably the swelling of the summit area of Kilauea, earthquakes, and the registration of harmonic tremors on seismographs, are indicative of an imminent outbreak of lava. A few Hawaiians believe that a *kahuna* (priest) or Pele (the volcano goddess) will tell them when an eruption is about to occur. In contrast to this belief in supranatural prediction, many haole (Caucasian) newcomers believe that the volcanologists with their instruments at the Hawaiian Volcano Observatory can predict the occurrence of eruptions.

Characteristics of eruptions

A simple semantic scale was used to gain insight into the way people perceive the characteristics of eruptions (Table 19–1). On four of the scales more than 70 percent of the respondents thought that volcanic eruptions were natural, irregular, unstable, and uncontrollable, all of which are objective physical characteristics of eruptions. On the other three scales there was a greater spread of response but in each case 59 to 60 percent of the respondents thought volcanic activity is localized, slow, and harmful. However, 27 percent of the respondents consider volcanic activity widespread, and on the slow-fast scale responses were spread.

Future occurrence

Seventy-nine percent of the respondents expected eruptions in the future, but 12 percent denied future occurrences. Furthermore, 78 percent believed that eruptions can happen anytime and that there are no cycles of sequences.

Experience with eruptions

Eighty-five percent of the respondents had experienced a number of eruptions. Unlike all Japanese, and most Hawaiians and Filipinos, many haoles were not resident in the area in 1955 and 1960. A majority of respondents (71 percent) thought that the 1960 eruption was by far the most destructive, with a small group (14 percent) considering the 1955 activity to be the worst. Eighty percent believed that since 1960 none of the eruptions had posed threats to human activities. These latter outpourings were well away from occupied areas, but it is pertinent to note that in contrast to popular belief, the 1955 eruption covered a larger area than did that of 1960, and destroyed a greater acreage of crops. The destruction of Kapoho in 1960 obviously has impressed itself on people's memories.

Personality factors

Factors of individual personality which might affect an individual's perception of volcanic hazard proved to be difficult to measure and even more troublesome to assess in terms of significance in the adjustment process. However, some information on how respondents view the future, how they report reacting to the stress caused by an eruption, and how individuals rationalize their responses to an eruption was gathered through the sentence completion test.

In terms of planning for the future, a majority of respondents claimed that they were not worried about it and had no plans. Another 29 percent claimed that they had plans laid out for the future. More important, however, is the information concerning views on how people "get ahead in the world" (Table 19–2). Interestingly enough, "cooperation" was seen to be the major factor, particularly among Hawaiians and Filipinos, and even among Japanese, where it ranked second. The dominance of this factor as the means to success reflects Hawaiian and Filipino, and to a lesser extent Japanese cultural characteristics, in that cooperation and helping neighbors in the community is of long-standing importance. Among the Japanese, however, hard work was the single most important response, and overall this factor ranked second, followed by education, and then by a number of lesser factors.

Natural hazards such as a volcanic eruption cause stress, particularly when people realize an eruption is imminent, and are uncertain about what is going to

Table 19–1. Perceived characteristics of eruptions[a]

	(percent)					
Natural	91	1	2	0	6	Unnatural
Regular	18	0	6	4	72	Irregular
Unstable	74	3	9	2	12	Stable
Widespread	27	1	8	4	60	Localized
Slow	61	9	19	3	8	Fast
Harmful	60	2	12	2	24	Unharmful
Controllable	7	1	3	3	86	Uncontrollable

[a]Respondents queried this scale and when answering stated that they were referring to property, not human life. N = 99.

Table 19–2. Assessment of fate control (percent; N = 101)

	Japanese	Hawaiian	Filipino	Haole	Other	Total
Cooperation	7	14	7	2		30
Hard work	10	7	1	3	2	23
Education	3	2	2	2		9
Faith	4			1	1	6
Competition	2		1	1		4
Good health	3					3
Other	4	6	1	13	1	25
Total	33	29	12	22	4	100

happen. Forty-one percent of the respondents claimed that they do not worry and remain calm when they hear an eruption is imminent. Those who report feeling scared or worried (43 percent in total) say that they do not fear for their lives, but for property and belongings. Regardless of how people respond to the news of an eruption, there are a number of ways by which they rationalize the feelings of uncertainty associated with volcanic activity. A large group (28 percent) believe nothing can be done to stop an eruption. They seem to accept the danger passively without denying or denigrating the existence of the hazard. A larger group (36 percent) say that because they have had past experience with eruptions and because they claim to know what to do, they feel certain how they would behave. In other words, while they realize that future occurrences of eruptions cannot be predicted, they are not bothered by this uncertainty because they feel sure about how they will respond.

Is the volcano perceived as hazardous?

Many people failed to perceive volcanic eruptions as hazardous. More than 90 percent of the respondents claimed that there are more advantages in living in the area than disadvantages, despite the fact that they (over 82 percent) realize that there are other places with little or no volcanic activity where they could earn as good a living. Even those who replied that they are aware of the hazards of volcanic activity (57 percent) emphasized that people are in no danger, only land and property. This dichotomy in people's minds between danger to human life and danger to land and property seems to be a fundamental one in the area. Intuitively speaking we believe that most of the remaining 43 percent of the respondents know full well of the dangers of eruptions to land and property, but do not think of such dangers as natural hazards as they do not threaten life.

Human adjustment to volcanic activity

Information about eruptions

Kilauea Volcano is constantly monitored by scientists from Hawaiian Volcano Observatory. They can tell from their instruments when magma is building up under the summit and when it is moving under one of the rift zones. Normally, just before magma breaks out onto the surface, seismographs record a pattern known as a harmonic tremor.

When the instruments indicate that an eruption has occurred or seems to be imminent a procedure is followed to alert the personnel of Hawaii Volcanoes National Park, Hawaii Civil Defense Agency, the police, various county agencies, and the mass media, all of which inform the general public. Depending on the location and magnitude of the eruption the administrator of the Hawaii Civil Defense Agency can (1) activate the emergency operations center; (2) mobilize civil de-

fense personnel; and (3) take whatever precautionary measures are deemed necessary (Hawaii, County of, 1969).

In actuality these procedures were only followed in 1955 and 1960. The eruptions along the East Rift Zone during the 1960s occurred within Hawaii Volcanoes National Park and did not threaten agricultural land and property. The park takes special precautions because of the enormous number of visitors that an eruption brings into the area.

Emergency adjustments

If the eruption is in an occupied area, the emergency procedure goes into effect. Evacuation of people and valuables is the standard method employed. This is carried out by the Civil Defense Agency, with the help of the Red Cross, Police Department, Fire Department, state and county governments, the National Guard, the Boy Scouts, and individual volunteers. The evacuees are supplied with shelter, food, food stamps, clothing, loans, transportation, and other immediate needs. During the 1955 eruption, people from Kalapana spent about a month at Pahoa and Keaau living in school buildings. When it appears that residences will be destroyed, as at Kapoho in 1960, valuables are removed.

Another kind of emergency adjustment involves appeals to the supranatural, mainly Pele, the volcano goddess. Many people, especially old Hawaiians, make offerings of liquor, candy, tobacco, flowers, food, and ti leaves to Pele. Further, it is possible to pray to dead relatives who have been dedicated to Pele to intercede with the goddess to direct a lava flow away from a particular piece of land. Pele beliefs also play a part in behavior associated with evacuation on the part of old Hawaiians. Such people may designate a flow as being Pele who "can take anything she wants when she will," and when a lava flow approaches their property they will do nothing, believing that to take any action would anger the goddess. During the 1955 eruption, an old Hawaiian man refused to move his belongings, including a new television set and refrigerator, because of his beliefs.

Various agencies give assistance to people affected by eruptions e.g., the Red Cross, the Civil Defense Agency, and a number of community associations. All of these groups were involved in providing postevent relief in 1955 and 1960. In addition to this type of relief, under Act 1973 of the state of Hawaii, taxes can be remitted, refunded, or forgiven equal to losses due to natural disasters. This act, passed in 1961, was in direct response to the natural disasters—volcanic and tsunami—of 1960, both of which badly hit the island of Hawaii.

Emergency adjustment experiments

Evacuation and prayer are the only adjustments practiced during an eruption. However, various experiments have been attempted in efforts to change the path of lava flows, to stop lava flows, and to protect property.

Walls. Walls are constructed during the 1960 Kapoho eruption in an endeavor to channel a lava flow away from a group of houses. Macdonald and Abbott (1970) suggest that walls are definitely useful in reducing damage, if the topography is suitable. It has been proposed to build permanent walls near Hilo to prevent lava from Mauna Loa reaching the city, though the proposal has never been taken seriously by people and government.

Bombing. Although this has not been done in the study area, in 1935 and 1942, lava flows from Mauna Loa which were threatening Hilo were bombed in an effort to divert them into different courses and collapse lava tubes and clog the flows. It is also claimed that bombing may cause violent stirrings of the lava and transform it from *pahoehoe* to *a'a* which is less liquid and congeals more rapidly (Macdonald, 1962). Bombing, of course, does not seem practical in inhabited areas.

Watering. The effect of water on flow margins was tested at Kapoho in 1960. Macdonald (1962, p. 277) reports the results:

> . . . in the case of a stationary margin it is possible to cool the flow surface enough to prevent radiant heat from igniting wooden structures only a few meters away. It was also possible to locally check the advance of a flow margin. Although the check is temporary, several hours only, it gives the time needed to remove furnishings, or even to remove the building itself.

Long-term adjustments

Given the almost uncontrollable nature of volcanic eruptions, all long-term adjustments involve modifications in the human use system. But it is difficult to assess whether some land-use changes are as much a result of volcanic activity as they are economically motivated.

Land use. Since 1955 volcanic activity has definitely been a factor in changing land use. Approximately 1,000 hectares (2,500 acres) of crops were destroyed or cut off and abandoned by the various lava flows from the 1955 eruption, and more crops were destroyed at Kapoho in 1960. More than 600 hectares (1,500 acres) of sugarcane were destroyed in 1955 and 1960. Although sugarcane continues to be an important use of land, as transportation and production costs increase more and more land in the Kapoho and rift zone area is shifting from cane production to other uses. Two new uses—subdivisions for housing and diversified agriculture—have become major factors in the changing land use of the study area. Lava flows were the destructive force influencing some of these changes, but economic factors also were significant.

Settlement relocation. Volcanic hazard differs from other natural hazards throughout the United States in that once a house or village is destroyed by lava people rarely resettle the same place. They move to nearby subdivisions. After the 1955 eruption 21 households were resettled in the new village of Kaniahiku near Pahoa. House lots were provided by the county, and the Red Cross assisted people to purchase houses. People from Kapoho still live close to the East Rift Zone and are subject to the same hazard as before.

Insurance. There seems, at the moment, to be little possibility of insurance against volcanic activity. Fire insurance was actually paid for damages to sugarcane in 1955 and 1960, but insurance companies no longer admit fire initiated by radiant heat from lava as a valid cause.

Awareness of adjustments

When asked what action could be taken when eruption occurred, 98 percent of the respondents replied "evacuate" (Table 19-3). This response and removal of valuables were the only answers volunteered, although upon further probing people did admit to knowledge of supranatural adjustments (prayer to God or Pele, offerings to Pele, intercession of dead relatives). A few were aware of the various experiments by experts carried out in 1960.

Evacuation is thus the only emergency adjustment with which residents are familiar. This is to be expected as to date it has proven to be the only practical measure that can be taken in the face of an eruption. At the moment there seems to be little motivation to search for new modes of adjustment. Indeed, the whole warning and emergency system is premised upon the belief that if an inhabited area is threatened by lava, evacuation will take place. Virtually the entire population knows about the evacuation procedure, which has been well publicized by the Hawaii Civil Defense Agency.

Evaluation of emergency adjustments

As evacuation is the only known adjustment apart from appeals to the supranatural, little choice is involved. When the hazard perception threshold reaches a certain

Table 19-3. Awareness of adjustments (percent; N = 101)

Adjustment	Mentioned by respondent	When asked, yes
Evacuate	98	98
Walls		7
Trenches		6
Bombing		4
Prayer		88
Offerings		31
Relatives to intercede		32
Remove valuables	2	2
Move house		2
Watering		1

value, people either move themselves, or the civil defense and related agencies are mobilized to assist. Given the nature of volcanic eruptions and the character of the human use system, evacuation seems most suitable for the environmental setting and is certainly technically feasible in response to the warning and emergency system.

Eruptions have occurred frequently along the East Rift Zone, but only two have occurred in inhabited areas within living memory. Although these two occurred only 5 years apart, the overall frequency is low. In areas of low frequency, other hazard studies have found that people adopt few, if any, adjustments (Kates, 1971). This certainly seems to be the case in eastern Puna in regard to emergency adjustments.

As for long-term adjustments, it is difficult to relate changes in land use to volcanic activity, apart from the land, crops, and property destroyed in 1955 and 1960. Changes in land use have been economically gainful, and it seems that eruptions have played but a small role, apart from the relocation of some settlements.

A potential problem relating to adjustments in the future arises from the housing subdivisions. While most lots are unoccupied there is little need to consider other types of adjustments such as volcano insurance. But what will happen if the area becomes more heavily settled and the degree of hazard from the volcano increases?

Summary: adjustment to volcanic hazard

In eastern Puna District a limited set of adjustments exist to mitigate the hazards of volcanic activity. The most common adjustment designed to modify the natural events system is an appeal to the supranatural, especially Pele, the volcano goddess. In addition, experiments have been made to discover whether lava flows can be channelized, directed, or stopped.

Evacuation, the most common adjustment, eliminates danger to human life, although not to property. The most widely used of all other adjustments is the bearing of losses when they occur. Although insurance is not now available, relief and tax benefits are, both of which spread the burden of loss. Relief also spreads the effects of an eruption.

Variables affecting perception and adjustment

Two sets of analyses, a chi-square test of association and a correlation analysis, were carried out on selected variables in order to test hypotheses concerning how people perceive, evaluate, and choose adjustments to hazard.

Tests of association

Chi-square analysis was used to test the association of four variables—severity of hazard, tenure, total known adjustments, and income—with a number of other variables relating to perception, knowledge, and evaluation of both volcanic activity and adjustments (Table 19–4).

The results of these tests were with three exceptions not significant at the 5 percent confidence level. This would indicate that perception of volcanic activity and its future occurrence, what people intend to do if another eruption occurs, and what they know of adjustments have little to do with their location in relationship to volcanic hazard (high, medium, and low zones), whether they own or rent, their knowledge of adjustments, and their income level.

Three associations were significant at the 5 percent level: severity of hazard with (1) prediction of future occurrence; (2) a different adjustment the next time an eruption occurs; and (3) ownership and movement out of the area. First, inhabitants of the zone of highest hazard had a higher than expected value concerning the total unpredictability of volcanic eruptions, which suggests that scientifically objective expectations of occurrance may be associated with proximity to hazard. Second, more people than expected in the low hazard zone said that they will pray in the future; third, fewer owners than expected intended to move—both reflec-

Table 19–4. Summary of chi-square analyses for selected variables with nominal data

	Level of Significance[a]
Vs. severity of hazard (high, medium, low)	
Number of perceived hazards	0.005
Predictions of future occurrence	NS
Expectation to move	—
Who discuss community problems with	—
Different future adjustment	—
Different future adjustment when warned	NS
Total number of adjustments	TS
Adjustments by neighbors	TS
Vs. tenure (own, rent)	
Effects of eruption	—
Number of perceived hazards	—
Predictions of future occurrence	—
Expectation to move	NS
Vs. total no. of known adjustments	
Effects of eruption	TS
Damage	TS
Number of perceived hazards	TS
Predictions of future occurrence	TS
Expectation to move	TS
Vs. income (high, medium, low)	
Number of perceived hazards	—
Prediction of future occurrence	—
Expectation to move	—
Who discusses community problems with	—
Different future adjustment	—
Different future adjustment when warned	—
Total number of adjustments	—
Adjustments by neighbors	—

[a]NS = not significant; TS = sample too small.

tions, we suggest, of changes in the population structure associated with development.

Results of correlation analysis

Several significant conclusions about perception of volcanic activity and knowledge of adjustments emerge from the correlation analysis. Age and length of residence are important in relationship to knowledge about volcanic activity and knowledge of adjustments. On this basis we suggest that while socioeconomic and locational variables (with the exceptions noted earlier) are not important, experience, reflected by age and length of residence, is important. The average length of residence among respondents was 20 years, and most had experienced the 1955 and 1960 eruptions. Thus long-time residents of the area know well the fickleness of Pele, but what of the more recent arrivals, and particularly those who have purchased their 1.2-hectare (3-acre) dream lot 1½–3 kilometers (a mile or two) from the East Rift Zone? Information on the characteristics of eruptions and emergency procedures should be provided to new purchasers. The question is where responsibility for providing such information lies: government, developer, or individual. Information is probably no substitute for experience, but if it is provided, and all three units—government, developers, and individuals—are willing to take risks and bear losses, there is little else that can be done other than forbidding residence in the area.

Conclusion

The simple frequency distributions and the results of the chi-square test and correlation analysis give us insight into the circumstances under which people in eastern Puna make decisions to cope with volcanic hazard. Most important in understanding how people perceive and evaluate the hazard, and evaluate and adopt adjustments, is experience as reflected by age, length of residence in the area, and personal encounters with the hazard. Factors of individual personality may also be important, but further analysis, particularly of interethnic variation, is necessary. In terms of awareness of adjustments, experience is again the critical factor. Apart from bearing losses, only one major adjustment, evacuation, is commonly used. This, we conclude, is related to the overwhelming nature of eruptions, though they have only occurred twice in inhabited areas. Thus perhaps the small number of adjustments adopted reflects the low frequency of volcanic activity in the inhabited areas. Socioeconomic and locational variables are not important in understanding perception, evaluation, and adjustment. The only exception is that people in close proximity to the Rift Zone appear to have a better understanding of the unpredictable nature of eruptions.

To people living within 1½–3 kilometers (a mile or two) from the East Rift Zone the volcano is constantly with them—jumbled flows, cones, steaming vents. They are obviously willing to take risks in utilizing land for agriculture, but it is noteworthy that many farmers do not live on their farms but around Pahoa, further from the Rift Zone. On the other hand, much land has been sold for house lots along the Rift Zone. If houses are ever built in that area the potential property losses will increase enormously.

The eruptions of 1955 and 1960 were disruptive to the economy of the island of Hawaii. Considerable losses were incurred by government, private companies, and individuals. People were forced to leave their homes and although there was no loss of lives, several communities were disrupted. Economic losses have been great enough for the state of Hawaii to enact special tax legislation. On a national scale, the effects of volcanic activity in Puna have not been important. Crop losses did not effect mainland marketing of sugar. Today, with the growth of papaya and flower farming, mainly for the mainland market, there could be some nationwide effects. These products are luxury items and their loss would not be critical. The federal government has not been involved directly in post-emergency activity, as the state and county have been able to handle matters thus far. Puna has never been declared a national disaster area.

There is little question that volcanic activity has physically affected the land-use pattern in eastern Puna although the growth of diversified agriculture, and the creation of the subdivisions, cannot be attributed to the eruptions of 1955 and 1960. We suggest that land-use change, particularly residential development, should be cognizant of the volcanic hazard if greater losses are to be avoided in the future.

References

Hawaii, County of, Civil Defense Agency. (1969) *Natural Disaster Instructions.* Hilo, Hawaii.
Honolulu Star Bulletin. (1960) October 16.
Kates, R. W. (1971) "Natural hazard in human ecological perspective: hypotheses and model." *Economic Geography* 47: 438–51.
Macdonald, G. A. (1962) "The 1959 and 1960 eruptions of Kilauea Volcano, Hawaii, and the construction of walls to restrict the spread of the lava flows." *Bulletin Volcanologique* 24:248–94.
_____, and Abbott, A. T. (1970) *Volcanoes in the Sea: The Geology of Hawaii.* Honolulu: University of Hawaii Press.
_____, and Eaton, J. P. (1964) "Hawaiian volcanoes during 1955." *Geological Survey Bulletin* 1171.
Peterson, D. W. (1971) Personal communication.
Puna Volcanic Fact Finding Committee, Report of the (1955). Mimeo.
Natural Hazards Research (1970) "Suggestions for comparative field observations on natural hazards." Boulder: Univ. of Colorado, Inst. of Behavioral Science, Working Paper No. 16.
Swanson, D. A., Jackson, D. B., Duffield, W. A., and Peterson, D. W. (1971) "Mauna Ulu eruption, Kilauea Volcano." *Geotimes* 1:12–16.

20. Human adjustment to the earthquake hazard of San Francisco, California

EDGAR L. JACKSON
University of Toronto

TAPAN MUKERJEE
University of the Pacific

Out of the estimated loss of assets from earthquakes in the United States in the period 1925–71, approximately $600 million of losses occurred in the state of California. California is part of the Pacific seismic belt, which is responsible for about 80 percent of the world's earthquakes. Approximately 90 percent of the seismic activity of the continental United States, including Alaska, occurs in California and western Nevada (Wood and Heck, 1966).

Several major earthquakes occurred in California between 1860 and 1972: Owens Valley (1872), magnitude 8.3R (R = Richter scale), 60 deaths; San Francisco (1906), 8.3R, 700 deaths, losses around 1.6 billion (all losses in 1958 prices); Long Beach (1933), 6.3R, 115 deaths, losses $89 million; Imperial Valley (1940), 7.1R, 9 deaths, losses at least $12 million; Kern County (1952), 7.7R, 12 deaths, losses around $63.0 million; San Fernando (1971), 6.6R, 65 deaths, estimated losses approximately $439 million (Wood and Heck, 1966; Joint Committee on Seismic Safety, 1971).

The city of San Francisco, which is the focus of this study, lies in seismic zone 3 (see Fig. 27–1, below), one of the seismically most hazardous parts of California (Algermissen, 1969).

History of settlement

San Francisco and the surrounding cities of the Bay Area owe their growth, in part, to their location on a natural harbor, the San Francisco Bay (Watson, 1963). The main West Coast port of the United States, San Francisco's 1970 population was 715,764 (information provided by San Francisco's Department of City Planning).

Figure 20–1 provides a useful picture of San Francisco's growth, with early development at the northeast of the peninsula in the 1850s. Expansion took place along the main northeast-southwest artery. Through the early years of the twentieth century, the city expanded westward and northward along the more accessible flat-

Fig. 20–1. The growth of San Francisco showing street along which property was developed. Prepared by the San Francisco Department of City Planning

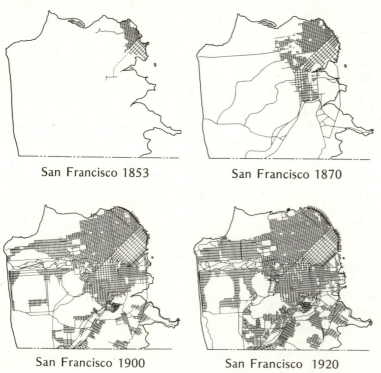

San Francisco 1853

San Francisco 1870

San Francisco 1900

San Francisco 1920

San Francisco 1966

ter land. The great earthquake and fire of 1906, which destroyed more than 500 downtown blocks, had little effect on the street pattern, since the city was rebuilt largely as it had been before (Bowden 1970). In spite of the extremely hilly terrain, the pattern of streets remained rectangular until about the Second World War (Bartz, 1968). Subsequent development consisted largely of in-filling within the built-up area, and can be recognized in Fig. 20–1 as irregular street patterns signifying expansion onto the steeper hills.

Exhaustion of available land for residential development had important implications for the local earthquake hazard: it fostered building on the less secure foundation of Bay "fill," or "made land."

Physical background to earthquake hazard

The hilly peninsula on which San Francisco lies rises to over 270 meters (900 feet), and is one of a series of nearly parallel ridges trending obliquely to the coast in a northwesterly direction. The bay itself, a deep depression formed by warping and faulting, and drowned by the sea, lies between the Berkeley Hills on the east and, on the west, the San Francisco Peninsula and its northern continuation across the Golden Gate Channel, the Marin Peninsula (Goldman, 1969) (Fig. 20–2).

The earthquake hazard is connected to local geology in two ways: stratigraphy and proximity to faults. The peninsula consists of solid outcrops of rock interspersed with areas of soft sandstone, marine sands, dune sands, river alluvium, as well as "fill" (Goldman, 1969). Houses and other buildings constructed on solid rock foundations are less liable to sustain damage in an earthquake than those on "fill" or other less consolidated foundations (Oakeshott, 1969; Steinbrugge, 1968).

Earthquakes in California are closely related to active faults, of which the most important are Garlock, White Wolf, Elsinore, San Gabriel, San Jacinto, Death Valley, Hayward, Calaveras, and San Andreas (U.S. Department of the Interior, 1969). The San Andreas has been referred to as the "master fault" of the network, and Oakeshott (1959) has called it "the greatest active fault of historic times—both the longest and the one on which earthquakes have originated most frequently." Movement along the San Andreas Fault was responsible for the great earthquake of 1906 and many others of lesser magnitude (U.S. Department of the Interior, 1970).

Complete instrumental observations have been made only since 1927, and for the period prior to that date, reliance has to be placed on historical records (Wood and Heck, 1966). Difficulties arise in comparing measures of intensity, since settlement areas and densities have changed, building practices improved, and so on (Tocher, 1959). Between 1800 and 1963, however, there are know to have been 59 earthquakes of intensity VII or greater, in California. Of these, 22 were of intensity VII, 19 of intensity VIII, 7 of intensity IX, and 2 of intensity X. There were, in addition, three

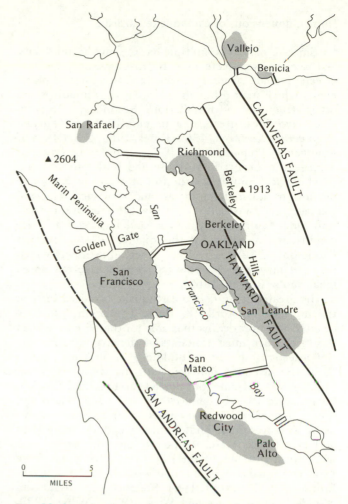

Fig. 20–2. Location of earthquake faults in the San Francisco area, elevations in Feet. Sources: Steinbrugge (1968); Paterson (1970)

larger shocks. (The intensities of a further six are not accurately known.) The five largest earthquakes affecting San Francisco took place in 1836, 1838, 1865, 1968, and 1906. There have been about 12 damaging earthquakes per century since the first settlement, though unevenly distributed in space, time, and magnitude (Steinbrugge, 1968).

It is not yet possible to predict exactly when or where future earthquakes will occur. Estimates have been made, however, of recurrence intervals for earthquakes of selected magnitudes on the San Andreas Fault (*California Geology,* 1971). At least two authorities have agreed that "on the basis of the historical record, and in view of the accumulating strains, *for planning purposes* it is reasonable to anticipate a major or great earthquake in the San Francisco Bay Area once every 60 to 100 years" (Steinbrugge, 1968, p. 10; Oakeshott, 1959, p. 14).

Human dimensions of earthquake hazard

Earthquake hazard is multiple hazard, and both primary and secondary effects may be distinguished. Primary effects include ground movements (shaking, displacement) which may result in the collapse of buildings and other structures and installations. Secondary effects include landslides, fires, tsunamis, and floods (*Earthquake Engineering Research,* 1969; Kates, 1960; U.S. Office of Emergency Preparedness, 1972). Four main types of damage can occur: loss of life and both physical and psychological injury; destruction of property; economic disruption and indirect losses; and ecological damages. Damages are caused not only by the magnitude, location, and depth of the shock, and local soil and site conditions, but also by the quality of construction (Hodgson, 1956), as well as other adjustments which may or may not be adopted by an affected area's inhabitants (Russell, 1969).

The great San Francisco earthquake occurred early in the morning of April 18, 1906. The timing probably resulted in fewer deaths than might otherwise have been the case, since most residents of the city were indoors and asleep at the time. Buildings in all parts of San Francisco sustained damage, with the worst effects on "fill," where pavements buckled, arched, and fissured, houses were severely damaged or destroyed, sewers and water mains broke, and streetcar tracks were bent. Many fires resulted, and within two days, more than 500 city blocks were completely gutted. Damage was widespread throughout the Bay Area. (Accounts of the 1906 earthquake may be found in Bowden, 1967, 1970; Bronson, 1959; Iacopi, 1964; Steinbrugge, 1968; Sutherland, 1971; Thomas and Witts, 1971; Tocher, 1959; Wood and Heck, 1966).

In contrast, the earthquake of March 22, 1957, of magnitude 5.3R and intensity VII, was a relatively minor event. Although the most damaging since 1906, no lives were lost, and only about 40 people suffered injuries, these of a minor nature. Total damages to roads, buildings, and a reservoir amounted to about $1 million (Brazee and Cloud, 1959; California, 1959; Wood and Heck, 1966).

The damage potential in 1972 is much higher. As far as magnitude of future earthquakes is concerned, it has been said that "the forces generated by the 1906 San Francisco earthquake appear to be a reasonable upper limit" (Steinbrugge, 1968, p. 10). It is estimated by the Joint Committee on Seismic Safety that under certain conditions, an earthquake of magnitude 8.0R of a minute's duration could result in damages to assets in the range of $30 billion in 1970 prices. This would amount to about 3 percent of the 1971 Gross National Product of the United States. In terms of the total personal income of the state of California in 1970 (Stockton *Record,* 1971), it would amount to 33.7 percent. The damage figures most likely do not include the losses due to decline in economic opportunity such as falls in wages, salaries, profits, rents, and taxes; or, alternatively, a decline in the level of production as a result of the disaster. If these losses are taken into account together with multiplier effects, the total economic impact of a future earthquake in San Francisco could indeed be catastrophic. Furthermore, casualties up to 350,000 dead and injured could result (Thomas and Witts, 1971).

Theoretical range of adjustment to earthquake hazard[1]

Expert opinions indicate the following range of possible adjustments.

Affect the cause

No such adjustment is presently available, although research is moving in this direction. Nor can earthquakes yet be predicted.

Modify the hazard

Such adjustments include, for the individual homeowner, choice of a stable site for home location, and purchase of an earthquake-resistant dwelling. Land-use zoning and risk zoning, as well as soil and slope stabilization, can be employed by the municipality. Home protection against fire is an adjustment practiced by most homeowners (but not necessarily with regard to earthquakes), though dependence tends to be placed on assistance from the local fire department. Municipal disaster operations plans allow for conflagrations following an earthquake.

Modify the loss potential

For both the individual homeowner and the city government, these adjustments include planning during the preimpact period for emergency action to be carried out in the postimpact situation. Also available are structural reinforcements, which are more likely to be adopted when enforced. Making new buildings earthquake-resistant is encouraged by the adoption and enforcement of building codes. Bond issues are made for reinforcing public buildings such as schools.

Adjust to the damages

Doing nothing and therefore bearing the loss probably constitutes the most widespread adjustment to the earthquake hazard. Loss sharing is a possibility, though insurance premiums are high: the premium for a San Francisco single frame dwelling is $2.00 per $1,000 of

---

1. The list of adjustments was compiled from several sources, the main ones being Clarke and Hauge (1971), Kates (1970), Mukerjee (1971), and Steinbrugge (1968).

insurance; for other structures the rate is $10.35 per $1,000. Finally there are local, state, and federal relief measures.

A reluctant sample

A sample of 120 respondents was drawn from nine local areas of the city of San Francisco, stratified by average income and damages in the 1906 earthquake. The interviewers experienced a 78 percent refusal rate. As a result, the conclusions drawn from the data cannot be said to represent the residents of San Francisco. The discussion, therefore, relates to the sample of 120 who agreed to talk about earthquakes.

Among possible reasons for the high refusal rate are (1) unwillingness to spend the required time in answering this questionnaire, which took about 30 minutes to administer; (2) unwillingness to take part in questionnaire surveys of any kind; (3) unwillingness to let strangers into the home—this may be a considerable problem in large cities like San Francisco, where there are high crime rates; (4) denial of the earthquake problem, which may be the most important reason. The introductory statement used to recruit the respondents mentioned the topic of earthquakes, which possibly discouraged a significant proportion of those who refused from participating. Common responses included "What do you want to know about that for?"; "Earthquakes don't bother me"; "Let it rock"; "We have no earthquake problem"; and so on. The tone of most such refusals suggested that many people *are* worried about the earthquake hazard but do not allow themselves to think about it. This may well be minimization of the earthquake threat through a process of dissonance reduction.

A subsequent study in Cornwall, Ontario, experienced a refusal rate of only 10–15 percent, and more recent research in Los Angeles, British Columbia, and Anchorage resulted in similarly low rates of refusal. In these studies, earthquakes were not mentioned until the interviews were under way, and the introduction included only a general statement about environmental problems. The authors feel, however, that had they employed this method in San Francisco, many interviews would have been terminated as soon as the topic of interest became apparent.

Perception of hazard

Attachment of San Franciscans to their city

This is a fundamental consideration in estimating the importance of the earthquake hazard to San Francisco residents. Asked about disadvantages and advantages of their place of residence and of San Francisco, 80.8 percent and 87.5 percent respectively stressed advantages. Only 10.0 percent and 8.3 percent emphasized disadvantages. Those mentioned tended to be social in nature, though a few respondents did mention such

natural phenomena as cold or fog. Not one respondent included earthquakes when listing disadvantages, or in a list of community problems where 44.2 percent could think of no such problems. Those of concern were mainly social in nature, with education and school busing for integration at the top of the list (16.7 percent), followed by problems concerning surrounding streets and their cleanliness (10.0 percent) and crime (8.3 percent). Other problems, mentioned by less than 6 percent of the respondents, included traffic, ecology, employment, and welfare. The earthquake hazard was absent.

The notion of close attachment to San Francisco is further supported by expressed intentions to remain in the city: 67.5 percent intended to do so; 9.2 percent would move to other places in the hazard area; 5.8 percent were not sure; only 13.3 percent would move out of the area altogether. Typical responses included "I *love* this city," which may partly account for the minimization of personal vulnerability from the earthquake threat.

Earthquake hazard and experience

Few of the respondents were unaware of San Francisco's earthquake hazard: asked if the residents of the city have trouble with earthquakes, only 13.3 percent replied they did not know, while 36.7 percent replied yes and 45.8 percent no. Many of these last respondents, however, said that while there had been earthquakes, they did not give any trouble to the people of San Francisco.

Only 14.2 percent of the sample had never experienced an earthquake. Five percent had experienced at least one earthquake, but in places other than San Francisco; 80.8 percent had experienced at least one earthquake in San Francisco.

The range of experience varied from only one up to "dozens." This date, however, should be treated with care, since the respondents did not describe their experience in detail. Perhaps those people claiming to have experienced only one or very few omitted the smallest tremors, which might have been included by those claiming to have experienced many earthquakes.

Although earthquakes were never suggested as a problem and many persons had refused to discuss them, those who agreed to talk were willing to admit the hazard when questioned. In total, 42.5 percent expected another earthquake within the next few years, that is, 15.8 percent "soon," and 26.7 percent "in a few years" (Table 20-1); 28.3 percent expected another but were unwilling to state a time period; 4.2 percent thought there would be another, but not for many years; and only 4.2 percent of the respondents thought an earthquake would not occur again.

As far as expected damages are concerned, 20.0 percent thought there would be no effects to them personally. Among those who believed there would be damages, structure and contents were mentioned most

Table 20–1. Prediction of future occurrence

Response	No.	%
No	5	4.2
Don't know	25	20.8
Yes, soon	19	15.8
Yes, in a few years	32	26.7
Yes, in many years	5	4.2
Yes, but don't know when	34	28.3
Total	120	100.0

often, by 36.6 percent and 13.3 percent respectively. Regarding the extent of these damages, only 6.7 percent thought that future damages would be "total," while 20.0 percent thought they would be substantial, 25.8 percent thought they would be slight or nonexistent, and 22.5 percent did not know.

From the preceding paragraphs it can be seen that the San Franciscans in the sample were aware of the earthquake hazard, and had a high rate of expectation of future events. There was a tendency, however, to dismiss the hazard as not troublesome and, in general, to minimize the damage that would result to respondents from future earthquakes.

The fact that scientists cannot predict earthquakes is reflected in responses to the questions concerning signs of future earthquakes: 67.8 percent believed no such signs exist; 12.7 percent put their faith in scientists' ability to predict. Belief in phenomena such as "earthquake weather" was expressed by 11.0 percent of the sample. The remaining 8.5 percent were not aware of any signs, but did not categorically state that none existed.

Adjustment to earthquake hazard

More than half the respondents (55.9 percent) believed something could be done to prevent damages. The rest either stated nothing could be done (33.3 percent) or they did not know of anything (10.8 percent). The respondents placed greatest faith in structural modifications and building reinforcements (20.8 percent). Responses to the question thus indicated that a total of 44.1 percent were unaware of the existence of measures to reduce damages. The next question, however, revealed that only 26.7 percent did not practice any of the adjustments included in the questionnaire. This suggests that certain courses of action are not thought of as loss-reducing.

Adjustments included in the questionnaire survey were: do nothing; pray; evacuate; protect home against fire and looters; structural changes to home; earthquake insurance; and impact and postimpact emergency action, for example, run outside to an open space or take shelter in a safe place. Of the respondents, 26.7 percent

had never done or would not do any of these things; one or two adjustments had been or would be adopted by 54.1 percent; three or more adjustments (to a maximum of five) had been carried out by only 19.2 percent.

Doing nothing or praying, each reported by 36.7 percent, were the two most practiced adjustments, closely followed by protection (35.8 percent). Most respondents employed day-to-day fire precautions in their homes, and the problem of looters was usually thought minimal or nonexistent. Also, 42.5 percent had carried out or would rely on emergency action in the form of shelter, while 16.8 percent preferred to get outside to an open space. Adjustments of a more substantial nature, requiring preparation, time, and investment, were adopted by few respondents. Such adjustments include insurance (7.5 percent), structural changes (7.5 percent), and evacuation (5.0 percent).

Taking shelter and home protection were most often evaluated as "good" adjustments, by 79.2 percent and 89.2 percent respectively (Table 20–2). Interestingly, structural changes, insurance, and evacuation were rated as "good" things to do by higher proportions of the respondents (53.3 percent, 40.8 percent, and 32.5 percent) than had in fact adopted such adjustments. There were greater differences of opinion over evacuation than over the other adjustments. The issue was also divided over insurance.

The authors believe that the adjustments adopted represent a poor adaptation to the damage potential. Reliance on emergency actions, on praying or doing nothing were the adjustments most widely adopted, with relatively little emphasis placed on modifying the loss potential or preparing for losses. This suboptimal level of adjustment cannot be explained by lack of knowledge of the range of choice. Table 20–3 shows that except in the case of earthquake insurance, lack of awareness of an adjustment was never given as a reason for not adopting it.

How, then, were adjustments evaluated? Reference to Table 20–3 shows that for each adjustment, between 12.5 percent and 27.4 percent could not verbalize their reasons for accepting or rejecting it (column 9). As far as doing nothing is concerned, 39.8 percent believed

Table 20–2. Evaluation of adjustments

Adjustment	Good		Bad		Don't know	
	No.	%	No.	%	No.	%
Nothing	15	12.5	68	56.7	37	30.8
Pray	80	66.7	18	15.0	22	18.3
Evacuate	39	32.5	58	48.3	23	19.2
Structural change	64	53.3	29	24.2	27	22.5
Protect	107	89.2	4	3.3	9	7.5
Insurance	49	40.8	41	34.2	29	24.2
Open space	43	35.8	61	50.8	16	13.3
Shelter	95	79.2	10	8.3	15	12.5

Table 20–3. Reasons for opinions (percent)

Adjustment	1. Not heard of adjustment	2. Hazard estimation environmental fit	3. Technology	4. Would or would not pay	5. Cannot afford	6. Spatial linkage	7. Won't work	8. Other	9. No reason given	Total
Do nothing	0.0	0.8	0.0	2.5	0.8	3.3	39.8	25.0	27.4	100.0
Pray	0.0	0.0	0.0	0.0	0.0	1.7	16.7	61.5	19.9	100.0
Evacuate	0.0	50.8	0.0	2.5	0.0	2.5	10.8	20.8	12.5	100.0
Structural change	0.0	29.8	10.0	6.7	7.5	0.8	5.0	20.0	20.0	100.0
Protect	0.0	5.8	2.5	43.3	0.8	0.0	0.8	30.7	15.0	100.0
Insurance	5.8	2.5	0.8	25.8	27.4	0.0	5.8	10.8	20.8	100.0
Open space	0.0	72.3	0.0	0.0	0.0	4.2	0.0	10.0	13.3	100.0
Shelter	0.0	53.5	0.0	1.7	0.0	10.8	3.3	18.3	12.5	100.0

that "you must do something" (column 7). Others replied "what *can* you do?" (column 8). Then, 61.5 percent gave a variety of reasons for praying, including "I always pray"; others (16.7 percent) considered prayer futile. The choice to evacuate is apparently made in the context of the damages: column 2 includes such responses as: "I would only evacuate if the damages were great" and "the damages would not be so great as to force me to evacuate." Potential damages also affected the evaluation of structural changes, although again there were differences of opinion. Column 2 includes both the respondents rejecting and those accepting structural changes in the context of assumed damages. Other reasons given for evaluation of structural changes were inability to carry out, and the costs involved. That it would pay to do so was the reason most often given for home protection. Insurance was most often rejected because of cost, either because it would not pay, or because the respondents could not afford it. A few thought insurance not worthwhile in the context of the perceived damage potential. Emergency actions were accepted or rejected in the context of hazard estimation: some respondents would run outside to get away from falling buildings; others would remain inside *because* of falling buildings, wires, etc. For the latter, indoors would be "the best place to be."

It appears that what neighbors would or would not do has little if any effect on the respondents' choice of alternative adjustments. More than half (52.5 percent) either did not know or could not remember what their neighbors would do. There was little variation in assumed adoption of adjustments by neighbors.

Of the 103 respondents who had experienced an earthquake, almost half (43.3 percent) would not do anything different next time, nor were they planning to make any alternative long-term adjustments. The major

changes would be with regard to emergency action (evacuation and shelter, 11.7 percent each.)

Suggestions have been made that warnings might encourage people to adopt more optimal adjustments to the earthquake hazard. However, almost half the respondents (49.2 percent) asserted that they would do nothing if a warning were given. Many stated a disbelief in such warnings. Of the remainder, the action to be taken by most would be evacuation (24.2 percent). Preparatory adjustments such as structural modification would be adopted by few (4.2 percent), which is surprising, since the question postulated a warning time of up to 1 year.

Variations in the adoption of adjustments are difficult to explain on the basis of this survey. Nevertheless, some tests of association have been carried out. The number of adjustments adopted was related to the number of earthquakes experienced in the past, but not to expectations of remaining in San Francisco. There was no significant relationship between adjustments adopted and future expectation of earthquakes and damages.

Some suggestions, however, may be made on the basis of the data at hand. First, the earthquake hazard was perceived as trivial. Second, while experience and awareness of the hazard were high, future damages were expected to be low. In other words, respondents tended to minimize their perceived personal vulnerability. This probably relates to the recent lack of damaging earthquake activity in San Francisco. Finally, the indeterminacy of the occurrence of the event may have been an important factor—earthquakes were considered an act of God, with no signs to tell when the next one is coming.

The authors believe that the relative lack of recent seismic activity in San Francisco explains the suboptimal level of adjustment to the earthquake hazard. Few respondents had experienced intense and damaging earthquakes, and thus dismissed the hazard as unworthy of consideration and preparation.

Conclusion

San Franciscans, represented by the sample, and self-selected as willing to talk about the hazard, give low priority to the earthquake hazard. They are highly attached to their city, and there appears to be a general hesitancy to recognize hazards of either a natural or social nature. The majority of respondents had experienced at least one earthquake either in San Francisco or elsewhere and expected an earthquake sometime in the future. Awareness of the hazard was also high, although about 50 percent of the respondents did not see past earthquakes as being troublesome. In general, future damages were expected to be slight. In this context, emphasis seems to be placed on loss bearing and emergency action, with relatively little attention or thought given to the predisaster type of adjustment requiring investment and preparation. There is no evidence as to how those who refused to talk might respond.

More recent research, however, conducted in Cornwall, Ontario; Los Angeles; Vancouver and Victoria, British Columbia; and Anchorage, Alaska, largely supported the tentative conclusions offered here. It was found, once more, that the relative intensity of experience with the hazard was most closely related to the adoption of adjustments. No relationship was found between adoption and future expectation of earthquakes and damage, although socioeconomic status had a minor effect. The majority of respondents in the later studies perceived the earthquake hazard as lacking in importance, and were unwilling to adopt preparatory adjustments unless they themselves had experienced an intense and damaging earthquake.

Acknowledgments

The interviews were carried out by Edgar Jackson and David Polster; the authors are indebted to the latter for his assistance.

References

Algermissen, S. T. (1969) "Seismic risk studies in the United States." Santiago, Chile: Fourth World Conference on Earthquake Engineering.

Bartz, Fritz. (1968) *San Francisco—Oakland Metropolitan Area.*

Bowden, Martyn J. (1967) "The dynamics of city growth: an historical geography of the San Francisco Central District, 1850–1931." Berkeley: University of California, unpublished Ph.D. thesis.

_____. (1970) "Reconstruction following catastrophe: the laissez-faire rebuilding of downtown San Francisco after the earthquake and fire of 1906." *Proceedings of the Association of American Geographers* 2:22–26.

Brazee, Rutlage J., and Cloud, William K. (1959) *United States Earthquakes 1957.* Washington, D.C.: U.S. Department of Commerce, Coast and Geodetic Survey.

Bronson, William. (1959) *The Earth Shook, the Sky Burned.* Garden City, N.Y.: Doubleday.

California, State of. (1959) *San Francisco Earthquakes of March 1957.* Division of Mines and Geology, Special Report No. 57.

California Geology. (1971) "Earthquake recurrence curves plotted for San Andreas Fault." 24:46.

Clarke, William B., and Hauge, Carl J. (1971) "When the earth shakes . . . you can reduce the damage." *California Geology* 24:203–16.

Earthquake Engineering Research. (1969) Washington, D.C.: National Academy of Sciences, Committee on Engineering Research, Division of Engineering, National Research Council.

Goldman, Harold B. (1969) "Geology of San Francisco Bay." In *Geologic and Engineering Aspects of San Francisco Bay Fill.* State of California, Division of Mines and Geology, Special Report 97.

Hodgson, John H. (1956) "A seismic probability map for Canada." *Canadian Underwriter.*

Iacopi, Robert. (1964) *Earthquake Country.* Menlo Park, Calif.: Lane Book Company.

Joint Committee on Seismic Safety. (1971) *Preliminary Report on the San Fernando Earthquake Study.* Sacramento: California State Legislature.

Kates, Robert W. (1970) "Human adjustment to earthquake hazard." In *The Great Alaska Earthquake of 1964: Human Ecology Volume.* Washington, D.C.: National Academy of Sciences.

Mukerjee, Tapan. (1971) "Economic analysis of natural hazards: a study of adjustments to earthquakes and their costs." Toronto: Univ. of Toronto, Natural Hazard Research, Working Paper No. 17.

Oakeshott, Gordon B. (1959) "San Andreas Fault in Marin and San Mateo counties." In *San Francisco Earthquakes of March 1957.* State of California, Division of Mines and Geology, Special Report No. 57.

_____. (1969) "Geologic features of earthquakes in the Bay Area." In *Geologic and Engineering Aspects of San Francisco Bay Fill.* State of California, Division of Mines and Geology, Special Report No. 97.

Paterson, J. H. (1970) *North America, a Geography of Canada and the United States.*

Russell, Clifford S. (1969) "Losses from natural hazards." Toronto: Natural Hazard Research Working Paper No. 10.

Steinbrugge, Karl V. (1968) *Earthquake Hazard in the San Francisco Bay Area: A Continuing Problem in Public Policy.* Berkeley: University of California, Institute of Governmental Studies.

Stockton *Record.* (1971) *California—Past, Present, and Future.* Stockton: California Almanac Company.

Sutherland, Monica. (1971) *The Damndest Finest Ruins.* New York: Ballantine Books.

Thomas, Gordon, and Witts, Max Morgan. (1971) *The San Francisco Earthquake.* New York: Stein & Day.

Tocher, Don. (1959) "Seismic history of the San Francisco region." In *San Francisco Earthquakes of March 1957.* State of California, Division of Mines and Geology, Special Report No. 57.

U.S. Department of the Interior. (1969) "The San Andreas Fault." Washington, D.C.: Geological Survey.

_____. (1970) "Active faults in California." Washington, D.C.: Geological Survey.

U.S. Office of Emergency Preparedness. (1972) *Disaster Preparedness,* vol. 3. Washington, D.C.: Executive Office of the President.

Watson, J. Wreford. (1963) *North America, Its Countries and Regions.* London: Longmans.

Wood, H. O., and Heck N. (1966) *Earthquake History of the United States. Part II. Stronger Earthquakes of California and Western Nevada.* Washington, D.C.: Environmental Science Services Administration.

21. Urban snow hazard: Marquette, Michigan

FILLMORE C. F. EARNEY
Northern Michigan University

BRIAN A. KNOWLES
University of Colorado

Snow as a hazard, the focus of this study, is examined within the limits of one urban setting—Marquette, Michigan. The present paper is based upon work carried out during the summer of 1971 as a comparative field observation.

Site and sequent occupance

Marquette is located on the northern shore of Michigan's "Upper Peninsula," approximately midway between Wisconsin's border on the west and Sault Ste. Marie, Michigan, on the east (Fig. 21–1). Situated on Lake Superior's south shore, the city's physiographic features are dominated by low sandy plains along the shoreline with a fairly rapid rise in elevation away from the lake, especially in the south portion of town where some parts of the city lie 150 meters (500 feet) above lake level. Soils belong mostly to the podzol group, lying on bedrock of schist, granite, and gneiss.

For more than a century after their arrival on the eastern seaboard, European settlers paid little attention to the upper Great Lakes region. The area remained a wilderness inhabited only by Indian tribes who hunted within its forests or fished along the shores of Lakes Michigan and Superior. In the early seventeenth century French explorers penetrated the region, first in quest of a Northwest Passage and, when this effort proved futile, for furs. The French, English, and Americans, in their turn, contributed to the development of the fur trade that reached its apex in the early nineteenth century and then declined to insignificance by 1850. Throughout this period, the present site of Marquette remained part of a familiar but undeveloped wilderness. A significant change was soon to come.

Entry of American pioneers into the area led to the discovery of what was to become the main economic base of Marquette and vicinity—iron ore. Surveyors accidentally discovered outcrops of ore in the hills approximately 21 kilometers (13 miles) to the west of Marquette. After a slow beginning in the mid-1840s, and in spite of difficult terrain, inadequate transport facilities, and distance from eastern markets, the industry grew rapidly. Marquette's position directly to the east of the iron range, along with a relatively sheltered harborage, played a significant role in shaping its economic function as a transport link between rail and ship. Various rail transport improvements and the opening of the St. Mary's Ship Canal in 1855 set the stage for rapid growth in the iron-ore industry and contributed to further expansion of the service functions of the city. As the

Fig. 21–1. Upper Peninsula-Michigan

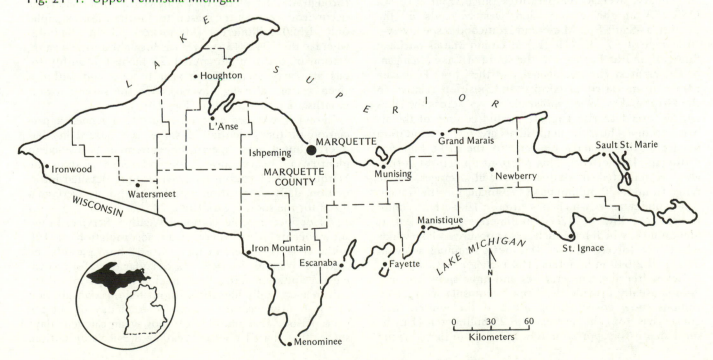

mining industry grew so did industries supplying timber, food, and mining machinery.

Prosperity in the iron-ore industry has been erratic and its immediate economic condition has dictated, to a large degree, the health of Marquette's economy. This condition, however, has changed considerably as a result of the establishment of a college in 1899; as this institution has grown, it has come to play a more significant role in the local economy, now employing approximately 1,000 people.

Land use within the city is mostly residential, although significant areas are occupied by the central business district, widely scattered recreational units, a large shopping complex on the west, and two ore terminals on the lakeshore—one (active) in the northeast and another (inactive) near the central business district.

Marquette and the Upper Peninsula region compared

Michigan's Upper Peninsula has a population of approximately 304,000; Marquette's total is 22,000, or 7.2 percent of the region's population. Marquette is in many respects a regional service center for shopping, medical, and educational needs.

Comparatively, the Upper Peninsula is rather homogeneous climatically. All parts receive 690–910 millimeters (27–36 inches) of precipitation annually, usually well distributed throughout the year; Marquette receives 790 millimeters (31 inches). Average temperatures for July vary from 17° C. (63° F.) at Grand Marais in the central Upper Peninsula on the shore of Lake Superior and at Houghton on the Keweenaw Peninsula to 21° C. (70° F.) along the Wisconsin border, inland from Lake Michigan. At Marquette the average is 19° C. (66° F.). In January, average temperatures range from - 10° C. (14° F.) at the eastern and western ends of the peninsula—Sault Ste. Marie and Ironwood respectively—to a high of -7° C. (19° F.) at Grand Marais on Lake Superior and at Fayette on the shore of Lake Michigan.

The greatest climatic diversity of the Upper Peninsula comes from varied proximity and position relative to the Great Lakes. When moist air moves across the lakes and is forced to rise abruptly, rapid cooling of the air mass occurs. Thus, the peninsula's greatest precipitation occurs in the uplands adjacent to the Lake Superior shore (the lake effect). This relationship is most noticeable as reflected in annual snowfall averages. These range from 4,470 millimeters (175 inches) in the Houghton-Calumet area, which has intense lake-effect conditions, to a low of 1,270 millimeters (50 inches) at Menominee, which has much less exposure to lake winds which are forced to rise. Marquette's annual snowfall average is 2,640 millimeters (104 inches).

Below-freezing temperatures and large snow accumulations usually provide the Upper Peninsula with a continuous snow cover during most of the winter. Marquette has, on the average, a 25-millimeter (1-inch) snow depth for approximately 145 days in the year; the

Keweenaw Peninsula has some areas that average nearly 160 days; Menominee, in the southernmost portion of the Upper Peninsula, has approximately 100 days. Each winter, on the average, Marquette experiences approximately 8 days with a snow depth of 790 millimeters (31 inches) or more and has had an absolute maximum of 1,250 millimeters (49 inches) on the ground. The Keweenaw, area, in the vicinity of Houghton, averages approximately 30 days with 790 millimeters (31 inches) on the ground and 11 with 910 millimeters (36 inches) or more and has had a maximum depth of 1,750 millimeters (69 inches).

The Marquette field study

The cooperative field study of the urban snow hazard in Marquette was limited to the legally defined incorporated limits of the city.

The observing station

The United States Weather Service in Marquette has been in continuous operation since 1881, although its location has been changed several times. The distance of removal was no more than two city blocks and the general exposure of the instruments has remained relatively constant. Therefore, data for the various locations should be comparable (Mueller, 1971).

Historical occurrence of the hazard

Marquette's annual "snow-year" average exceeds 2,540 millimeters (100 inches) in 7 of every 10 years. (A snow year is a period from September 1 of a given year through May 31 of the following year.) Great variations occur from one winter season to another. For example, only 1,520 millimeters (60 inches) fell in 1941–42, whereas the 1942–43 snow–year brought a total accumulation of 3,190 millimeters (125 inches). Monthly totals also vary greatly from year to year and within a given season, although averages for the several winter months are rather uniform (Fig. 21–2).

Snowfall averages can contribute to a general appreciation of the potential of snow as a hazard in Marquette, but a more specific measurement is needed to identify a "heavy snow," the reference used in the Marquette field study. To provide more detailed information, daily observation data for Marquette were analyzed for the snow–years 1950–70.

As defined by the National Weather Service, heavy snow occurs when there is an accumulation of 100 millimeters (4 inches) or more in a 12-hour period. The National Weather Service uses calendar days for recording precipitation data, thus making these periods difficult to specifically identify without tedious consultation of the hourly weather record for each day of the 20-year period. Examination of snow-year calendar days shows that, based on the Weather Service's definition,

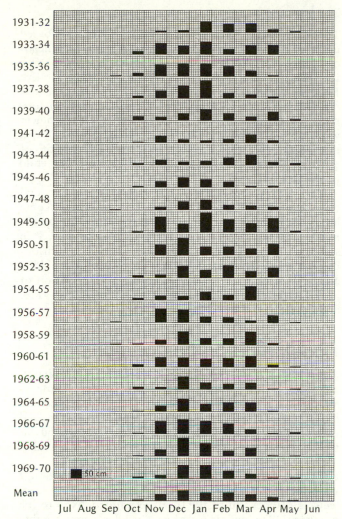

Fig. 21-2. Monthly snowfall at Marquette, 1931-70

heavy snowfalls occurred 115 times during the 20-year period, with the greatest frequency (30) occurring in December. If the magnitude is increased to 250 millimeters (10 inches) for one calendar day, a total of 8 such snowfalls occurred during the 20 years (Table 21-1).

Tabulations for heavy snowfalls that utilize only calendar days fail to include 100 millimeter (4-inch) magnitudes occurring in a time period encompassing part of two calendar days. When the two-calendar-day criterion is used, an additional 68 occurrences can be identified for the 20-year period, giving a total of 183. If the same criterion is applied to the 250-millimeter (10-inch) magnitude, then another 19 occurrences show up for a total of 27 during the 20-year period (Table 21-1). Of course, not all these would have occurred in a calendar-day overlap and in a consecutive 12-hour period.

Although Marquette is not noted for extremely heavy snowfall in a 24-hour period (the record fall is 432 millimeters or 17 inches in February 1890), a few rather exceptional storms of more than 24 hours' duration have occurred. These storms range from a fall of 590 millimeters (23 inches) to several occasions when about 400 millimeters (16 inches) fell. One of these heavy storms occurred in 1938. The year 1938 is of special significance to this study because it was so frequently mentioned by interviewees in response to the question "What was the worst year for heavy snowfalls?" A total of 10 of the 120 interviewees chose 1938 as the worst year for snow; and 7 others, based on their identification of a proximate year or other statements, probably had this snow-year in mind when they responded to the question. These figures take on added significance when it is realized that only 70 of the respondents were born before that winter and only 38 of those were as old as 10. Also, it is unlikely that all of them lived in Marquette at that time.

Table 21-1. Number of occurrences of snowfall for selected magnitudes based on one and two calendar days, 1950-70

| | One calendar day | | | | Two calendar days | |
	25 mm or more	75 mm or more	100 mm or more	250 mm or more	100 mm or more	250 mm or more
Total	658	195	115	8	68	19
Mean	32.9	9.8	5.8	0.4	3.4	0.95
Maximum in 1 year	52	21	14	2	0.7	2
Minimum in 1 year	22	3	2	0	0	0

Source: Tabulated from monthly record of observations at Marquette, Michigan, 1950-70. Records obtained at the National Weather Service Office in Marquette, Michigan, July 1971.

Most of these 17 respondents identified the 1937–38 snow year in relation to one particular storm—that of January 24–27. Because of this storm's overall impact on the community and its relevance to this study, it merits a closer look. On January 23, prior to the storm, a total of 270 millimeters (11 inches) of snow was on the ground. Snow began to fall between 5:00 and 6:00 A.M. on the 24th with wind velocities increasing from 10 kilometers (6 miles) per hour at 12:00 A.M. to 40 kilometers (25 miles) per hour at 12:00 midnight. By 6:30 P.M. on the 24th a total of 220 millimeters (9 inches) had fallen. Snow and high winds (averaging 40 kilometers or 25 miles per hour) continued through the 25th. Snow and winds tapered off on the 26th, but snowfall did not end entirely until late afternoon on the 27th. During the storm a total of only 480 millimeters (19 inches) of snow fell (390 millimeters or 15 inches during the 24th and 25th), but strong winds created extreme drifting. At the end of the storm 710 millimeters (28 inches) of snow were on the ground. Here are the notes of the weather observer as he recorded them during the storm:

Jan. 24

Special observation at noon for .16 [inches] fall in pressure in 3 hrs.

Snow became heavy during the forenoon and by evening became a blizzard. The snow was quite wet and stuck to trees and wires. By 8 p.m. streets were almost impassable. Many cars were stalled and abandoned for the night. Schools were dismissed and county pupils sent home at noon. About 5:10 p.m. the new 188-foot antenna of WBEO crashes.

Jan. 25

Blizzard continued unabated. The strong wind has made giant drifts. Auto traffic is at a standstill and pedestrians have a struggle. Snowshoes are at a premium. There have been no trains in or out all day and no deliveries of mail or food except in emergency cases. Fire, starting early this morning completely destroyed the Masonic Temple, all the offices it contained and the four first floor stores; Woolworth, Scott, Jean's Jewelry, and the Nightingale Cafe.

The three large stacks of the Big Bay Lumber Co. blew down at 12:30 a.m.

City electric service suffered several short interruptions; due to falling trees.

Many trees down all the way to Big Bay. [Marquette, Mich., 1938]

Based upon the information presented here and the findings of research studies focused upon the urban snow hazard for other areas of the United States, it would seem that Marquette may indeed have a relatively significant snow hazard (Rooney, 1967). Marquette is also subject to several other hazards but these are, for the most part, infrequent. Recently, local shoreline erosion has become a problem because of high water levels and man-created current alterations. A seiche sometimes occurs; the most recent seiche was in the summer of 1968. Only one documented case of a tornado within the city limits has been recorded (1888).

The interview sample

The urban snow hazard study in Marquette was based upon responses to an oral interview given to 120 heads of household. The interview was patterned closely after the General Questionnaire used in the Comparative Field Observations on Natural Hazards Program. The respondents were selected by: (1) random door-to-door requests, (2) personal acquaintance of the interviewer with the interviewee, and (3) random meetings with individuals who were involved in their daily activities of work or leisure. All households had at least one child below working age.

The sample was stratified into high-, middle-, and low-income levels (relative to local conditions). Income levels used for the study interviews were low, $0–5,999; middle, $6,000–8,999; and high, $9,000 or above. In spite of efforts to obtain a balance among the various income strata, the sample is skewed toward the middle- and high-income levels. The percentage distribution of interviewees at the three income levels were low, 18.6, middle, 39.8, and high, 41.5.

The hazard's impact: costs and adjustments

A concerted effort was made to provide a meaningful analysis of the economic and social costs of snow to Marquette. These efforts were not completely successful but some data were obtained through interviews and tabulated data provided by city and state officials as well as through the administration of the questionnaire.

Economic and social costs. Some individuals do not view snow exclusively as a hazard. A total of 12 interviewees (10 percent of the total) responded "No" to the question: "Do people of this place have any trouble with heavy snowfalls?" An additional 6 respondents replied "Don't know" to this question. Although they may recognize snow as a hazard in some ways, many look upon it as an aesthetic or challenging experience. Many adjustments identified such as "watch it," "sit and enjoy it," "take a walk," and "get the snow machine ready for riding" exemplify these attitudes.

Although most respondents seem not to consider snow a physically dangerous hazard, they do recognize its adverse effects in relation to their personal activity. This statement seems justified in that 43 percent perceived snow as having its main effects on their activities but only 7 percent felt that, if damages should occur as a result of heavy snow, they would be substantial. Another 91 percent indicated that damages would be only slight or nonexistent.

Interviewees were asked to estimate their annual expenditures for coping with snow. They were requested to exclude heating costs but told to include such things as snow tires, snow removal equipment and tools, clothing designed specifically for snow, extra gasoline, and other similar purchase items. Some respondents seemed not to have previously thought of snow as contributing to living costs; others appeared to recognize snow as a costly economic factor. Of those responding, expenditures for snow as a hazard ranged from an estimated low of $5 annually to a high of $1,000. The average estimate was approximately $285.

Costs for public snow plowing and removal and ice control of streets and walks represent a significant capital outlay. Marquette during fiscal year 1970–71 spent more than $160,000; this was 35 percent of total city maintenance expenditures (McNabb, 1971). In addition, over $31,000 was spent in capital outlay for snow removal equipment. As of 1971, city investment in snow removal equipment totaled $294,052, equal to 50 percent of total investment in all city equipment, including fire, police, and general maintenance vehicles and associated equipment. Other service groups such as the public schools, city hospital, and a medical center expend relatively large portions of their annual maintenance budget for snow removal. Outlays range from 27 percent of the total maintenance budget being spent for snow removal by the hospital to only 2.5 percent by Northern Michigan University. The university's figure fails to include all expenditures; therefore, it should be viewed only as indicative of total removal costs (Table 21–2).

Michigan taxpayers provide an important source of maintenance revenue for Marquette's streets. The state maintains a general motor vehicle highway fund financed by total revenues generated from annual license plate fees and gasoline taxes. Twenty percent of the fund is set aside for street repair and maintenance in cities and villages; the amount each receives is prorated by using a weighting factor based on total population. During the period 1960–70 Marquette received from this fund an annual average of $172,168 (Sherman, 1971). Because these funds are inadequate to cover all maintenance costs—including snow removal—some Upper Peninsula cities that experience especially heavy snow accumulations, prior to the Marquette study, sponsored a bill in the state legislature that would have provided additional money specifically for snow removal. The proposal was to earmark for snow maintenance 0.8 percent of the cities' 20 percent allocation of the general fund. The bill, unfortunately for Marquette taxpayers, was defeated.

The city of Marquette also receives state funds for maintaining the state trunk-line highway, U.S. 41, that runs through the central business district. Reimbursement is made by the Department of State Highways for the cost of labor, materials, and equipment used. During the 1966–71 period Marquette's average annual reimbursement was $33,700 (Dempsey, 1971). These funds are not part of the general highway fund discussed previously.

As mentioned earlier, "activity" was frequently cited as a significant area of household experience that was affected by heavy snow. This problem is closely asso-

Table 21–2. Maintenance expenditures of selected institutional groups in Marquette for the 1970–71 snow year ($)

	Total maintenance	Snow removal	Snow removal as a percent of total maintenance expenditures	Capital investment in snow removal equipment
Marquette public school system[a]	98,262	10,000	10.2	5,000
Northern Michigan University[b]	1,518,673	35,617[c]	2.5	27,500
St. Luke's Hospital and Medical Center[d]	18,500	5,000	27.0	10,000[e]
City of Marquette[f]	453,156	160,288	35.0	294,052[g]

[a]Tiziani (1971).

[b]Neumann (1971).

[c]Data do not include some expenses incurred for outside help in snow removal.

[d]Steen (1971).

[e]This figure represents the cost of installing heated sidewalks at the Marquette Medical Center; it does not include other snow removal equipment investments as data for these items were not available.

[f]McNabb (1971).

[g]Does not include capital investment in equipment that is seasonally used (such as end loaders and graders) for snow removal.

ciated with road and highway conditions. Reaching the place of work or social participation was often identified as a problem associated with snow. Failure of the individual to reach work may mean loss of income; snowbound shoppers or recreationists contribute little to the business community; empty classrooms waste the public tax dollar. Each day Marquette public schools are forced to close it costs the taxpayers $12,500. During the 1970–71 school year 5 days were lost because of snow; the cost: $62,500. From 1966 to 1971 the public schools had a total of 9 "snow days" (Tiziani, 1971).

An effort was made to determine costs of snow to individuals and to the public by evaluating several years' automobile traffic accident reports which might list snow as a "contributing" cause. This proved impossible because the city police department's reporting system does not always identify specifically the contributing causes to an accident. Similar efforts were directed toward associating automobile insurance claims with snow-caused accidents. This too proved impracticable; insurance companies contend the data would be too difficult to obtain under present claim adjustment systems.

Adjustments. The number of adjustments to snow as a hazard in Marquette, Michigan, is considerable. At times, however, evaluation of their overall relationship to snow presents difficulties. This stems from ambiguity in interviewee responses where they mention adjustments to cold-season hazards in general rather than to snow only. Snow-related adjustments only will be discussed here.

A total of 53 different adjustments were mentioned by respondents in answer to the question "When a heavy snowfall comes, what do you do?" These adjustments and the frequency with which they were mentioned are given in Table 21–3. Although there are several adjustments which were mentioned by a number of respondents, not one was mentioned by more than 26 percent of those interviewed. Furthermore, a large number of adjustments were mentioned by only one or two respondents. Many of these personal or idiosyncratic responses revealed a tendency to treat snow not as a hazard but simply as an incidental event to be disregarded for practical purposes. Table 21–4 lists an additional set of individual and community adjustments identified by the research team as practiced but not mentioned by respondents. Two adjustments are included which, although practiced in the region, are not known to be practiced in Marquette.

In an attempt to determine factors which might explain differences in the number and kind of adjustments mentioned by respondents, a set of scales was developed from the interview. These scales measured a variety of characteristics including demographic variables, perceptions of the snow hazard and its effects, and knowledge of possible adjustments and sources of relief. For each respondent, a value was derived for each of the scales.

Table 21–3. Individual adjustments mentioned in order of frequency

Adjustment	Frequency
Prepare	31
Snow tires	19
Shovel snow	18
Extra groceries	16
Clothing	15
Snow removal equipment	14
Protect	10
Check fuel supplies	7
Plow driveway	7
Emergency heating supply	5
Stay home	5
Storm windows	4
Extra gas	3
Take only necessary trips	3
Sit inside and watch it	3
Walk instead of drive	3
Things to pass time	2
Stay in	2
Wood in fireplace	2
Charge or heat car	2
Listen to radio and TV weather reports	2
Proper medicines	2
Make sure family is safe	2
Put car in garage	2
Carry on as usual	1
Take a drive	1
Recreational skiing	1
Stay home and enjoy it	1
"I'm ready"	1
Wait till it stops	1
Go to classes	1
Go to library	1
Play cards	1
Visit neighbors	1
Recover items outside	1
Find a way to work	1
Shelter for shrubs	1
No travel	1
Snow blowers used	1
Worry about roof	1
Hope I can get home	1
Get off highways	1
Take care of business	1
Shelter	1
Extra lighting material	1
Wind up windows in car	1
Bank house with snow	1
Store recreation equipment	1
Leave early for work	1
Buy a snow shovel	1
Head home early from work	1
Rearrange schedule	1
Check house	1

Each of these scales was then correlated with the number of adjustments mentioned by the respondents.

Next, each of the respondents was placed in one of five groups based upon the type of adjustments mentioned. The groups were defined as follows:

1. Persons who prepare for heavy snow (acquiring sup-

Table 21—4. Individual and community adjustments
practiced but not mentioned

Automobile

Tire chains   Sand supply in car
Colored ball or flag on radio antenna   Move car off street
Iceproof windshield wipers   Reduce driving speed
Snow-blade windshield wipers[a]   Overnight parking ban
Full tank of gasoline   Public warned to stay off roads
Back car into drive   Highway snowplow convoy for cars
Carry weight in back of car or   Highway closed to traffic
   pickup truck

Home and Other Structures

Sheet metal on edge of eaves   Metal or wooden stakes placed over
Entire sheet-metal roof   objects
Anchor bars on roof (especially   Special braces and closer placement
   above doorways)   of studs
Electric heating wires on eaves   Minimum roof pitch
Houses built high off ground   Roof guttering made removable or not used
Plank walks from street to front   Wide eaves
   porch[a]   Bracing of fence posts
                                             Heated walks

Cleanup and Control

Use of large sheet-metal or wooden   Sanding of streets
   snow scoops   Plowing of public roads and walks
Jeeps and trucks fitted with plows   Loading snow on trucks and hauling away
Chemicals used for melting snow on   Shovel off roof
   walks   Snow fences
Salting of streets

Personal

Walk in snow   Look after domestic animals and pets
Check on elderly neighbors and   Baby carriages fitted with runners
   those living alone   Tracked snow vehicles (e.g., snowmobiles)
Watch TV or do other things for   used for travel needs
   entertainment   Use skis for traveling short distances
Check on canceled events   Use snowshoes for traveling
                                             Arrange for more time needed for daily
                                                activities
                                             Administrative decisions for closures

---

[a]Not known to be practiced in Marquette.

plies of food, snow removal equipment, etc.) and who also mentioned other adjustments aimed at clearing up after snow, carrying on as usual, reducing the inconvenience and so on ($N = 32$).

2. Persons who prepare for snow as in the first group but who claim no other adjustment ($N = 20$).

3. Persons who indicate activities of clearing up after heavy snow (shoveling or plowing) but who do not prepare for heavy snow in any way. They may, however, claim other adjustments ($N = 21$).

4. Persons who mentioned only miscellaneous adjustments involving carrying on as usual, reducing the inconvenience, or others, but did not mention preparing for heavy snow or clearing up afterward ($N = 23$).

5. Persons who mentioned no adjustments whatsoever ($N = 24$).

These five groups were compared on each of the scales by means of an analysis of variance of the scale scores.

These analyses yielded the following significant results:

1. Those in adjustment group 3 were most likely to perceive heavy snow as the principal hazard of the area; those in group 5 were least likely to do so: $F(4, 59) = 2.93$, $p < 0.05$. In other words, persons who report the most physical activity resulting from snow

(mentioning plowing and shoveling) find snow the most salient local hazard.

2. Reliance on official or formal sources of storm prediction was related to both number and type of adjustment mentioned. The greater an individual's reliance on formal sources, the more adjustments he mentioned: $r = 0.22$, $p<0.05$. Also, those in adjustment groups 1 and 3 indicated the greatest reliance on formal sources while those in groups 2, 4, and 5 indicated the least: $F(4, 115) = 5.39$, $p<0.001$. That is, persons who indicated that they make the most extensive adjustments for heavy snow take most cognizance of available formal warnings.

3. The greater the degree of damage estimated as resulting from heavy snowfall, the more adjustments an individual mentioned: $r = 0.23$, $p<0.05$. In addition, respondents in groups 1, 2, and 3 estimated the greatest damage from heavy snow; those in adjustment groups 4 and 5 estimated the least: $F(4, 115) = 2.74$, $p<0.05$. In short, the more damage perceived, the more likely a person was to make several adjustments and the less likely he was to simply carry on as usual, or ignore the snowfall.

4. The more permanent or long-lived the type of effects attributed to heavy snow, the greater the number of adjustments mentioned: $r = 0.27$, $p<0.01$. Furthermore, those who prepared for heavy snow in various ways (groups 1 and 2) indicated the most permanent kinds of damage while the remaining groups claimed the least: $F(4, 103) = 3.03$, $p<0.05$.

5. Finally, the greater a respondent's knowledge of organized sources of help or relief following a heavy snowfall, the more adjustments he gave: $r = 0.28$, $p<0.01$. Also, those in adjustment group 1, who both prepare for snow and attempt to carry on in spite of it, were more familiar with organized sources of help: $F(4, 115) = 4.59$, $p<0.01$.

In short, there is some question of how seriously the urban snow hazard is regarded in Marquette. Many people simply do not think of it as a hazard. Further, many respondents indicated they do nothing special in response to heavy snow or they may simply do what is necessary to carry on as usual. The evidence for this conclusion may be deceptive, however. A wide range of adjustments was found and many of these are more frequently practiced than was apparent from responses to the questionnaire. In addition, many adjustments have become so routinized that, although they are known to be widely used in Marquette, no one thought to mention them. It appears that heavy snow is a frequent enough occurrence in Marquette that many people have simply adapted and their responses indicate adaptations as a part of their normal activities.

It is important to note, however, that both the number and kind of adjustments which were mentioned are related to a variety of other factors. These include perceptions of snow as a hazard, reliance on formal storm prediction, the kind and extent of damage observed, and knowledge of organized sources of help and relief following heavy snowfall.

Acknowledgments

Investigators in the original data collection program included a basic research team of Fillmore C. F. Earney (director), William Hanson, Richard Kierzek, and Marvin Ruspakka. In addition, some interviews were made by Joe Riepe, Daniel Lehman, and Martin Eskelinen. The preliminary draft of Fig. 21–1 was made by Robert Sommerwell, Cartographic Laboratory, Northern Michigan Univ.

References

Dempsey, John T. (1971) Director, Bureau of Programs and Budget, State of Michigan, Lansing. Letter, August 13.

Marquette, Mich. (1938) Daily weather observation records.

McNabb, Thomas R. (1971) City Manager, Marquette, Mich. Interviews, June 28, August 3.

Mueller, Fred. (1971) Chief Meteorologist, National Weather Service, Marquette, Mich. Interview, August 5.

Neumann, Edward L. (1971) Chief Engineer of Operations and Maintenance, Northern Michigan University, Marquette, Mich. Interview, August 2.

Rooney, John F. (1967) "The urban snow hazard in the United States: an appraisal of disruption." *Geographical Review* 57:538–59.

Sherman, A. C. (1971) Engineer of Local Government, Department of Highways, State of Michigan, Lansing. Letter, August 17.

Steen, Duane J. (1971) Comptroller for St. Luke's Hospital and the Marquette Medical Center, Marquette, Mich. Interview, August 2.

Tiziani, Julius J. (1971) Assistant Business Manager, Marquette Public Schools, Marquette, Mich. Interview, August 2.

22. Avalanche problems in Norway

GUNNAR RAMSLI
University of Oslo

Among the various forms of rapid mass movements, snow avalanches are those that cause the most damage and accidents in Norway. The number of deaths caused by avalanches between 1836 and the present day is approximately 1,600 or an average of 12 per year. Switzerland, with the highest accident rate of this type, has by comparison an average of 25 per year. It should, however, be noted that the greater share of avalanche accidents in Switzerland involves skiers, whereas in Norway the permanent local population is mainly affected.

Losses of life and property

The worst known year for avalanche accidents in Norway was 1679, when there is reason to believe that between 400 and 500 people died. The year 1755 is also notable, when presumably 200 deaths occurred. In 1886, snow avalanches claimed 161 lives. Other years with relatively high death rates were 1880 with 20 victims, 1881 with 60, 1895 with 24, 1906 with 29, and 1918 and 1919 with 29 and 31 respectively. More recently, 30 people were killed in the winter of 1955–56.

These figures do not, however, indicate the extent of other damage: permanent injuries to people, loss of cattle, houses completely or partly destroyed, damage to agricultural land and forest, inconvenience to road and rail traffic and to industry and installations in danger areas.

Some idea of how much avalanches cost Norway in

Fig. 22–1. Example of an avalanche map, 1953

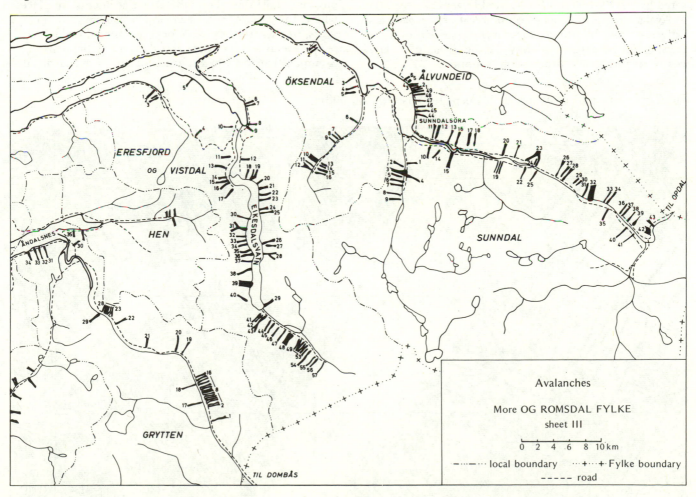

monetary terms can be gained from the annual reports of the State Fund for Natural Hazards for the period 1962–71. During these years, material damage to private property was valued at a total of approximately N.kr. 8.6 million (U.S. $1.6 million). At the same time, approximately N.kr. 4.2 million (U.S. $600,000) was contributed to protective measures, mostly for private property.

Additional damage to and protection of public property, larger industries, and other installations which are not eligible for compensation under the rules of the fund is not accounted for by the annual reports. The Department of Roads, in particular, has spent large sums on clearing roads blocked by avalanches, and on protective measures in the form of tunnels and snowsheds. Repairs and protective measures constitute a considerable economic burden to industry and other installations. In the course of a 10-year period one mining company in north Norway had an outlay of N.kr. 5–6 million (U.S. $0.9–1.1 million) in connection with avalanche problems.

It is possible, but not usual, to insure against avalanche damage. Nor is insurance necessary for private property, as damage will, for the most part, be compensated by the State Fund for Natural Hazards.

Avalanches also hinder communications. Roads and railways may be closed for varying periods of time, causing great cost to society in the form of delays, costly rerouting, and production stoppages.

Psychological aspects

In addition to the economic aspects of the problem, many people live under considerable psychological pressure caused by avalanche threats during parts of the winter. In many cases, people have had no alternative but to evacuate their homes during the most dangerous periods. In some areas, children are sent away from home, while the grown-ups remain as long as possible to do the daily work. When an accident is first reported, anxiety and fear often spread through several districts. Those who live or travel beneath steep snow-laden slopes live daily in fear of their lives.

Distribution and occurrence

Snow avalanches are most widespread in west and north Norway, where the fjords and valleys are steep-sided. The timberline is low, precipitation is high, and conditions of temperature and wind fluctuate often throughout the winter. In more central areas, the landscape is less steep, precipitation is less, the timberlines are higher, and meteorological conditions are more stable.

The "occasional" avalanche is often the most dangerous and can bring about the worst accidents. In places where there have not been any avalanches for many years, but where they were once an annual occurrence, people tend to overlook the danger. It happens, for example, that areas in the track of an old avalanche are

Fig. 22–2. Example of an avalanche map, 1971 (scale 1:50,000)

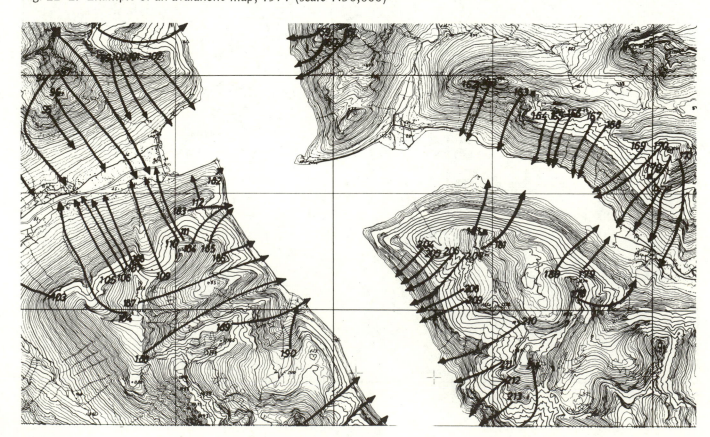

developed in the belief that avalanches have ceased there. In reality, any winter with unusual snow, wind, and temperature conditions provides the perfect situation for an accident.

Mapping of hazard areas

It is of great importance to register and map avalanches. Between 1948 and 1960 mapping was partly organized under the auspices of the Ministry of Agriculture. The purpose of this investigation of avalanches in Norway was to decide what preventive measures could be taken to reduce damage and accidents. During that time, approximately 1,500 avalanches were mapped and recorded. Attention was directed primarily to west Norway, but information was also gathered from north and south Norway. Usually, only the most important avalanches were taken into account, especially those which presented a threat to settlement, agricultural land, and communications (Fig. 22–1). As mapping proceeded, as

much information as possible was gathered on each individual avalanche. This included periodicity, favorable weather conditions, damage, and inconvenience caused. Emphasis was placed, furthermore, on previous accidents and damage; and in many cases the terrain was described.

Mapping was resumed in 1970 at the Department of Geography, University of Oslo, and is partly financed by the State Fund for Natural Hazards, and partly by interested local authorities. The avalanches are drawn on the maps at the scale of 1:50,000 (Fig. 22–2) and on economic maps at 1:5,000 (Fig. 22–3).

These maps together with the additional information are, in many cases, of value in areal planning and the choice of new lines of communication or the rerouting of old roads. Unfortunate and sometimes dangerous location of housing, industry, and installations may thereby be avoided; and in road planning it is possible in the preparatory stage to give far more attention than before to avalanche problems and associated inconveniences.

Fig. 22–3. Example of a large-scale avalanche map, 1971. These maps are drawn at the scale of 1:5,000. The figure, however, is reduced

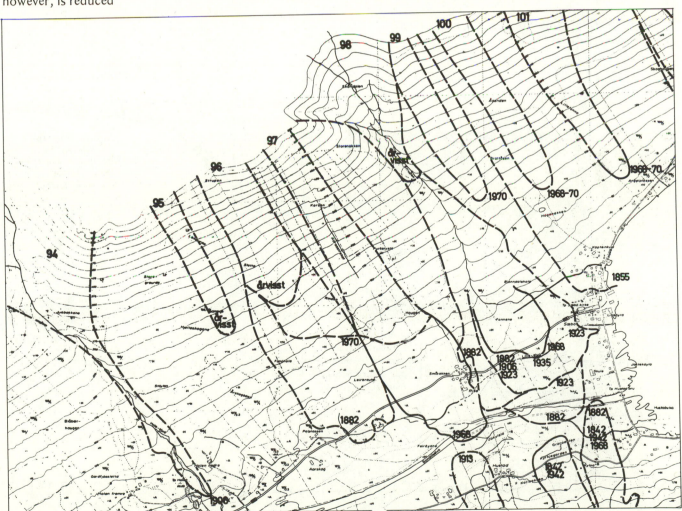

Public regulation of the areas mapped is prescribed by the Statute on Building Development. This allows building only when there exists sufficient security against snow avalanches, earth- and rockslides, and floods. Where such security does not exist, the law demands that threatened areas be classified as dangerous in area development plans.

The maps will, in themselves, constitute an indirect protective measure. Mapping, however, does not solve all problems. In many cases, protective installations will continue to be necessary, and presumably to a greater extent than at present.

Protective measures

Throughout history, the population in localities prone to avalanches has attempted to protect itself. In many places along the fjords and valleys of west Norway it is possible to find houses built beneath rocky protections, or in the lee or on top of natural rises in the ground. The reasons for such locations may, of course, vary; but in many cases, the sites were chosen out of consideration for avalanche danger.

In places where natural obstacles are insufficient, protective installations may be used (Fig. 22–4). These may be divided into three categories: (1) installations in the depositional area or track of the avalanche, (2) installations in the rupture zone of the avalanche, and (3) installations to the windward of the rupture zone. The purpose of the first category is either to stop the snow masses before they reach the threatened area, or to deflect them or carry them over the exposed objects. The second and third categories are intended to prevent the formation of avalanches either directly or indirectly.

Once an avalanche has begun to move, it is usually impossible to stop it in its tracks. The only solution liable to succeed is to dam up the snow when it has reached the valley floor and has lost much of its velocity. This may be effected by transverse walls or so-called arresting dams.

Installations intended to divide or divert the snow masses are of more practical importance. The simplest forms are earth or stone heaps on the exposed side of the object requiring protection. Deflecting walls or splitting wedges are a further development. By making the

Fig. 22–4. Methods and types of defense works.

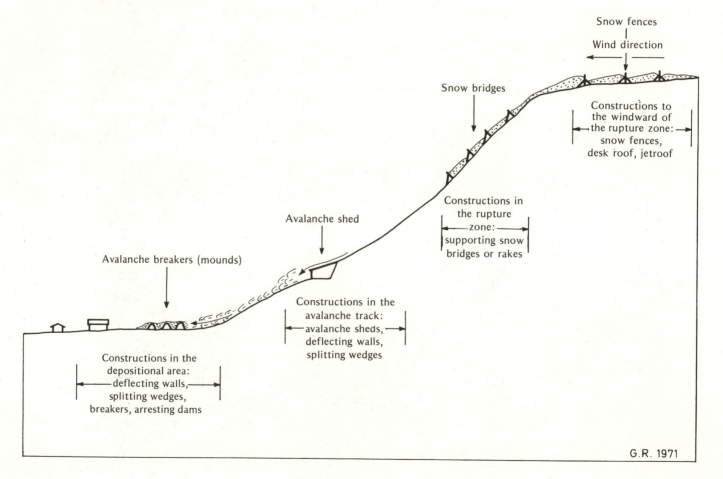

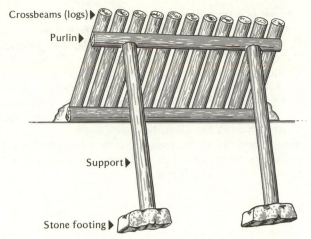

Simple snow rake (cross-beams upright) built from round timber

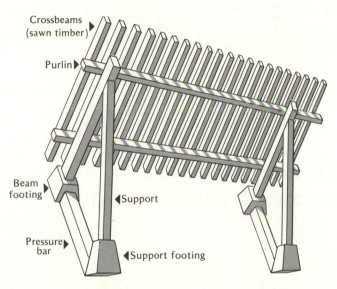

Snow rake built from sawn timber on a concrete foundation

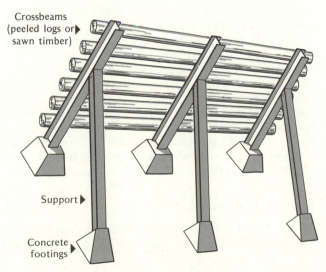

Snow bridge (crossbeams horizontal) with concrete foundation, steel frame, and timber crossbeams

wall wedge- or plow-shaped with its apex toward the avalanche, the efficiency of the installation is increased, and the snow masses slide along its sides.

In more recent years, the principle of dividing the snow masses after they have reached more gentle slopes has been used in the construction of so-called avalanche breakers. The method involves the erection of several earth-splitting wedges or mounds in the same area and according to a certain system. In this way, the snow masses are progressively divided and lose much of their energy, so that the avalanche stops prematurely.

If the exposed object lies in the periphery of an avalanche, one or several deflecting walls will often provide effective protection. The purpose of these is to confine the avalanche to certain limits, and direct the snow in the most suitable direction. A combination of deflection walls and breakers is often advantageous.

In the case of roads and railways, where it is not possible to build tunnels, it is usual to construct avalanche sheds. Such installations, of course, can be used to protect other objects, e.g., buildings which for some reason must be sited in dangerous places.

In Norway, the most common types of installation used in the rupture zone are steel and wooden fences erected along the valley side. In a few cases, aluminum is also used. (See Fig. 22–5.)

Drifted snow constitutes the greater part of the snow masses in the rupture zones of many avalanches. The danger may therefore be considerably reduced in many places by partial or complete prevention of this type of accumulation. The feasibility of this is, however, strongly dependent upon local ground and wind conditions. As a rule, an effective solution will only be possible in localities where the snow is supplied to the rupture zone from relatively wide, flat plateaus, and where the main wind direction is fairly constant. Ordinary snow fences on the plateau to the windward of the rupture zone often give good results. In many areas of Norway conditions are well suited to this method.

Finally it may be mentioned that explosives have also been used as a protective device, either hand-placed or fired from guns. This method has mainly been used to release snow slabs and cornices along roads, to ensure that the snow can be taken down under full control before the snow accumulation in the rupture zone becomes excessive.

Until about 20 years ago little had been done in Norway on the technical aspects of protective installations. Of course, a number of old splitting wedges and deflecting walls existed, and some snowsheds were used on railways and roads. In later years, however, a considerable number of protective measures have been taken, and in most cases with success. Practically all known methods and types of installations have been tried. Work has been concentrated in west and north Norway, and in several places larger new projects are being undertaken or are in the planning stage.

Fig. 22–5. Supporting structures for avalanche starting zones. Source: Mellor (1968)

Some of the protective installations are very expensive. For avalanche sheds in concrete the price per meter (yard) is approximately N. kr. 10,000 (U.S. $1,800), and for snow bridges in aluminum, approximately N. kr. 2,500 (U.S. $450). For this reason, constructions in the rupture zone are only used to protect major investments, such as, for example, villages and industrial plants, while separate houses or small farms will be protected by more simple constructions, such as deflection walls, splitting wedges and earth mounds.

Coordination of public action

Experience has shown that the possibilities of preventing or reducing avalanche damage are excellent. Much can be done through research, mapping, public information, and direct protective measures. Until recently, these activities have been organized, for the most part, by the State Fund for Natural Hazards in cooperation with the Norwegian Snow Research Council (a body consisting of representatives from the Norwegian State Railways, the Department of Roads, the Norwegian Water and Electricity Board, and the State Fund of Natural Hazards). In December 1972, however, a recommendation by a government commission to transfer the work to the Norwegian Geotechnical Institute was passed by Parliament. This institute was already engaged in similar work on earthslides and rock avalanches; and now, since the recommendation has been followed, all problems connected with rapid mass movements will be handled by one institution.

References

"Lawinenschutz in der Schweiz." (1972) *Bunderwald, Zeitschrift des Bundnerischen Forstvereins und der Selva*, Chur, Special issue No. 9.

Mellor, Malcolm. (1968) *Avalanches, Cold Regions Science and Engineering.* Hanover, N.H.: U.S. Army Material Command.

23. Problems in the use of a standardized questionnaire for cross-cultural research on perception of natural hazards

THOMAS F. SAARINEN
University of Arizona

Collaborators in the current cross-cultural study of natural hazards are participating in an interesting test of methodology. Their studies should indicate whether a standardized questionnaire which has evolved over time in a series of studies in North America can be effectively transferred to a much broader range of study sites around the world. Given such a test it seems appropriate that at least one paper should be devoted to an examination of what happened to the questionnaire (see p. 6) as it was transferred to other areas and to the investigation of a broad range of hazards.

In a project of such scope there are many possible sources of error. These were compounded by limitations in time and money and the necessity of cooperation with many collaborators who were engaging in such survey research for the first time. Besides the usual problems which result from inexperienced interviewers and a limited number of pilot studies, there were the additional ones of adapting the questionnaire to other cultures with specific local problems and translating the questions into other languages. All of these problems will not be investigated here.

The present study is limited to an examination of what happens to the individual questions as the questionnaire is adapted for use in a new area. The method adopted is the question-by-question examination of a series of 20 adapted questionnaires obtained from a variety of collaborators in many different world areas (Table 23-1). The questionnaires analyzed are those which were sent to the author following a request to all participants for a copy of the final form used in their survey. In addition comments on why certain questions were included, modified, or omitted were solicited. Final sources of data were interviews with a number of collaborators in sites outside of North America.

Selection of questions

The methods employed lead to certain limitations for the study. It is assumed that questions were modified

Table 23–1. Questionnaires analyzed listed according to hazard, area, and investigator

Hazard	Area	Investigator
Air pollution	Illinois	N. Moline
	Sheffield, England	G. Wall
	Edinburgh	D. Billingsley
	Ljubljana, Yugoslavia	D. Kromm and
		S. Vizjak
	Christchurch,	
	New Zealand	J. Hay and
		R. Johnston
	South Yorkshire, England	G. Wall
	Budapest	F. Probald
Drought	Northeast Brazil	R. Brooks
	Northwest Nigeria	W. Roder
	Sukumaland, Tanzania	R. Kates and
		T. Hankins
	Yucatan, Mexico	C. Parra
Earthquake	Huaylas, Peru	C. Penaherrera
	Cornwall, Canada	E. Jackson
	San Francisco	E. Jackson
Flood	Kyoto	M. Kusaka
	Osaka	M. Oya
	Illinois	N. Moline
Frost	Utah	R. Jackson
Volcano	Hawaii	B. Murton and
		S. Shimabukuro
Weather		
(to agriculture)	Florida	R. Ward

and omitted for good reasons and those which remained were perceived as useful at least by the director of the local survey. However, little information was available on the quality of replies elicited by the questions finally selected. When such information was provided in comments it was used but unfortunately not all who sent in questionnaires included comments. Thus the ideas developed here would have to be weighed against the results to further probe the strengths and weaknesses of the questions. Neither sampling methods nor any of the other pitfalls of survey research were considered. Not even the sample of questionnaires used was systematic but it does include a majority of those used in the study. A weakness is the large number derived from air pollution surveys, and pains were taken to try to avoid giving them disproportionate weight. Another way to approach the problem of assessing the strengths and weaknesses of the questionnaire would be to analyze the types of questions added. However, time limitations ruled it out in this study.

Table 23–2 lists all the questions in the general questionnaire indicating the proportions which were included verbatim, modified somewhat in wording, and omitted entirely in the 20 questionnaires analyzed. The percentages can be considered as a first rough guide to items which could pose problems in adaptation of the questionnaire for surveys of many different hazards in various areas of the world. For the convenience of the reader the questions are grouped into four main cate-

gories in Table 23–3, which takes into account the proportion of surveys using each question unchanged and the degree of modification of those changed, and corrects for the excessively strong weight of the 7 questionnaires on perception of air pollution. Certain questions were omitted from all 7 of these surveys and the unadjusted percentages of Table 23–1 tend to exaggerate this effect. Each of the four groups of questions is discussed in turn.

Generally noncontroversial

Questions found in group 1 were all used in practically every survey with only minor modifications to fit the particular contexts. It is not surprising that many· of these pose no problems since the preliminary identifying characteristics and final comments may be provided directly by the interviewer regardless of the reaction of the respondent. They are strictly factual, simple, noncontroversial data. Those that require the cooperation of the respondent tend to be less universally applied in the original form. Thus the sometimes touchy question of religion was omitted in some cases, as in the two surveys in the Communist bloc countries; and the items on occupation and educational level were considerably modified to fit local circumstances. The item on predominant language spoken was often omitted in surveys conducted in monolingual areas where the answer was obvious. In several cases, the background information was solicited after the main body of questions. With this type of question, such a change in ordering is not serious. One can conclude that in cross-cultural surveys no difficulty is perceived in noting the date and the location of the interview and obtaining such simple facts as the sex, age (the ages of children are not always accurately recalled in certain nonliterate societies but in such cases they can be estimated close enough for the purposes of analysis), occupation, and general educational level of the respondent. However, these data, while essential for full analysis, are not the most crucial for furthering our understanding of hazard perception so they will be dismissed from further discussion.

Minor problems

The second group of questions, though less universally applied than the preceding group, were still retained in unchanged or slightly modified form in the great majority of surveys. They generally appeared in two-thirds or more of the questionnaires. These items can be characterized as simple, factual, relevant, short-answer questions which can be handled by the respondent with a minimum of thought or reflection. Thus, for example, questions 21 and 22 ask for the worst and most recent occurrences of the hazard in question (these questions worked well where hazards were discrete events but were not as useful in the case of air pollution where the problem is more continuous in nature), items 18 and 33

Table 23–2. Fate of general questionnaire items in later surveys (N = 20)

Question number	% included verbatim	% modified	% omitted	Question number	% included verbatim	% modified	% omitted
1	95	0	5	27	45	35	20
2	100	0	0	28[a]	45	25	30
3	100	0	0	29[a]	50	5	45
4	95	0	5	30[a]	50	5	45
5	55	0	45	31[a]	55	5	40
6	95	0	5	32[a]	45	5	55
7	95	0	5	33[a]	65	0	35
8	75	0	25	34[a]	40	5	55
9	60	40	0	35(1)	65	35	0
10	75	25	0	35(2)	80	10	10
11	65	5	30	36[a]	60	10	30
12	65	5	30	37	45	15	40
13	95	0	5	38(1)[a]	25	15	60
14	95	5	0	38(2)[a]	25	10	65
15	40	55	5	39[a]	60	5	35
16	40	25	35	40[a]	35	20	45
17	60	35	5	41[a]	50	10	40
18	70	25	5	42	45	15	40
19[a]	70	25	5	Comment 1	80	5	15
19[b]	45	20	35	Comment 2	85	0	15
19c	60	30	10	Comment 3	90	0	10
20[a]	40	40	20	1[b]	65	10	25
21	70	5	25	2[b]	40	30	30
22	80	0	20	3[b]	35	15	50
23[a]	50	40	10	4[b]	50	0	50
24[a]	30	15	55	5[b]	45	5	50
25[a]	30	0	60	6[b]	40	25	35
26	85	5	10	7[b]	55	15	30
				8[b]	60	0	40
				9[b]	60	0	40
				10[b]	65	5	30
				11[b]	60	10	30

[a]Omitted or modified in the 7 questionnaires on perception of air pollution.
[b]S.C.T. = sentence completion test.

Table 23–3. Classification of questions according to adaptability for use in multihazard cross-cultural context

Group 1: No problems (N = 17)	Group 2: Minor problems (N = 13)	Group 3: Some difficulties (N = 14)	Group 4: Most difficulties (N = 14)
1, 2, 3, 4, 6, 7, 8, 9, 10, 11, 12, 13, 14, 15; interviewer comments 1, 2, 3	18, 19a, 21, 22, 26, 27, 30, 31, 32, 33, 35; sentence completion test items 1, 10	5, 17, 19b, 19c, 23, 34, 36, 39, 41; sentence completion test items 4, 7, 8, 9, 11	16, 20, 24, 25, 28, 29, 37, 38, 40, 42; sentence completion test items 2, 3, 5, 6

may be answered by a simple yes or no, and items 27, 30, 31, and 32 require only a few words in reply. Somewhat more complex were items 19(a), 26, and 35, which are considered in more detail. The only other items included in group 2 were numbers 1 and 10 of the sentence completion test. It is clear that these two items are the most obviously relevant of the sentence comple-

tion stems for the subject of hazard perception and thus the ones most frequently selected for inclusion in the surveys.

Item 19(a) is a more open-ended question than most in group 2. However, since most of the respondents have experienced the problems involved the answers should be, essentially, straightforward factual accounts of their previous actions. The hypothetical wording of the question led to problems in some areas where the majority of respondents were illiterate and not accustomed to such modes of thought. In the drought surveys in Africa the wording was modified and specific questions added to probe more directly the exact effects of the hazard on the individual household. An additional source of confusion was introduced by the b and c portions of the question, whose purposes were not always understood particularly by less experienced interviewers.

Item 26 is a multiple-choice question more complex than most of those in group 2. Perhaps it was so often included because of its easy anecdotal style and simple selection of an answer. But due to its complexity and hypothetical character it was not always successful. Again negative comments were most frequent in areas where most respondents were illiterate or generally lacking in formal education. In such areas the question was sometimes described as too complicated.

Question 35 consists of two parts. The first asks the adjustments made in response to the hazard in an open-ended way while the second part provides the respondent with a checklist from which he selects the adjustments he has used. Both received favorable comments as very good questions and since they were also included in almost all the questionnaires one might conclude that they are among the most successful items. Since the information they elicit is very important in understanding how people perceive and react to various hazards this is good news. However, even here a problem of comparability arises, for in several surveys the wording was changed and in some cases the checklist was used alone since it seemed simpler and easier to administer. The result in such cases is that the respondent may have been guided toward certain choices while other possible ones were omitted. However, if the checklist is exhaustive only a reinforcement of the answer may have been lost. Part of the problem stems from the complicated nature of the table suggested in the standard questionnaire for recording the responses; this will be commented on further.

Significant difficulties

The third group of questions differs considerably from the preceding ones. Judged by the hard criterion of inclusion in an unaltered form, these questions are less successful. Although all were retained essentially unaltered in at least half of the surveys, none was included verbatim in as many as two-thirds of the questionnaires. Clearly they did not as successfully slip by the scrutiny of those adapting the questionnaire as the previous items. Why? The questions in group 3 are generally harder to answer than those already considered and this might militate against their inclusion. They are generally open-ended or require evaluative responses or expressions of opinion. Neither the interviewer nor the respondent can be sure what the correct answer is and the

resulting uncertainty makes these less comfortable questions. Items 17, 34, and 41, for example, resemble item 19(a). They are all open-ended but the items in group 3 are somewhat vaguer, more open, and less likely to have obvious answers. Questions 19(b), 19(c), 36, and 39 all require the respondent to make judgments on a series of alternatives. This is considerably more difficult than merely listing categories when an evaluation is required. The same is true of question 23 and the first part of 39, which demand opinions on topics about which the respondent presumably should know something. A special case is item 5, the only introductory question which was not included in group 1. It may have been often left out because of a lack of basic information on the physical properties of various hazards. The bulk of the sentence completion stems are included in group 3 but discussion will be delayed until later when the special problems of this type of question are considered.

Major difficulties

The final group of questions were not included in an unaltered form in at least half of the surveys. No doubt a wide variety of reasons are needed to fully explain why. Only the most obvious will be considered here. Group 4 includes the most complex and ambiguous questions, many seemingly redundant, and others requiring a long-range numerical memory as well as a broad perspective to answer accurately. The latter qualities may be present in the potential respondents but those adapting the questionnaires often indicated their doubt by leaving out the questions which might prove it. A certain redundancy is seen in item 16. It asks essentially the same thing as the more successful question 17 which tended to be retained when one of the two was dropped. Similarly, two of the least satisfactory of the sentence completion stems, numbers 3 and 6, also seem redundant. Items 20 and 29 demand numerical answers or even the division of a total series into proportions. Such questions were described by one investigator as "hopeless" for areas where the bulk of the respondents are illiterate. His sentiment was seconded by several others in similar areas who either modified such questions drastically, omitted them, or found them unsatisfactory. Apparently investigators balk at the prospect of asking questions about the respondent's impressions of other people's actions or other places, for questions 24, 25, and 38, which seek such information, are among the least used items from the general questionnaire. Furthermore, the interviewers' hesitation may be well founded, for among those who did use the questions there were many negative comments. The term "other places" was too vague for some investigators.

The most frequently omitted question was also the most complicated one. The prize for the most frequently omitted item would have to be awarded to the second part of item 38. This should not come as a surprise, for it is nearly the last item in a long series and seems redundant since preceding parts are identical except for the substitution of a word or two. The total set begins with question 35, already discussed, and continues through question 39. The series appears to be too complicated in format for interviewers and respondents alike. It was described as "repetitious," "too long," "tedious," "irritating to respondents," and "too complicated" by various investigators. The complexity led to various modifications. Some surveys used only a checklist to investigate adjustments to the hazards. Others opted to combine questions 35 and 38 so that the adjustments of the respondent plus those of his friends and neighbors were considered at the same time. With such modifications the results were described as interesting and useful.

The remaining questions in group 4 are stems 2 and 5 of the sentence completion test and items 40 and 42. Item 40 is an example of a question which does not apply to the full range of hazards. It seems most useful in agricultural areas and for drought or other fairly frequent weather-related events but is less clearly relevant for such hazards as air pollution, earthquakes, and floods. Another problem which makes it less satisfactory is the subsections requiring numerical estimates or proportions. Question 42 is a special case. Asking people their income is never easy and many surveys omitted the question, relying instead on the interviewer's estimate which was included in the final comments.

Sentence stems

The purpose of the sentence completion test was not well understood by the interviewers. Less than half of the surveys retained it in an unaltered form. Five omitted it entirely. In some cases it was placed in a different position in the sequence of questions, which probably would change the endings considerably, especially when the test followed exhaustive questioning on various aspects of the hazards. This ordering problem is one of the hazards of having surveys directed by people inexperienced in questionnaire construction. Another problem not really considered here should at least be mentioned; that is, the difficulty of translating various terms into other languages. Many Japanese geographers, for example, found it exceedingly difficult to translate exactly several stems of the sentence completion test since such basic concepts as "bad," "luck," and "emotion" have different connotations in English and Japanese. In general, the shorter, more concrete and direct stems were preferred. The longer, less specific ones and those centering on emotions and feelings were more frequently omitted. The sentence completion test is, of course, the most experimental portion of the questionnaire as there are many difficult problems of interpretation even after

the responses have been collected. Some of these are discussed elsewhere. They are beyond the scope of this paper.

Conclusion

Having examined which questions from the general questionnaire were retained, modified, and omitted it is now time to come to a few general conclusions. It is clear that the most simple direct and concrete questions are most easily adaptable to all surveys. As one increasingly probes into people's mental processes the degree of success diminishes. Thus, the open-ended and opinion questions are less likely to be used unaltered than the simple factual types. Hypothetical questions and those requiring precise numerical answers were of limited utility in areas where most of the respondents were illiterate. Complex or ambiguous questions lead to problems. The extra precision which is the aim of similar but slightly different questions seems to jeopardize the broader aim. The intention is not always clear and misunderstanding leads to major modifications which reduce comparability.

It is probably better for the time being at least to seek the broad though rough comparisons. There is a dilemma here, for the most interesting questions are often indirect and open to misunderstanding. While a wide range of information is obtainable through simple short-answer questions they have the disadvantage of depending entirely on the researcher's ideas and provide no means of tapping the cognitive world of the respondent on his own terms.

This, in the final analysis, is the major weakness of using any formal questionnaire for cross-cultural research. For the conceptual ordering of the data remains in the terms of the researcher which may bear little relationship to the respondent's structuring of the world into significant and insignificant events.[1] The present analysis cannot assess how successfully the questions did the job they were designed for. This depends on the analysis of the content of the answers. However, it is to be hoped that a greater awareness of some of the weaknesses of the individual questions noted here may aid in that final analysis and in the future design of similar studies.

References

Kirkby, Anne V. (1972) "Perception of air pollution as a hazard and individual adjustment to it in three British cities." Paper No. 39 presented at the 22d International Geographical Congress, Calgary, Canada.

1. This argument is used with great force by researchers in the new field of cognitive anthropology. See for example the introductory book of readings, Stephen A. Tyler, ed., *Cognitive Anthropology*, New York: Holt, Rinehart & Winston, 1969.

III Decision processes

24. Decision processes, rationality, and adjustment to natural hazards

PAUL SLOVIC
Oregon Research Institute

HOWARD KUNREUTHER
University of Pennsylvania

GILBERT F. WHITE
University of Colorado

The distress and disruption caused by extreme natural events has stimulated considerable interest in understanding and improving the decision-making processes that determine a manager's adjustment to natural hazards. Technological solutions to the problem of coping with hazards have typically been justified by a computation of benefits and costs that assume the people involved will behave in what the policy maker considers to be an economically rational way. However, it has slowly become evident that technological solutions, by themselves, are inadequate without knowledge of how they will affect decision making. In reviewing the wide range of adjustments to Gangetic floods or Nigerian drought or Norwegian avalanche, it has been observed that attempts to control nature and determine government policy will not succeed without a better understanding of the interplay among psychological, economic, and environmental factors as they determine the adjustment process.

Throughout the paper we shall use the terms "decision maker" and "manager" synonymously. Managers are defined as individuals who act in the management of an establishment. An establishment is defined as a single residence, agricultural, commercial, or public-service organization that has distinct use of an area. The term "adjustments" refers to the many courses of action available to the manager for coping with natural hazards. For example, in the case of floods, potential adjustments include bearing the loss, insurance, land elevation, structural works (such as dams), and public relief.

By assuming that managers are rational and that they act according to the same decision criteria that public agencies prescribe, government programs to reduce hazards have been based upon predictions that often failed to materialize. These failures have been attributed to the ignorance and irrationality of the occupants of a hazard zone; but recent work suggests that adjustments to hazards may be understandable and, in a sense, reasonable, within the framework of decision models different from the traditional optimization models. One such model is that of "bounded rationality," which takes into account limitations of the decision maker's perceptual and cognitive capabilities. The need for an improved understanding of the decision-making process is urgent, and is at the heart of systematic improvement of public policy. Such improvement becomes increasingly significant as man intervenes still further in natural processes and thereby opens himself to further hazard from their variability and uncertainty.

Aims and organization of the paper

This paper focuses on cognitive elements of decision making under risk that are important for understanding adjustment to natural hazards in a modern technological society. It includes such topics as human understanding of probabilistic events, perception of hazards, and the processes involved in balancing risks and benefits when choosing among alternative modes of adjustment to hazard. The phenomena to be reviewed are, for the most part, likely to generalize across cultures and across individuals and are likely to increase understanding of adjustments to man-made hazards as well as natural ones. This cognitive emphasis is not meant to deny the obvious importance of personality, cultural, and social factors in determining adjustments to natural hazards. The influence of culturally ingrained attitudes toward nature and toward fate is illustrated by Burton and Kates (1964), Baumann and Sims (1972), Parra (1971), and others. The effects of community organization are reviewed by Barton (1970). Individual personality factors that influence adjustment to hazards are discussed by Schiff (1970) and Burton (1972). The reader interested in a model of the interrelationships among cognitive, personality, and cultural factors in the context of adjustment to natural hazards should see Kates (1970).

The organization of the chapter is as follows. It begins with a brief overview of the leading normative and descriptive theories of decision making. Particular emphasis is given to a comparison between a decision theory which espouses maximization of expected utility as a normative guideline and a conceptualization of bounded rationality which has both normative and descriptive intent. The next section presents evidence from the psychological laboratory and data from field observations of adjustment to natural hazards to document the usefulness of the notion of bounded rationality as a framework for understanding adjustment to hazards. Whenever possible, related data from laboratory and field are juxtaposed to highlight the generality and importance of these phenomena. The picture that emerges from this work illustrates some rather startling limitations in the ability of the decision maker to think in probabilistic terms and to bring relevant information to bear on his judgments. However, the knowledge gained about human cognitive limitations has implications for improving the decision-making process. These implications are discussed in the latter part of the chapter.

Theories of decision making under risk

Maximization of expected utility

The objective of decision theory is to provide a rationale for making wise decisions under conditions of risk and uncertainty. It is normative in intent, concerned with prescribing the course of action that will conform most fully to the decision maker's own goals, expectations, and values. Since good expositions of decision theory are available elsewhere (Coombs, Dawes, and Tversky, 1970; Dillon, 1971; Luce and Raiffa, 1957; Savage, 1954), the coverage here will be quite brief.

Decisons under uncertainty are typically represented by a payoff matrix, in which the rows correspond to alternative acts that the decision maker can select and the columns correspond to possible states of nature. In the cells of the payoff matrix are one set of consequences contingent upon the joint occurrence of a decision and a state of nature. A simple illustration for a traveler is given in Table 24–1.

Since it is impossible to make a decision that will turn out best in any eventuality, decision theorists view choice alternatives as gambles and try to choose according to the "best bet." In 1738, Bernoulli defined the notion of a best bet as one that maximizes the "expected utility" of the decision. That is, it maximizes the quantity

$$EU(A) = \sum_{i=1}^{n} P(E_i)\, U(X_i)$$

where $EU(A)$ represents the expected utility of a course of action which has consequences $X_1, X_2, \ldots, X_n$ depending on events $E_1, E_2, \ldots, E_n$, $P(E_i)$ represents the probability of the ith outcome of that action, and $U(X_i)$ represents the subjective value or utility of that outcome. If we assume that the parenthesized values in the cells of Table 24–1 represent the traveler's utilities for the various consequences, and if the probability of sun and rain are taken to be 0.6 and 0.4, respectively, we can compute the expected utility for each action as follows:

$$EU(A_1) = 0.6(+1) + 0.4\,(+1) = 1.0$$
$$EU(A_2) = 0.6(+2) + 0.4(0) = 1.2$$

In this situation, leaving the umbrella has greater expected utility than taking it along. The same form of analysis can be applied to computing the expected utility of heeding a flood warning in Shrewsbury (chap. 6 above), planting drought-resistent maize in Kenya (chap. 11), or protecting against frost in Florida (chap. 17).

A major advance in decision theory came when von Neumann and Morgenstern (1947) developed a formal justification for the expected utility criterion. They showed that, if an individual's preferences satisfied certain basic axioms of rational behavior, then his decisions could be described as the maximization of expected utility. Savage (1954) later generalized the theory to

Table 24–1. Example of a payoff matrix

		State of nature	
		sun (E_1)	rain (E_2)
	A_1 carry umbrella	(+1) stay dry carrying umbrella	(+1) stay dry carrying umbrella
Alternatives			
	A_2 leave umbrella	(+2) dry and unburdened	(0) wet and unburdened

allow the $P(E_i)$ values to represent subjective or personal probabilities.

Maximization of expected utility commands respect as a guideline for wise behavior because it is deduced from axiomatic principles that presumably would be accepted by any rational man. One such principle, that of *transitivity*, is usually defined on outcomes but applies equally well to actions or probabilities. It asserts that, if a decision maker prefers outcome A to outcome B and outcome B to outcome C, it would be irrational for him to prefer outcome C to outcome A. Any individual who is deliberately and systematically intransitive can be used as a "money pump." You can say to him: "I'll give you C. Now, for a penny, I'll take back C and give you B." Since he prefers B to C, he accepts. Next you offer to replace B with A for another penny and again he accepts. The cycle is completed by offering to replace A by C for another penny; he accepts and is 3¢ poorer, back where he started, and ready for another round.

A second important tenet of rationality, known as *the extended sure-thing principle*, states that, if an outcome X_i is the same for two risky actions, then the value of X_i should be disregarded in choosing between the two options. Another way to state this principle is that outcomes that are not affected by your choice should not influence your decision.

These two principles, combined with several others of technical importance, imply a rather powerful conclusion—namely that the wise decision maker chooses that act whose expected utility is greatest. To do otherwise would violate one or more basic tenets of rationality.

Applied decision theory assumes that the rational decision maker wishes to select an action that is logically consistent with his basic preferences for outcomes and his feelings about the likelihoods of the events upon which those outcomes depend. Given this assumption, the practical problem becomes one of listing the alternatives and scaling the subjective values of outcomes and their likelihoods so that subjective expected utility can be calculated for each alternative. Another problem in application arises from the fact that the range of theoretically possible alternatives is often quite large. In addition to carrying an umbrella, the risk-taking traveler in our earlier example may have the options of carrying a raincoat, getting a ride, waiting for the rain to stop, and many others. Likewise, the outcomes are considerably more complex than in our simple example. For example, the consequences of building a dam are multiple, involving effects on flood potential, hydroelectric power, recreation, and local ecology. Some specific approaches that have been developed for dealing with the additional complexities of any real decision situation will be discussed later.

Descriptive decision theory and bounded rationality

Although the maximization theory described above grew primarily out of normative concerns, a good deal of debate and empirical research has centered around the question of whether this theory could also describe both the goals that motivate actual decision makers and the processes they employ when reaching their decisions. The leading critic of utility maximization as a descriptive theory has been Simon, who observed:

> The classical theory is a theory of a man choosing among fixed and known alternatives, to each of which is attached known consequences. But when perception and cognition intervene between the decision-maker and his objective environment, this model no longer proves adequate. We need a description of the choice process that recognizes that alternatives are not given but must be sought; and a description that takes into account the arduous task of determining what consequences will follow on each alternative. [Simon, 1959; p. 272].

As an alternative to the maximization hypothesis, Simon introduced the theory of "bounded rationality," which asserts that the cognitive limitations of the decision maker force him to construct a simplified model of the world to deal with it. The key principle of bounded rationality is the notion of "satisficing," whereby an organism strives to attain some satisfactory, though not necessarily maximal, level of achievement. Simon conjectured that " . . . however adaptive the behavior of organisms in learning and choice situations, this adaptiveness falls far short of the ideal of 'maximizing' postulated in economic theory. Evidently organisms adapt well enough to 'satisfice'; they do not, in general, optimize" (Simon, 1956; p. 129).

The "behavioral theory of the firm" proposed by Cyert and March (1963) elaborated the workings of bounded rationality in business organizations. Cyert and March argued that to understand decision making in the firm we must recognize that there are multiple goals and we must understand the development of these goals, the manner in which the firm acts to satisfy them, and the procedures the firm employs to reduce uncertainty. They described how uncertainty is avoided by following fixed decision rules (standard operating procedures) whenever possible and reacting to short-run feedback rather than trying to forecast the future (which is too uncertain). Firms avoid the uncertainties of depending on other persons by negotiating implicit and explicit arrangements with suppliers, competitors, and customers. A firm's search for new alternatives is triggered by a failure to satisfy one or more goals; thus a crisis is often required to spur corrective action. Cyert and March claim this short-run behavior is adaptive, given the complexity of the environment and the decision maker's cognitive limitations.

At about the same time that Simon and Cyert and March were developing their ideas, Lindblom (1964) was coming to a similar conclusion on the basis of his analysis of governmental policy making. Lindblom argued that administrators avoid the difficult task of taking all important factors into consideration and weighing their relative merits and drawbacks comprehensively by employing what he calls *"the method of successive limited comparisons"*. This method drastically simplifies decisions by comparing only those policies that differ in relatively small degree from policies already in effect. Thus, it is not necessary to undertake fundamental inquiry into an alternative and its consequences; one need study only those respects in which the proposed alternative and its consequences differ from the status quo. As an example, Lindblom cites the similarity between the major political parties in the United States. They agree on fundamentals and offer only a few small points of difference. Lindblom refers to this conservative method as "muddling through" and defends it as efficient and effective, although he admits that its use may cause good new policies to be overlooked, or worse—never even formulated.

Just as Cyert and March's business firms act and react on the basis of short-term feedback, Lindblom's policy maker recognizes his inability to avoid error in predicting the consequences of policy moves. He thus attempts to proceed through a succession of small changes, oriented toward remedying a negatively perceived situation, rather than attaining a preconceived goal:

> His decision is only one step, one that if successful can quickly be followed by another. . . . he is in effect able to test his previous predictions as he moves on to each further step. Lastly, he often can remedy a past error fairly quickly—more quickly than if policy proceeded through distinct steps widely spaced in time. [Lindblom, 1964, p. 166]

Comparison of the two theories

Although utility theory is primarily normative in intent and bounded rationality has a descriptive character, this distinction is not completely accurate. There are those who argue that utility theory has some relevance for describing how decisions are actually made and, as we shall see later, the notion of bounded rationality has normative as well as descriptive implications.

Utility theory is concerned with probabilities, payoffs, and the merger of these factors—expectation. The problem of comparing the worth of one consequence with the worth of another consequence is faced directly by translating both into a common scale of utility. The theory of bounded rationality, on the other hand, postulates that decision makers do not think probabilistically and that they try to avoid the necessity of facing uncertainty directly. Likewise they avoid the problems of evaluating utilities and comparing incommensurable features. The goal of the decision maker is assumed to be the achievement of a satisfactory, rather than a maximum, outcome. Because he is constrained by limitations of perception and intelligence, the boundedly rational decision maker is forced to proceed by trial and error, modifying plans that don't yield satisfactory outcomes and maintaining those that do until they fail.

Bounded rationality and adjustment to natural hazards

Several lines of evidence illustrate the workings of bounded rationality in the context of adjustment to natural hazards.

Limited range of alternatives. It is clear that the resource manager never has available the full range of alternatives from which to make a decision (White, 1961, 1964, 1970). Local regulations or cultural traditions eliminate some alternatives from consideration, and lack of awareness eliminates the others. Early studies of individual and public decisions regarding flood damage reduction, for example, revealed that the traditional choice for users of floodlands in the United States has been simply to bear the loss or to encourage the government to construct engineering works to protect against flooding. Other adjustments such as structural changes in buildings and land-use changes were practiced, until recently, by relatively few managers, and were typically ignored in public action.

Misperception of risks and denial of uncertainty. There are extensive data indicating that the risks of natural hazards are misjudged. For example, Burton and Kates (1964) pointed out that the estimates of hazards made by technical experts often fail to agree. As an illustration of this, they noted that three highly regarded methods of flood frequency analysis placed the long-run average return period of the largest flood on record in the Lehigh Valley as 27, 45, and 75 years.

Misperception of hazards by resource users is further illustrated by an extensive study of flood perception by Kates (1962), who interviewed occupants of locations for which detailed records of flood occurrences were available. The major findings related to the difficulties these floodplain dwellers had in interpreting the hazard within a probabilistic framework. Unlike the technical personnel, who never entirely discounted the possibility of a flood recurring in a previously flooded location, 84 out of 216 floodplain dwellers indicated they did not expect to be flooded in the future.

Close examination of the residents' views illustrated several systematic mechanisms for dispelling uncertainty. The most common of these was to view floods as repetitive, and even cyclical, phenomena. In this way, the randomness that characterizes the occurrence of the hazard is replaced by a determinate order in which history is seen as repeating itself at regular intervals (Burton and Kates, 1964). Another common view was the "law of averages" approach, in which the occurrence of a severe flood in one year made it unlikely to recur the following year. Other occupants reduced uncertainty by means of various forms of denial. Some thought that new protective devices made them 100 percent safe. Others attributed previous floods to a freak combination of circumstances unlikely to recur. Still others denied that past events were floods, viewing them instead as "high water." Another mechanism was to deny the determinability of natural phenomena. For these people, all was in the hands of a higher power (God or the government). Thus, they did not need to trouble themselves with the problem of dealing with the uncertainty.

Crisis orientation. Just as Cyert and March's business firms and Lindblom's policy analysts appear to need direct experience with misfortune as a stimulus to action, so do resource managers. It has been observed: "National catastrophes have led to insistent demands for national action, and the timing of the legislative process has been set by the tempo of destructive floods" (White, 1945, p. 24). Burton and Kates (1964) commented that, despite the self-image of the conservation movement as a conscious and rational attempt at long-range policy and planning, most of the major policy changes have arisen out of crises generated by catastrophic natural hazards. After interviewing floodplain residents, Kates (1962) concluded that it is only in areas where elaborate adjustments have *evolved* by repeated experiences that experience has been a teacher rather than a prison. He added: "Floods need to be experienced, not only in magnitude, but in frequency as well. Without repeated experiences, the process whereby managers evolve emergency measures of coping with floods does not take place" (Kates, 1962, p. 140).

Individual vs. collective management. It is tempting to draw generalizations embracing individuals and groups such as firms and community organizations, but the evidence for doing so is slim. The situational factors are quite different and the methods of handling information may be different for individuals than for groups with corporate memories and organized analysis. Nevertheless, it appears that there are a number of parallels among the bounded rationality of business firms, the behavior of political policy makers, the responses of organizations under stress, and the behavior of individuals with regard to the hazards in their environment. Specifically, decision makers in all these settings exhibit limited awareness of alternatives; they tend to misperceive probabilistic events and employ numerous mechanisms to reduce uncertainty and avoid dealing with it. Finally, they exhibit a short-run, crisis-oriented approach to adaptation.

Psychological research: further evidence for bounded rationality

Thus far the evidence for a theory of bounded rationality within the contexts of business, policy making, and adjustment to natural hazards has been reviewed. In doing so, little that is new or that has not been reviewed by others has been noted. However, most of the evidence has been anecdotal in nature, coming from close observation of behavior and interviews in natural settings. Although this type of analysis has the benefit of realism and relevance, it lacks rigor. Moreover, most of the evidence for bounded rationality in hazard adjustment comes from studies of floodplain residents, and the generality of these conclusions as applied to other types of hazards has not been fully demonstrated. Current efforts are underway to extend the studies of floods in the United States to other cultures and to other hazards. As the detailed analyses of that additional evidence about avalanches, droughts, earthquakes, frost, snow, tropical storms, volcanoes, and winds proceed, it is helpful to ask how the evidence from the field may be compared with that from the laboratory. This question is vital to understanding the complex processes by which man comes to terms with risk in nature, and it has wider implications. It bears upon the degree to which experience with natural hazards can be extrapolated to other sectors of behavior. Should comparable evidence be obtained from laboratory and field research, the validity and importance of both endeavors will be enhanced.

In keeping with these points, the recent psychological literature is examined here for evidence bearing upon man's information processing limitations as it may relate to hazard adjustment. This work differs from that previously discussed in that it comes primarily from controlled, laboratory research. Also, its relevance to hazard adjustment has not been reviewed before. Although Burton and Kates (1964) contended that the artificiality of the laboratory seemed to provide only limited insights into decision strategies and the perception of probabilities, it is our belief that the results of many recent laboratory studies merge nicely with the observations of geographers in their field studies, and help provide a more complete understanding of bounded rationality as it impinges upon adjustments to natural hazards.

The format to be followed in this section is as follows. The psychological principles and data will be reviewed, followed by speculations about the relevance of this work for understanding adjustment to natural hazards. Wherever possible, evidence from field surveys of human response to natural hazards will be presented to further highlight the relevance and generality of these phenomena.

The description of research on information processing is organized around several basic issues of concern to a decision maker. First, he wonders what will happen or how likely it is to happen, and his use of information to answer these questions involves him in probabilistic tasks such as inference, prediction, probability estimation, and diagnosis. He must also evaluate the worth of objects or consequences, and this often requires him to combine information from several components into an overall judgment. Finally, he is called upon to integrate his opinions about probabilities and values into the selection of some course of action. What is referred to as "weighing risks against benefits" is an example of the latter combinatorial process.

Studies of probabilistic information processing

As anyone who has ever planned a picnic understands, nature epitomizes uncertainty. The uncertainties of natural hazards are compounded by the need to plan for periods of time far longer than those considered in other endeavors. The usual damaging flood in urban areas, for example, has a recurrence interval of 50 to 100 years, and the great floods are less frequent.

Efficient adjustment to natural hazards demands an understanding of the probabilistic character of natural events and an ability to think in probabilistic terms. Because of the importance of probabilistic reasoning to decision making in general, a great deal of recent experimental effort has been devoted to understanding how people perceive, process, and evaluate the probabilities of uncertain events. Although no systematic theory about the psychology of uncertainty has emerged from this literature, several empirical generalizations have been established. Perhaps the most widespread conclusion is that people do not follow the principles of probability theory in judging the likelihood of uncertain events. Indeed, the distortions of subjective probabilities are often large, consistent, and difficult to eliminate. Instead of applying the correct rules for estimating probabilities, people replace the laws of chance by intuitive heuristics. These sometimes produce good estimates, but all too often yield large systematic biases. Given these findings, Kates's observations that individuals refuse to deal with natural hazards as probabilistic events are not surprising. To do otherwise may be beyond human cognitive abilities.

The law of small numbers. A series of recent studies of subjective probability has been reported by Tversky and Kahneman (1971), who analyzed the kinds of decisions psychologists make when planning their scientific experiments. Despite extensive formal training in statistics, psychologists usually rely upon their educated intuitions when they make their decisions about how large a sample of data to collect or whether they should repeat an experiment to make sure their results are reliable.

After questioning a number of psychologists about their research practices, and after studying the designs of experiments reported in psychological journals, Tversky and Kahneman concluded that these scientists had seriously incorrect notions about the amount of error and unreliability inherent in small samples of data. They found that the typical psychologist gambles his research hypotheses on small samples, without realizing that the odds against his obtaining accurate results are unreasonably high; second, he has undue confidence in early trends from the first few data points and in the stability of observed patterns of data. In addition, he has unreasonably high expectations about the replicability of statistically significant results. Finally, he rarely attributes a deviation of results from his expectations to sampling variability because he finds a causal explanation for any discrepancy.

Tversky and Kahneman summarized these results by asserting that people's intuitions seemed to satisfy a "law of small numbers," which means that the "law of large numbers" applies to small samples as well as to large ones. The "law of large numbers" says that very large samples will be highly representative of the population from which they are drawn. For the scientists in this study, small samples were also expected to be highly representative of the population. Since his acquaintance with logic or probability theory did not make the scientist any less susceptible to these cognitive biases, Tversky and Kahneman concluded that the only effective precaution is the use of formal statistical procedures, rather than intuition, to design experiments and evaluate data. People are not always incautious when drawing inferences from samples of data. Under somewhat different circumstances they become quite conservative, responding as though data are much less diagnostic than they truly are (Edwards, 1968).

In a related study, this time using Stanford University undergraduates as subjects, Kahneman and Tversky (1972) found that many of these subjects did not understand the fundamental principle of sampling—namely, the notion that the error in a sample becomes smaller as the sample size gets larger. To illustrate, consider one of the questions used in this study.

> A certain town is served by two hospitals. In the larger hospital about 45 babies are born each day, and in the smaller hospital about 15 babies are born each day. As you know, about 50 percent of all babies are boys. The exact percentage of baby boys, however, varies from day to day. Sometimes it may be higher than 50 percent, sometimes lower.
>
> For a period of one year, each hospital recorded the days on which more than 60 percent of the babies born were boys. Which hospital do you think recorded more such days?

Check one:
a) The larger hospital
b) The smaller hospital
c) About the same (i.e., number of days were within 5 percent of each other

About 24 percent of the subjects chose answer (a), 20 percent chose (b), and 56 percent selected (c). The correct answer is, of course, (b). A deviation of 10 percent or more from the 50 percent proportion in the population is more likely when the sample size is small.

From these and other results, Kahneman and Tversky (1972, pp. 444–45) concluded that "the notion that sampling variance decreases in proportion to sample size is apparently not part of man's repertoire of intuitions. . . . For anyone who would wish to view man as a reasonable intuitive statistician, such results are discouraging."

What are the implications of this work for adjustment to natural hazards? We hypothesize that those in charge of collecting data for the purposes of making inferences about degree of hazard would fall prey to the same tendency to overgeneralize on the basis of small samples as do the research psychologists, unless they employ formal statistical procedures to hold their intuitions in check. Although it is common for scientists concerned with recurrence intervals of extreme natural events to lament the short periods of recorded data, we suspect that once the computation is made, whether it is on the basis of data from 20 years or from 70 years, the results will be treated with equal confidence. One rather dramatic example of overgeneralization on the basis of a ridiculously small amount of evidence is given by Burton and Kates (1964), who describe how the occurrence of two earthquakes in London in 1750, exactly one lunar month apart (28 days), with the second more severe than the first, led to predictions that a third and more terrible earthquake would occur 28 days after the second. A contagious panic spread through the city, and which led to its being almost completely evacuated.

Perception of Randomness. A number of experiments bear ample testimony to the fact that people have a very poor conception of randomness; they don't recognize it when they see it and they cannot produce it when they try (Cohen and Hansel, 1956; Jarvik, 1951; Chapanis, 1953). The latter conclusion is illustrated in a study by Bakan (1960) where subjects were asked to generate a series of outcomes representing the tosses of a coin. Subjects' sequences showed more alternation than would be expected in a truly random sequence. Thus triples of responses, HHH and TTT, occurred less often than expected; alternating sequences, HHT, TTH, HTH, and THT, were produced too often. Ross and Levy (1958) found that subjects could not behave randomly even when warned of the types of biases to expect in their responses. The tendency to expect a tail to be more likely after a head, or series of heads, and vice versa is a common finding, and is known as the negative recency effect. Others call it the gambler's fallacy. This basic result is found also in the views of some of the floodplain residents interviewed by Kates (1962). These individuals believed that a flood was less likely to occur in year $x + 1$ if one had occurred in year x.

Judgments of correlation and causality

Another important facet of intuitive thinking is the perception of correlational relationships between pairs of variables that are related probabilistically. Correlation between two such variables implies that knowledge of one will help you to predict the value of the other, although perfect prediction may not be possible.

Chapman and Chapman (1969), studying a phenomenon they have labeled illusory correlation, have shown how one's prior expectations of probabilistic relationships can lead him to perceive relationships in data where they do not really exist. They presented naive subjects with human-figure drawings, each of which was paired with a statement about the personality of the patients who allegedly drew the figures. These statements were randomly paired with the figure drawings so that the figure cues were unrelated to the personality of the drawer. They found that most subjects learned to see what they expected to see. In fact, naive subjects discovered the same relationships between drawings and personality that expert clinicians report observing in clinical practice, although these relationships were absent in the experimental materials. The illusory correlates corresponded to commonly held expectations, such as figures with big eyes being drawn by suspicious people and muscular figures being drawn by individuals who worried about their manliness.

The Chapmans noted that in clinical practice the observer is reinforced in his observation of illusory correlates by the reports of his fellow clinicians, who themselves are subject to the same illusions. Such agreement among experts is, unfortunately, often mistaken as evidence for the truth of the observation. The Chapmans concluded that the clinician's cognitive task may exceed the capacity of the human intellect. They suggested that subjective intuition may need to be replaced, at least partly, by statistical methods of prediction.

Will illusory correlation influence perceptions of natural hazards? Remarks by Kates (1962, p. 141) suggest it may:

> For some managers, a belief that floods come in cycles reduces an uncertain world into a more predictable one. They might be expected to develop interpretive mechanisms that would enable them to transform any hazard information by selective abstraction into a buttress for their existing belief. Managers in LaFollette appear to do this with their observed experience and might find it even easier to do so with information conveyed by maps or printed word.

Several studies have investigated people's perceptions of correlation or causality in simple probabilistic situations involving two binary variables. Consider a 2 x 2 table of frequencies in which variable X is the antecedent or input variable and Y is the consequent or output variable (Table 24–2). The small letters are the frequencies with which the levels of these variables occur together, thus X_1 is followed by Y_1 on a occasions, is followed by Y_2 on b occasions, etc. A correlation exists between X and Y to the extent that the probability of Y_1, given X_1, differs from the probability of Y_1, given X_2—that is, to the extent that $a/(a + b)$ differs from $c/(c + d)$. In other words, if Y_1 is as likely to occur after X_2 as it is after X_1, there is no correlation between X and Y. Causal relationships can sometimes be inferred from tables such as these. If X_1 causes Y_1, we would expect the occurrence of Y_1 to be more probable after X_1 had occurred than after X_2 had occurred, other considerations being equal.

Research indicates that subjects' judgments of correlation and causality are not based on a comparison of $a/(a + b)$ versus $c/(c + d)$. For example, Smedslund (1963) had students of nursing judge the relation between a symptom and the occurrence of a particular disease across a series of trials where the symptom was either present or absent and the disease was either present or absent. He found that the judgments were based mainly on the frequency of joint occurrence of symptom and disease (cell a in the matrix), without taking the frequency of the other three event combinations into account. As a result, the judgments were unrelated to statistical correlation. Similar results were obtained by Jenkins and Ward (1965) and by Ward and Jenkins (1965).

The tendency for people to misperceive the degree to which causation is present in a probabilistic environment has important implications for decisions regarding natural hazards. For example, Boyd et al. (1971), in discussing the decision to modify a hurricane by cloud seeding, pointed out that observed changes in seeded hurricanes can result from both the effect of seeding and from the natural variability of the storm. Suppose

Table 24–2. Frequency with which various levels of Y occur jointly with levels of X

Output		
	Y_1	Y_2
X_1	a	b
Input		
X_2	c	d

that a seeded hurricane intensifies, changes course, and causes damage to a point not on the apparent trajectory before seeding. Would the public react to the joint occurrence of seeding and this unfortunate outcome by assuming that the seeding caused the unfortunate result? Would they conclude that meteorologists are irresponsible? The research described above strongly implies that the initiators of a cloud-seeding enterprise would be blamed for any unfavorable change in a hurricane, even though such changes would occur as frequently, or even more frequently, in the absence of human intervention. As Boyd et al. indicated, the government must be prepared to be held liable for damages occurring from a seeded hurricane, and this possibility must be weighed carefully in it's decisions concerning the general feasibility of hurricane modification programs. This clearly is a situation in which it will be imperative to educate the public on the uncertainty involved in such circumstances, lest a bad outcome be equated with a bad decision.

Judgment of probability by availability. Tversky and Kahneman (1973a) have proposed that people estimate probability and frequency by a number of heuristics, or mental strategies, which allow them to reduce these difficult tasks to simpler judgments. One such heuristic is that of availability, according to which one judges the probability of an event (e.g., snow in November) by the ease with which relevant instances are imagined or by the number of such instances that are readily retrieved from memory. Our everyday experience has taught us that instances of frequent events are easier to recall than instances of less frequent events, and that likely occurrences are easier to imagine than unlikely ones; thus mental availability will often be a valid cue for the assessment of frequency and probability. However, availability is also affected by recency, emotional saliency, and other subtle factors, which may be unrelated to actual frequency. If the availability heuristic is applied, then factors that increase the availability of instances should correspondingly increase the perceived frequency and subjective probability of the events under consideration. Thus, use of the availability heuristic results in predictable systematic biases in judgment.

Consider, for example, sampling a word (containing three or more letters) from an English text. Is it more likely that the word starts with a *k*, or that it has a *k* in the third position? To answer such a question, people often try to think of words beginning with a *k* (e.g., "key") and words that have a *k* in third position (e.g., "like"), and then compare the frequency or the ease with which the two types of words come to mind. It is easier to think of words that start with a *k* than of words with a *k* in the third position. As a result, the majority of people judge the former event more likely despite the fact that English text contains about twice as many words with a *k* in the third position. This example, and many other examples, are presented by

Tversky and Kahneman to document the pervasive effects of availability.

The notion of availability is potentially one of the most important ideas for helping us understand the distortions likely to occur in our perceptions of natural hazards. For example, Kates (1962, p. 140) writes:

> A major limitation to human ability to use improved flood hazard information is a basic reliance on experience. Men on flood plains appear to be very much prisoners of their experience.... Recently experienced floods appear to set an upward bound to the size of loss with which managers believe they ought to be concerned.

Kates further attributes much of the difficulty in achieving better flood control to the "inability of individuals to conceptualize floods that have never occurred" (p. 92). He observes that, in making forecasts of future flood potential, individuals "are strongly conditioned by their immediate past and limit their extrapolation to simplified constructs, seeing the future as a mirror of that past" (p. 88). In this regard, it is interesting to observe how the purchase of earthquake insurance increases sharply after a quake, but decreases steadily thereafter, as the memories become less vivid (Steinbrugge, McClure and Snow, 1969).

Some hazards may be inherently more memorable than others. For example, one would expect drought, with its gradual onset and offset, to be much less memorable, and thus less accurately perceived, than flooding. Kirkby (1972) provides some evidence for this hypothesis in her study of Oaxacan farmers. Kirkby also found that memory of salient natural events seems to begin with an extreme event, which effectively blots out recall of earlier events and acts as a fixed point against which to calibrate later points. A similar result was obtained by Parra (1971), studying farmers in the Yucatan. Parra found that perception of a lesser drought was obscured if it had been followed by a more severe drought. He also observed that droughts were perceived as greater in severity if they were recent and thus easier to remember.

Natural catastrophes are, typically, rare events. For example, Holmes (1961) found that 50 percent of the damage due to major floods was caused by floods whose probability of occurrence at that place in any year was less than .01. The city of Skopje was leveled by earthquakes in 518, 1555, and 1963. The mudflow that took 25,000 lives in Yungay, Peru, had similarly swept across the valley between 1,000 and 10,000 years before. Adequate decision making regarding natural hazards obviously requires a realistic appreciation of the likelihood of these rare events, yet such appreciation is likely to be especially sensitive to the effects of mental availability. For example, ease of imagination almost certainly plays an important role in the public's perception of the risks of injury and death from attack by a grizzly bear in the national parks of North America. In view of the widespread public concern over the dangerousness of these bears, it is indeed surprising that the rate of injury is only 1 per 2 million visitors and the rate of death is even smaller (Herrero, 1970). Imaginability of death by the claws of an enraged grizzly is heightened by newspaper stories and movies which only portray attacks, while the multitude of favorable public experiences go unpublicized.

The availability hypothesis implies that any factor which makes a hazard highly memorable or imaginable—such as a recent disaster or a vivid film or lecture—could considerably increase the perceived risk of that hazard. The Tennessee Valley Authority (TVA) apparently recognizes this, at least at an intuitive level. Kates (1962) noted that the TVA goes to considerable lengths to try to bring home the graphic reality of potential floods. It plots potential floods on easily read maps, and shows flood heights on photographs of familiar buildings. In a similar vein, a recent film entitled "The City That Waits to Die" depicts the vast death and destruction that would occur in San Francisco's next major earthquake. The film was promoted by a group attempting to prohibit the building of new skyscrapers in the city, but was initially banned from public showings. As Kates noted, there is a great need for well-designed studies investigating the effects of such graphic presentations on hazard perception. A decade after Kates's remarks, the need remains unmet.

One additional comment on availability seems warranted. Subtle changes in an individual's mental set are likely to alter the images and memories he brings to bear on the evaluation of hazard, with profound influence on his judgments. For example, an analyst who attempts to evaluate the likelihood of a flood of given magnitude may do so by recalling hydrologic conditions similar to those of the present or by recalling previous floods. The latter are easier to remember because they are more sharply defined, whereas hydrologic states are more difficult to characterize and, therefore, harder to recall. The resulting probability estimate is likely to be greatly dependent upon which of these two sets the analyst adopts. Even the form of the question may be important. Consider the following questions:

1. How likely is it that there will be a flood this season?
2. How likely is it that, given the present hydrologic state, there will be a flood this season?

The first question may focus attention on past instances of flood, whereas the latter may cause the analyst to think about previous hydrologic conditions. The answers to the two questions may be quite different.

Anchoring and adjustment in quantifying uncertainty. Another heuristic that seems useful in describing how humans ease the strain of integrating information is a process called anchoring and adjustment. In this process, a natural starting point is used as a first approximation to the judgment, an anchor, so to speak. This anchor is then adjusted to accommodate the implica-

tions of the additional information. Typically, the adjustment is a crude and imprecise one which fails to do justice to the importance of additional information.

Application of the anchoring and adjustment heuristic is hypothesized to produce an interesting bias that occurs when people attempt to calibrate the degree to which they are uncertain about an estimate or prediction. Specifically, in studies by Alpert and Raiffa (1968) and Tversky and Kahneman (1973b), subjects were given almanac questions such as the following:

> How many foreign cars were imported into the U.S. in 1968?
> a) Make a high estimate such that you feel there is only a 1% probability the true answer would exceed your estimate.
> b) Make a low estimate such that you feel there is only a 1% probability the true answer would be below this estimate.

In essence, the subject is being asked to estimate an interval such that he believes there is a 98 percent chance that the true answer will fall within the interval. The spacing between his high and low estimates is his expression of his uncertainty about the quantity in question. We cannot say that this single pair of estimates is right or wrong. However, if he were to make many such estimates, or if a large number of persons were to answer this question, we should expect the range between upper and lower estimates to include the truth about 98 percent of the time—if the subjective probabilities were unbiased. What is typically found, however, by Alpert and Raiffa and by Tversky and Kahneman, is that the 98 percent confidence range fails to include the true value from 40 to 50 percent of the time, across many subjects answering many kinds of almanac questions. In other words, subjects' confidence bands are much too narrow, given their state of knowledge. Alpert and Raiffa observed that this bias persisted even when subjects were given feedback about their overly narrow confidence bands and urged to widen the bands on a new set of estimation problems.

These studies indicate that people believe they have a much better picture of the truth than they really do. Why this happens is not entirely clear. Tversky and Kahneman tentatively hypothesize that people approach these problems by searching for a calculational scheme or algorithm by which to make a best estimate. They then adjust this estimate up and down to get a 98 percent confidence range. For example, in answering the above question, one might proceed as follows:

> I think there were about 180 million people in the U.S. in 1968; there is about one car for every three people thus there would have been about 60 million cars; the lifetime of a car is about 10 years, this suggests that there should be about 6 million new cars in a year but since the population and the number of cars is increasing let's make that 9 million for 1968; foreign cars make up about 10% of the U.S. market, thus there were probably about 900,000 foreign imports; to set my 98% confidence band, I'll add and subtract a few hundred thousand cars from my estimate of 900,000.

Tversky and Kahneman argue that people's estimates assume that their computational algorithms are 100 percent correct. However, there are two sources of uncertainty that plague these algorithms. First, there is uncertainty associated with every step in the algorithm and there is uncertainty about the algorithm itself. That is, the whole calculational scheme may be incorrect. It is apparently quite difficult to carry along these several sources of uncertainty and translate them intuitively into a confidence band. Once the "best guess" is arrived at as an anchor (e.g., the 900,000 figure above), the adjustments are insufficient in magnitude, failing to do justice to the many ways in which the estimate can be in error.

The research just described implies that our estimates may be grossly in error—even when we attempt to acknowledge our uncertainty. This may have profound implications for many kinds of judgments about the risks associated with natural hazards or the benefits of plans for coping with those hazards. It is likely, for example, that an individual's intuitive estimates of the size of a flood that would be exceeded only one time in one hundred will be conservative (i.e., too close to his estimate of the "most likely" flood magnitude) and he thus would allow too small a margin of safety in his protective adjustments.

Problems in integrating information from multiple sources

Thus far, our review of laboratory research has been concerned with the assessment of risks and estimation of uncertain quantities. A somewhat different problem is as follows. Suppose that we have good information about both risks and benefits. How capable is the decision maker of balancing these several factors and coming up with an optimal decision? By optimal, we don't mean a decision that will, necessarily, turn out well. Some good decisions work out poorly and vice versa. We're thinking of optimal decisions in the sense that such decisions faithfully reflect the decision maker's personal values and opinions.

Information processing biases in risk-taking judgments. As if we didn't have enough problems with our tendencies to bias probability judgments, there is some evidence to the effect that difficulties in integrating information may often lead people to make judgments that are inconsistent with their underlying values. An example of this within a risk-benefit context comes from two experiments (Lichtenstein and Slovic, 1971, 1972), one of which was conducted on the floor of the Four Queens Casino in Las Vegas. Consider the following pair of gambles used in the Las Vegas experiment:

Bet A

11/12 chance to win 12 chips
1/12 chance to win 24 chips

Bet B

2/12 chance to win 79 chips
10/12 chance to lose 5 chips

where the value of each chip has been previously fixed at, say, 25¢. Notice that bet A has a much better chance of winning, but bet B offers a higher winning payoff. Subjects were shown many such pairs of bets. They were asked to indicate, in two ways, how much they would like to play each bet in a pair. First they made a simple choice, A or B. Later they were asked to assume they owned a ticket to play each bet, and they were to state the lowest price for which they would sell this ticket.

Presumably, these selling prices and choices are both governed by the same underlying quality, the subjective attractiveness of each gamble. Therefore, the subject should state a higher selling price for the gamble that he prefers in the choice situation. However, the results indicated that subjects often chose one gamble, yet stated a higher selling price for the other gamble. For the particular pair of gambles shown above, bets A and B were chosen about equally often. However, bet B received a higher selling price about 88 percent of the time. Of the subjects who chose bet A, 87 percent gave a higher selling price to bet B, thus exhibiting an inconsistent preference pattern.

What accounts for the inconsistent pattern of preferences? Lichtenstein and Slovic conclude that subjects use different cognitive strategies for setting prices than for making choices. Subjects choose bet A because of its good odds, but they set a higher price for B because of its large winning payoff. Specifically, it was found that, when making pricing judgments, people who find a gamble basically attractive use the amount to win as a natural starting point. They then adjust the amount to win downward to take into account the less-than-perfect chance of winning and the fact that there is some amount to lose as well. Typically, this adjustment is insufficient and that is why large winning payoffs lead people to set prices that are inconsistent with their choices. Because the pricing and choice responses are inconsistent, it is obvious that at least one of these responses does not accurately reflect what the decision maker believes to be the most important attribute in a gamble.

A "compatibility" effect seems to be operating here. Since a selling price is expressed in terms of monetary units, subjects apparently found it easier to use the monetary aspects of the gamble to produce this type of response. Such a bias did not exist with the choices, since each attribute of one gamble could be directly compared with the same attribute of the other gamble. With no reason to use payoffs as a starting point, subjects were free to use any number of strategies to determine their choices.

Compatibility bias. The overdependence on payoff cues when pricing a gamble suggests a general hypothesis to the effect that the compatibility or commensurability between a dimension of information and the required response affects the importance of that information in determining the response. This hypothesis was tested further in an experiment by Slovic and MacPhillamy (in press), who predicted that dimensions common to each alternative in a choice situation would have greater influence upon decisions than would dimensions that were unique to a particular alternative. They asked subjects to compare pairs of students and predict which would get the higher college grade-point average. The subjects were given each student's scores on two cue dimensions (tests) on which to base their judgments. One dimension was common to both students, and the other was unique. For example, student A might be described in terms of his scores on Need for Achievement and Quantitative Ability, while student B might be described by his scores on Need for Achievement and English Skill.

In this example, since Need for Achievement is a dimension common to both students, the compatibility hypothesis suggests it will be weighted particularly heavily. The rationale for this prediction is as follows: A comparison between two students along the same dimension should be easier, cognitively, than a comparison between different dimensions, and this ease of use should lead to greater reliance on the common dimension. The data strongly confirmed this hypothesis. Dimensions were weighted more heavily when common than when they were unique. Interrogation of the subjects after the experiment indicated that most did not wish to give more weight to the common dimension and were unaware that they had done so.

The message in these experiments is that the amalgamation of different types of information and different types of values into an overall judgment or decision is a difficult cognitive process and, in our attempts to ease the strain of processing information, we often resort to judgmental strategies that may do an injustice to our underlying values. In other words, even when the risks and benefits are known and made explicit, as in the gambling situation, subtle aspects of the decision we have to make, acting in combination with our intellectual limitations, may bias the balance we strike among the many relevant attributes.

Relevance to decisions regarding natural hazards. The research on information integration described above suggests that simplified strategies for easing the strain of making decisions about natural hazards may be used by experts and laymen alike. Although this hypothesis has not been studied systematically, a few relevant examples exist. Perhaps the simplest way to minimize the strain of integrating information is to avoid making decisions. Kates (1962) found that many floodplain managers wanted to abdicate their responsibilities and leave the decision making to the experts. White (1966) noted that, when attention turned to the possibility of setting aside floodplains for open space, some municipalities adopted the blanket policy of buying up valley bottoms for recreational use without even attempting to weigh the alternatives in any given instance. And Kates (1962) observed that three structures in different sites in a town were each elevated by 0.3 meter (1 foot), despite a wide variation in hazard among the sites. One foot is a convenient number, and these decisions suggest that a crude approximation rule was used to determine the elevation changes, much as such approximations were used in the risk-taking studies described above. One wonders also about the depth of analysis that led to the selection of the "100-year flood" as a standard criterion in the design of flood-protection structures.

It is interesting to compare these observations with another example of how simplistic thinking can influence even the most important decisions. With regard to the decision to place a 1.0-megaton (1,000,000-ton) nuclear warhead atop the first Atlas missile, physicist Herbert York commented:

> . . . why 1.0 megaton? The answer is because and only because one million is a particularly round number in our culture. We picked a one-megaton yield for the Atlas warhead for the same reason that everyone speaks of rich men as being millionaires and never as being tenmillionaires or one-hundred thousandaires. It really was that mystical, and I was one of the mystics. Thus, the actual physical size of the first Atlas warhead and the number of people it would kill were determined by the fact that human beings have two hands with five fingers each and therefore count by tens. [York, 1970, pp. 89–90]

Even technical persons, whose job is to aid the decision-making process, can be accused of grossly oversimplified use of information. Their chief tool, cost-benefit analysis, has focused primarily on the dollar values of various adjustments, presumably because these are readily measured and commensurable. The tendency is to ignore noneconomic considerations such as aesthetic and recreational values, or the emotional costs of leaving friends and familiar surroundings when moving to a less hazardous location.

One mechanism that is useful for bringing disparate considerations to bear upon a decision without actually attempting to make them commensurable is to employ a lexicographic decision rule in which one dimension of information is considered at a time. The most important dimension is considered first. Only if this first dimension does not lead to a clearly preferred alternative is the next most important dimension considered. An example of lexicographic behavior in the laboratory that led people to be systematically intransitive in their preferences is presented by Tversky (1969). A natural hazards example is provided in a study of how people, drawing water for household use in a rural area, choose among alternative sources (White, Bradley, and White, 1972). It was found that the users classified the sources as good or bad solely on the basis of health effects. If more than one source was satisfactory on this primary dimension, the remaining "good" sources were then discriminated on the basis of the economic costs of transporting the water. There was little indication that they were willing to "trade off" lower-quality water with accompanying health hazard for lower economic costs. The two dimensions were simply not compensatory.

Another noncompensatory mode of processing diverse dimensions of information is to set a criterion level on one or more of these dimensions. Alternatives that do not promise to meet that criterion are rejected. For those alternatives that remain, another dimension can then be employed as a basis for discrimination. This sort of mechanism has been observed in a laboratory study of risk taking by Lichtenstein, Slovic, and Zink (1969). A natural hazards example is given by Kunreuther (1972), who hypothesized that peasant farmers seek reasonable assurance of survival when deciding how to allocate their resources among crops varying in risk and expected yield. Only for those allocation plans in which survival needs are likely to be met is it likely that maximizing expected yield becomes a consideration. What happens when none of the alternatives meet all of the decision maker's requirements? Something must be sacrificed, and Kunreuther (chap. 25 below) hypothesizes that this sacrifice occurs by means of a lexicographic process whereby the decision maker proceeds sequentially, trying always to satisfy his more important goals, while relaxing those of lesser importance. Kunreuther again uses a crop-allocation decision in the face of natural disasters to illustrate the process.

Investigating bounded rationality in field settings

On the preceding pages we have described a number of aspects of bounded rationality that have been demonstrated in laboratory experiments. Some of the results have close parallel with findings from field studies of floodplain residents. However, most field work has not been oriented toward cognitive processes, and, therefore, has not provided data relevant to the phenomena described above. We believe it would be profitable to look for illustrations of bounded rationality in future field surveys, much as one would examine personality, cultural, or institutional influences upon behavior in the face of natural disaster. The following serves as an overview of the research described above and a brief guide to the kinds of phenomena one might wish to examine in the field.

The law of small numbers. Do individuals overgeneralize on the basis of small samples of evidence? Do they fail to discriminate between short and long periods of record when evaluating evidence or making decisions? Do they take conclusions on faith without questioning the amount of data upon which those conclusions were based?

Judgments of causality and correlation. Do people attribute a bad outcome to a bad decision and a good outcome to a good decision? Do they interpret evidence as supporting a preconceived hypothesis when it does not (illusory correlation)? That is, do they perceive relationships that they expect to see in the data, even when these relationships are not present?

Availability. Do factors of imaginability or memorability influence perception of hazards or actions regarding the hazard? Does rephrasing a question about hazard likelihood to influence memorability also influence the answer? Do vivid films, lectures, or newspaper articles influence perception of rare events? Hazards differ in characteristics that may affect their memorability or imaginability. Some have more sudden onset and offset than others. Duration varies. Contrast a flash flood, for example, with a drought. Do these characteristics systematically affect perception of the hazard? Are people prisoners of their experience, seeing the future as a mirror of the past? Do they predict the future by describing the past?

Anchoring and insufficient adjustment. Do individuals use simple starting-point and adjustment mechanisms when making estimates about quantities? When they attempt to calibrate their uncertainty by placing confidence bounds on their estimates, are those bounds too narrow, thus resulting in rare events occurring more often than they were expected to occur?

Information processing shortcuts. Is there evidence for simple decision strategies that avoid weighing of multiple considerations? Do people avoid making decisions by relying on experts, authority, fate, custom, etc.? Is there evidence for lexicographic processes or other noncompensatory decision modes in the evaluation of adjustments?

Additional needs for research. Finally, there are a number of important situational factors about which we have neither laboratory nor field data. For example, we need to better understand the effects of savings and reserves, time horizon, and amount of diversification upon perception of alternatives and efficiency of adjustment. Will larger amounts of reserves make it more likely that an individual will consider alternatives that have greater risk but also greater expected payoffs? Similarly, will diversification of farming activity reduce the risk of failing to meet one's goals of subsistence and thus permit the farmer to consider risky but profitable alternatives?

We need to know more about the condition of decision as it affects behavior. Will individuals become aware of a wider range of perceived alternatives if they are required to make a decision with respect to a given risk as opposed to conditions where the decision is voluntary?

Finally, we need theoretical models of boundedly rational behavior which, from reasonable assumptions about the constraints pertinent to a given natural-hazards decision, yield testable hypotheses about the effects of income reserves, insurance, time horizon, etc., upon factors such as range of perceived alternatives, criteria for choice, and level of aspiration.

Comment

The experimental work described in this section documents man's difficulties in weighing information and judging uncertainty. Yet this work is quite recent in origin and still very much in the exploratory stage. In addition, its implications do not fit with the high level of confidence that we typically accord our higher mental processes. Consider, for example, the statement by a famed economist: "We are so built that what seems reasonable to us is likely to be confirmed by experience or we could not live in the world at all" (Knight, 1965, p. 227). Since the laboratory results greatly contradict our self-image, it is reasonable to question whether the observed information processing difficulties will persist outside the confines of the laboratory in situations where the decision maker uses familiar sources of information to make decisions that are personally important to him.

In light of this natural skepticism, and since our coverage of the psychological experiments was necessarily rather brief, we should like to point out that evidence for cognitive limitations pervades a wide variety of tasks where intelligent individuals served as decision makers, often under conditions designed to maximize motivation and involvement. For example, the subjects studied by Tversky and Kahneman (1971) were scientists, highly trained in statistics, evaluating problems similar to those they faced in their work. Likewise, Alpert and Raiffa (1968) found it extremely difficult to reduce the biased confidence judgments in their subjects, who were students in the advanced management program at a leading graduate school.

In many of the experiments reported above, extreme measures were taken to maximize the subjects' motivation to be unbiased. When Lichtenstein and Slovic (1971) observed inconsistent patterns of choices and prices among college-student subjects gambling for relatively small stakes, they repeated the study, with identical results, on the floor of a Las Vegas casino. It should also be noted that their experiments involving selling-price responses employed a rather elaborate procedure devised by Becker, De Groot, and Marschak (1964) to persuade the subject to report his true subjective value of the bet as his lowest selling price; any deviations from this strategy, any efforts to "beat the game," necessarily resulted in a game of lesser value to the subject than the game resulting when he honestly reported his subjective valuations. Tversky and Kahneman have also resorted to extreme

measures to motivate their subjects to behave in an unbiased manner.

Finally, the laboratory conclusions are congruent with many observations of nonoptimal decision making outside the laboratory—in business, governmental policy setting, and adjustment to natural hazards. The belief that man can behave optimally when it is worthwhile for him to do so gains little support from these studies. The sources of judgmental bias appear to be cognitive, nor motivational. They have a persistent quality not unlike that of perceptual illusions.

It is interesting to speculate about why we have such great confidence in our intuitive judgments, in the light of the deficiencies that emerge when they are exposed to scientific scrutiny. For one thing, our basic perceptual motor skills are remarkably good, the product of a long period of evolution, and thus we can process *sensory* information with remarkable ease. This may fool us into thinking that we can process *conceptual* information with similar facility. Anyone who tries to predict where a baseball will land by calculating its impact against the bat, trajectory of flight, etc., will quickly realize that his analytic skills are inferior to his perceptual motor abilities. Another reason for our confidence is that the world is structured in such a complex, multiply-determined way that we can usually find some reason for our failures, other than our inherent inadequacies—bad luck is a particularly good excuse in a probabilistic world. In many situations, we get little or no feedback about the results of our decisions and, in other instances, the criterion for judging our decisions is sufficiently vague that we can't tell how poorly we are actually doing. Finally, when we do make a mistake and recognize it as such, we often have the opportunity to take corrective action—thus we may move from crisis to crisis but, in between crises, we have periods of fairly effective functioning. When we have the opportunity to learn from our mistakes, and can afford to do so, this may be a satisfactory method of proceeding. When we cannot, we must look toward whatever decision aids are available to help us minimize errors of judgment.

Implications for future policy: how can we improve adjustment to natural hazards?

Research, in both natural and laboratory settings, strongly supports the view of decision processes as boundedly rational. Given this awareness of our cognitive limitations, how are we to maximize our capability for making intelligent decisions about natural hazards?

Two answers to this question are considered here. The first is primarily nonanalytic in character and works within the framework of bounded rationality. The second is an analytic approach that accepts the notion that human beings are fallible in processing information, but strives to help them come as close as possible to an ideal conception of rational decision making.

Implications of bounded rationality

Knowledge of the workings of bounded rationality forms a basis for understanding constraints on decision making and suggests methods for helping the decision maker improve as an adapting system. For example, Cyert and March (1963) describe how policy inputs can trigger a search for new alternatives by introducing constraints which make old habits of adjustment unacceptable. Within the context of business decision making, Cyert and March point to three ways in which the firm's decision-making behavior can be altered via policy changes. The first changes the inputs to standard decision rules as exemplified by changes in the product specifications or work regulations. The second use of policy is to force a failure in meeting some valued goal by setting explicit constraints on costs, prices, profits, or the like. The third use of policy is to modify the consequences of potential solutions to problems to enhance the attractiveness of solutions that would otherwise be unacceptable.

How might we apply knowledge of bounded rationality to improving adjustment to natural hazards? Consider two key aspects of the problem—the need to make the decision maker's perceptions of the hazard more accurate and the need to make him aware of a more complete set of alternative courses of action.

To improve probabilistic perception of hazards, it is essential that complete historical records be kept, analyzed, and made available in understandable form to all resource managers. Technical experts should be taught how to express hazards probabilistically and their opinions should be made available in a format which attempts to be comprehensible to individuals not particularly skilled in probabilistic thinking. Records should be continually updated and, when a new development occurs that might render the historical data invalid, the technical expert should estimate the effect of this change on the hazard.

There has been a small amount of experimentation with physical formats for expressing probabilities of natural extremes: the U.S. Geological Survey and the Corps of Engineers have tried several ways of presenting flood frequencies, including historical summaries, graphs of recurrent intervals, eyewitness accounts, photographs, and maps. An elaborate set of maps showing susceptibility to geophysical hazards such as earthquake and landslide are under preparation by the USGS and the Department of Housing and Urban Development for the San Francisco Bay region. However, there has been no serious effort to find out what effect, if any, the different formats have upon understanding of probability. Perhaps the only relatively searching attempt has been in connection with public interpretation of weather forecasts which use probability estimates (Murphy and Winkler, 1971).

Of course, given our limited ability to comprehend probabilistic information, imaginative presentations of records may not be enough. Creative new devices will be

necessary to facilitate imaginability and to break through the "prison of experience" that shackles probabilistic thinking. One procedure worth exploration is that of informing decision makers of the biases that are likely to distort their interpretation and use of information. Another device is simulation, which might be particularly effective in conveying an appreciation of sampling variability and probabilities. Consider the important practical situation where a farmer in a frequently drought-stricken area must decide whether or not to plant drought-resistant maize. Such maize will provide greater yield than regular maize if a drought occurs, but will do worse in the event of normal rainfall. The farmer can be shown a historical record of rainfall for the past 50 years, but from what we know of the Tversky and Kahneman experiments and geographical surveys by Kates, it is unlikely that he will be able to use this information properly. The farmer's problem is increased by the difficulty of taking into account the utilities of various yields, as well as their probabilities. It is here that simulation can be particularly valuable. A farming game can give the decision maker realistic and appropriate experience with this type of decision and its consequences. The farmer begins with a specific amount of cash. In year 1 he makes the decision about what percent of his maize crop to plant with drought-resistant seed. Nature runs its course and the farmer receives an appropriate bounty. Our subject plays against nature and quickly gains experience that would ordinarily accrue only over many years. Simulations such as this have already been introduced as teaching aids in high-school and college geography courses (Patton, 1970; High and Richards, 1972). Kates (1962, p. 140) observed: "Without frequent experience, learned adjustments wither and atrophy with time." Simulation might be a quick and painless way to provide the concrete experiences needed to produce adjustments that are maximally adaptive.

With regard to widening the range of perceived alternatives, several possibilities exist. For example, since we know that perception is typically incomplete, we can take special measures to inform resource managers of the range of available options. Although there have been frequent pleas for encouraging people to consider a wider range of alternatives in coping with hazards (NAS-NRC, 1966; WRC, 1971) the means of doing so have been explored only casually. Thus, the National Environmental Protection Act of 1969 specifies that environmental impact statements shall indicate alternative measures for resource allocation, but it does not indicate how this should be done. The principal measures now being tried are survey reports, public hearings, public discussions, and informational brochures. With the exception of the studies of Corps of Engineers' public consultations (Borton et al., 1970), these have not been evaluated.

Another way to widen the range of perceived choice is to employ policy to modify the potential consequences of an alternative, thus making a previously unattractive alternative worthy of consideration. Compulsory insurance is a good example of a policy that can play a role in improving hazard perception and widening the range of alternatives. By guaranteeing individuals a minimum level of income if they adopt an innovative adjustment, insurance can decrease the risk entailed by the innovation, thereby enchancing its attractiveness. Probably the most significant role that may be played by insurance is in requiring explicit, conscious attention to risk by the individual concerned. He is faced with an estimate of risk expressed by an annual premium charge, and in some cases may also be provided with a schedule of reduced premiums contingent upon his taking certain actions, such as flood proofing his home. A government scheme which fails to base premiums on risk may have undesirable effects (chap. 26 below). For a more detailed discussion of the role of insurance in the context of natural hazards, see chapter 25 below and Kunreuther (1968), Dacy and Kunreuther (1969), and Lave (1968).

How safe is safe enough? There are some who believe that, given a static environment, man learns by trial, error, and subsequent corrective actions to arrive at a reasonably optimal balance between the benefit from an activity and its risk. One such individual is Starr (1969, 1972), who has developed a quantitative measure of the acceptable risk-benefit ratio for an activity, based on this belief. Starr assumes that historical national-accident records are adequate for revealing consistent patterns in risk-benefit ratios and that these historically revealed social preferences are sufficiently enduring to permit their use for predicting what risk levels will be acceptable to society when setting policies or introducing new technologies. Implicit in this approach is yet another assumption, that what is best for society is approximately equivalent to what is traditionally acceptable.

Starr distinguishes between voluntary activities, which the individual can evaluate via his own value system, and involuntary activities, where the options and criteria of evaluation are determined for the individual by some controlling body. His measure of risk is the statistical expectation of fatalities per hour of exposure to the activity under consideration. For voluntary activities, his measure of benefit is assumed to be approximately equal to the amount of money spent on an activity by the average individual. For involuntary activities, benefit is assumed proportional to the contribution that activity makes to an individual's annual income.

Analysis of a number of natural and man-made risks, according to these considerations, points to several important conclusions: (1) the public seems willing to accept voluntary risks roughly 1,000 times greater than involuntary risks at a given level of benefit; (2) the acceptability of a risk is roughly proportional to the real and perceived benefits; and (3) the acceptable level of risk is inversely related to the number of persons participating in an activity.

Starr's assessment technique falls within the purview of bounded rationality approaches because, rather than assuming that individuals can indicate directly an optimal risk-benefit trade-off, it merely assumes that across a large group of individuals, given an opportunity to learn from their mistakes, a satisfactory level will emerge.

The importance of knowing the acceptable risk level for an activity cannot be overestimated. The Starr technique thus promises to be a valuable aid for decisions regarding natural hazards and, in fact, a similar approach has already been used to guide the development of a new earthquake building code for the city of Long Beach, California (Wiggins, 1972). However, several reservations bear mentioning. First, the psychological research described above points to the prevalence of systematic biases in risk-taking decisions. It is unlikely that all such biases will be eliminated as a result of experience. Therefore, just as an individual's decisions may not accurately reflect his "true preferences," historical data may not necessarily reflect the underlying preferences of a group of people. Second, the validity of historical data as an indicant of preference assumes that the public has available a wide selection of alternatives and, furthermore, that these alternatives are perceived as being available. Can we really assume, for example, that the public will demand automobiles that are as safe as they would wish, given the available benefits? Unless the public really knows what is possible from a design standpoint, and unless the automobile industry cooperates in making available information that may not necessarily serve its own profit maximization interests, the answer is likely to be no. (For a more detailed discussion of the limitations of the public's "market" behavior as an indicant of its risk values see Schelling, 1968.) Finally, the Starr approach does not consider the question of who should bear the costs of damages from natural hazards. As Kunreuther (1973) has pointed out, individuals and communities might adopt quite different risk levels if they were forced to bear the costs of disasters rather than relying on liberal government relief.

With these qualifications, the Starr approach would seem to merit serious consideration as a method for designing and evaluating adjustments to natural hazards.

The analytic approach to improving adjustments

Bounded rationality, with its emphasis on short-run feedback and adaptation triggered by crises, may work satisfactorily in some settings, particularly in static environments where the same decision is made repeatedly and the consequences of a poor decision are not too disastrous. However, where natural hazards are concerned, we may prefer not to rely upon learning from experience. First, relevant experiences may be few and far between, and second, mistakes are likely to be too costly. With so much at stake, it is important to search for methods other than the clever ways of "muddling through" and "satisficing" advanced by the advocates of bounded rationality.

The alternative to muddling through is the application of scientific methods and formal analysis to problems of decision making. The analytic approach originated during World War II from the need to solve strategic and tactical problems in situations where experience was either costly or impossible to acquire. It was first labeled "operations analysis," and later became known as "operations research." Operations research is an interdisciplinary effort, bringing together the talents of mathematicians, statisticians, economists, engineers, and others. Since the war, its sphere of application has been extended primarily to business, but its potential is equally great for all areas of decision making.

Simon (1960) outlined the stages of an operations research analysis as follows: The first step is to construct a mathematical model that mirrors the important factors in the situation of interest. Among the mathematical tools that have been particularly useful in this regard are linear programming, dynamic programming, and probability theory. The second step is to define a criterion function which is to be used to compare the relative merits of the possible alternative actions. Next, empirical estimates are obtained for the numerical parameters in the model for the specific situation under study. Finally, mathematical analysis is applied to determine the course of action that maximizes the criterion function.

During recent years, a number of closely related offshoots of operations research have been applied to decision problems. These include systems analysis and cost-benefit analysis. Systems analysis is a branch of engineering, whose objective is capturing the interactions and dynamic behavior of complex systems. Cost-benefit analysis attempts to quantify the prospective gains and losses from some proposed action, usually in terms of dollars. If the calculated gain from an act or project is positive, it is said that the benefits outweigh the costs, and its acceptance is recommended (see, for example, the application of cost-benefit analysis to the study of auto-safety features by Lave and Weber, 1970).

Decision analysis. What systems analysis and operations research approaches lacked for many years was an effective normative framework for dealing either with the uncertainty in the world or with the subjectivity of decision makers' values and expectations. The emergence of decision theory provided the general normative rationale missing from these early analytic approaches. By the same token, systems analysis and operations research had something to offer applied decision theory. There is an awesome gap between the simple decisions that are typically used to illustrate decision theoretic principles (e.g., whether or not to carry an umbrella) and the complex real-world problems one wishes to address. Systems analysis attempts to provide the sophisticated modeling of the decision situation needed to bridge the gap. The result of the natural merger between decision theory and engineering approaches has been labeled "decision analysis." Our review of decision anal-

ysis will be brief. For further details, see the tutorial papers by Howard (1968a, 1968b) and Matheson (1969–70), and the books by Raiffa (1968) and Schlaifer (1969).

A thorough decision analysis takes a great deal of time and effort and thus should be applied only to important problems. Typically, these problems involve a complex structure where many interrelated factors affect the decision, and where uncertainty, long-run implications, and complex trade-offs among outcomes further complicate matters.

A key element of decision analysis is its emphasis upon *structuring* the decision problem and *decomposing* it into a number of more elementary problems. In this sense, it attempts a simplification process that, unlike the potentially detrimental simplifications the unaided decision maker might employ, maintains all the essential ingredients that are necessary to make the decision and ensures that they are used in a manner logically consistent with the decision maker's basic preferences. Raiffa (1968, p. 271) expresses this attitude well in the following statement:

> The spirit of decision analysis is divide and conquer: Decompose a complex problem into simpler problems, get your thinking straight in these simpler problems, paste these analyses together with a logical glue, and come out with a program for action for the complex problem. Experts are not asked complicated, fuzzy questions, but crystal clear, unambiguous, elemental hypothetical questions.

Decision analysis of hurricane modification. The technique of decision analysis is best communicated via a specific example. Fortunately, there is a detailed example available in the analysis of hurricane modification prepared by the decision analysis group of Stanford Research Institute (SRI) on behalf of the National Oceanic and Atmospheric Administration (Boyd et al., 1971; Howard, Matheson, and North, 1972). An overview of this analysis is presented below.

In the case of hurricane modification, one important decision is strategic: "Should cloud seeding ever be performed?" If the answer is yes, then tactical decisions concerning which hurricanes are to be seeded become important. The SRI analysis focuses on the strategic decision.

The basic approach is to consider a representative severe hurricane bearing down on a coastal area and to analyze the decision to seed or not to seed this hurricane. Maximum sustained surface wind speed is used as the measure of the storm's intensity, since it is this characteristic (which is the primary cause of destruction) that seeding is expected to influence. The analysis assumes that the direct consequence of a decision on seeding is the property damage caused by the hurricane.

However, property damage alone is insufficient to describe the consequences of hurricane seeding. There are indirect social and legal effects that arise from the fact of human intervention; thus, the government might have some legal responsibility for the damage from a seeded hurricane. The trade-off between accepting the responsibility for seeding and accepting higher probabilities of severe property damage is viewed as the crucial issue in this decision.

The first step in the SRI analysis was to merge current experimental evidence with the best prior scientific opinion to obtain a probability distribution over changes in the intensity of the representative hurricane as measured by its maximum surface wind speed. This was done for both alternatives—seeding and not seeding. Then, data from past hurricanes were used to infer the relationship between wind speed and property damage. On the basis of this information, the expected loss in terms of property damage was calculated to be about 20 percent less if the hurricane is seeded. Varying the assumptions of the analysis over a wide range of values caused this reduction to vacillate between 10 percent and 30 percent, but did not change the preferred alternative.

The above analysis favors seeding but does not take the negative utility of government responsibility into account. The assessment of responsibility costs entails considerable introspective effort on the part of the decision maker, who must make judgments such as "Estimate x such that the government would be indifferent between a seeded hurricane that intensifies 16 percent between time of seeding and landfall and an unseeded hurricane that produces x percent more damage than that of the seeded hurricane."

On the basis of estimates such as the above, it was inferred that the responsibility costs needed to change the decision were a substantial fraction (about 20 percent) of the property damage caused by the hurricane. This, and further analyses, led to the conclusion that on the basis of present information, the probability of severe damage is less if a hurricane is seeded, and that seeding should be permitted on an emergency basis and encouraged on an experimental basis.

Critique of decision analysis. It is difficult to convey in a summary such as that given above the depth of thinking and the logic underlying the decision analysis. The brief description necessarily simplifies the analysis and highlights a chief objection to decision analysis in general—the claim that it oversimplifies the situation and thus misleads. Nevertheless, even those who read the complete analysis may have concerns over its validity. They may note that Howard, Matheson, and North (1972) have constrained their analysis to ignore the beneficial and detrimental aspects of hurricanes in their major contribution to the water balance of the areas affected. The analysis also ignores the possibility that knowledge of an operational seeding program will give residents a false sense of security, thus inviting greater damages than might occur without seeding. The critics argue that such decision analyses are inevitably constrained by time, effort, and imagination, and must systematically exclude many considerations.

A second major objection to decision analysis is the possibility that it may be used to justify, and give a gloss of respectability to, decisions made on other, and perhaps less rational, grounds.

Decision analysts counter these attacks by invoking one of their basic tenets—namely, that any alternative must be considered in the context of other alternatives. What, they ask, are the alternatives to decision analysis, and are they any more immune to the criticisms raised above? The analysts point out that traditional modes of decision making are equally constrained by limits of time, effort, and imagination, and are even more likely to induce systematic biases (as illustrated earlier in this chapter). Such biases are much harder to detect and minimize than the deficiencies in the explicit inputs to decision analysis. Furthermore, they argue, if some factors are unknown or poorly understood, can traditional methods deal with them more adequately than decision analysis does? Traditional methods also are susceptible to the "gloss of respectability" criticism noted above. We often resort to expertise to buttress our decisions without really knowing the assumptions and logic underlying the experts' judgments. Decision analysis makes these assumptions explicit. Such explicit data are easy for knowledgeable persons to criticize and the explicitness thus focuses debate on the right issues.

Decision analysts would agree that their craft is no panacea, that incomplete or poorly designed analyses may be worse than no analyses at all, and that analysis may be used to "overwhelm the opposition." It seems clear, however, that the main task for the future is not so much to criticize decision analysis but rather to see how it can be used most appropriately.

Conclusion

In coping with the hazard of natural events, man enlarges the social costs of those events and tends to make himself more vulnerable to the consequences of the great extremes. His response to uncertainty in the timing and magnitude of droughts, earthquakes, floods, and similar unusual events has led to increases in the toll of life and property which they take.

Understanding why this is so is essential to the wise design of new policies. Much of the improved policy necessarily will depend upon action by individuals within public constraints. Here it is important to recognize how people make their choices in the face of uncertainty in nature and how they might be expected to respond within a different set of constraints.

Enough is known about the process of choice to be sure that it cannot be accurately described as a simple effort to maximize net marginal returns. Nor can it be explained solely in terms of the culture or the personality of the decision makers. It is not easily predicted as a product of particular environmental or organizational conditions. It is a complex, multidetermined phenomenon. The need for relatively clear analysis of its essential elements is urgent and is at the heart of systematic improvement of public policy.

The present paper attempts to show that: (1) convergent evidence from psychology, business, governmental policy making, and geography documents the usefulness of bounded rationality as a framework for conceptualizing decision processes; (2) an understanding of the workings of bounded rationality can be exploited to improve adjustment to natural hazards; and (3) decision analysis, though still in an early stage of development, promises to be a valuable aid for the important decisions man must make regarding natural hazards.

Measures needed to increase understanding of the decision process and provide opportunities for improving it require a combination of theoretical, laboratory, and empirical approaches. Determining a rationale for optimal behavior in the face of a capricious nature requires theoretical development. The basic modes of assessing probabilities of rare natural events and of assigning values to consequences involve cognitive processes that may be discerned most clearly in controlled laboratory experiments. The recognition of ways in which cultural and situational factors may influence decisions calls for observation in field settings.

Acknowledgments

Principal support for this paper came from Grant 6S-2882X from the National Science Foundation to the University of Colorado. Additional support was provided by grants to the Oregon Research Institute from The National Science Foundation (Grant GS-32505) and the U.S. Public Health Service (Grant MH-21216).

References

Alpert, M., and Raiffa, H. (1968) "A progress report on the training of probability assessors." Cambridge, Mass.: Harvard University, unpublished manuscript.

Bakan, P. (1960) "Response tendencies in attempts to generate random binary series." *American Journal of Psychology* 73:127–31.

Barton, Allen H. (1970) *Communities in Disaster*. New York: Anchor Books.

Baumann, D. D., and Sims, J. H. (1972) "The tornado threat: coping styles of the North and South." *Science* 176:1386–92.

Becker, G. M., De Groot, M. H., and Marschak, J. (1964) "Measuring utility by a single-response sequential method." *Behavioral Science* 9:226–32.

Borton, T. E. et. al. (1970) *The Susquehanna Communication-Participation Study*. Springfield, Virginia: FSTI.

Boyd, D. W., et al. (1971) *Decision Analysis of Hurricane Modification*. Menlo Park, California: Stanford Research Institute. Final Report: Project 8503. (This report is available through the National Technical Information Service, U.S. Department of Commerce, Washington, D.C., accession number COM-71-00784.)

Burton, I. (1972) "Cultural and personality variables in the perception of natural hazards." In J. F. Wohlwill and D. H. Carson, eds., *Environment and the Social Sciences: Perspectives and Applications*. Washington, D.C.: American Psychological Association.

———. and Kates, R. W. (1964) "The perception of natural hazards in resource management." *Natural Resources Journal* 3:412–41.

Chapanis, A. (1953) "Random number guessing behavior." *American Psychologist* 8:332. Abstract.

Chapman, L. J., and Chapman, J. P. (1969) "Illusory correlation as an obstacle to the use of valid psychodiagnostic signs." *Journal of Abnormal Psychology* 74:271–80.

Cohen, J., and Hansel, C. E. M. (1956) *Risk and Gambling*. London: Longmans Green.

Coombs, C. H., Dawes, R. M., and Tversky, A. (1970) *Mathematical Psychology*. Englewood Cliffs, N.J.: Prentice-Hall.

Cyert, R. M., and March, J. G. (1963) *A Behavioral Theory of the Firm*. Englewood Cliffs, N.J.: Prentice-Hall.

Dacy, D. C., and Kunreuther, H. (1969) *The Economics of Natural Disasters*. New York: Free Press.

Dillon, J. L. (1971) "An expository review of Bernoullian decision theory in agriculture: is utility futility?" *Review of Marketing and Agricultural Economics* 39:3–80.

Edwards, W. (1968) "Conservatism in human information processing." In B. Kleinmuntz, ed., *Formal Representation of Human Judgment*. New York: Wiley.

Herrero, S. (1970) "Human injury inflicted by grizzly bears." *Science* 170:593–97.

High, C., and Richards, P. (1972) "The random walk drainage simulation model as a teaching exercise." *Journal of Geography* 71:41–51.

Holmes, R. C. (1961) "Composition and size of flood losses." In G. F. White, ed., *Papers on Flood Problems*. Chicago: University of Chicago, Department of Geography, Research Paper No. 70.

Howard, R. A. (1968a) "The foundations of decision analysis." *IEEE Transactions on Systems Science and Cybernetics* 4:211–19.

———. (1968b) "Decision analysis: applied decision theory. In D. B. Hertz and J. Melese, eds., *Proceedings of the Fourth International Conference on Operational Research*. New York: Wiley.

———, Matheson, J. E., and North, D. W. (1972) "The decision to seed hurricanes." *Science* 176:1191–202.

Jarvik, M. E. (1951) "Probability learning and a negative recency effect in the serial anticipation of alternative symbols." *Journal of Experimental Psychology* 41:291–97.

Jenkins, H. M., and Ward, W. C. (1965) "Judgment of contingency between responses and outcomes. *Psychological Monographs* 79, whole no. 594.

Kahneman, D., and Tversky, A. (1972) "Subjective probability: a judgment of representativeness. *Cognitive Psychology* 3:430–54.

Kates, R. W. (1962) *Hazard and Choice Perception in Flood Plain Management*. Chicago: University of Chicago, Department of Geography, Research Paper No. 78.

———. (1970) *Natural Hazard in Ecological Perspective: Hypotheses and Models*. Natural Hazards Research Working Paper No. 14. Univ. of Toronto, Dept. of Geography.

Kirkby, A. V. (1972) *Perception of Rainfall Variability and Agricultural and Social Adaptation to Hazard by Peasant Cultivators in the Valley of Oaxaca, Mexico*. Calgary, Alta.: paper presented at the 22d International Geographical Congress.

Knight, F. H. (1965) *Risk, Uncertainty, and Profit*. New York: Harper & Row.

Kunreuther, H. (1968) "The case for comprehensive disaster insurance." *Journal of Law and Economics* 11:133–63.

———. (1972) *Risk-Taking and Farmer's Crop Growing Decisions*. Chicago: University of Chicago: Center for Mathematical Studies in Business and Economics, Report No. 7219.

———, (1973) "Values and costs." In *Building Practices for Disaster Mitigation*. Washington, D.C.: U.S. Department of Commerce, National Bureau of Standards.

Lave, L. B. (1968) "Safety in transportation: the role of government." *Law and Contemporary Problems* 33:512–35.

———, and Weber, W. E. (1970) "A benefit-cost analysis of auto safety features." *Applied Economics* 2:265–75.

Lichtenstein, S., and Slovic, P. (1971) "Reversals of preference between bids and choices in gambling decisions." *Journal of Experimental Psychology* 89:46–55.

———, and Slovic, P. (1972) "Response-induced reversals of preference in gambling: an extended replication in Las Vegas." *Oregon Research Institute Research Bulletin* 12, no. 6.

———, Slovic, P., and Zink, D. (1969) "Effect of instruction in expected value on optimality of gambling decisions." *Journal of Experimental Psychology* 79:236–40.

Lindblom, C. E. (1964) "The science of muddling through." In W. J. Gore and J. W. Dyson, eds., *The Making of Decisions*. New York: Free Press.

Luce, D., and Raiffa, H. (1957) *Games and Decisions*. New York: Wiley.

Matheson, J. E. (1969–70) "Decision analysis practice: examples and insights." In J. Lawrence, ed., *Proceedings of the Fifth International Conference on Operational Research*. London: Tavistock.

Murphy, A. H., and Winkler, R. H. (1971) "Forecasters and probability forecasts: some current problems." *Bulletin of the American Meteorological Society* 52:239–47.

National Academy of Sciences–National Research Council (1966) *Alternatives in Water Management*. Publication 1408.

Parra, C. G. (1971) "Perception of past droughts in Ticul, Yucatan." *Proceedings* of the Great Plains—Rocky Mountain Meeting of the American Association of Geographers, Colorado Springs.

Patton, D. J., ed. (1970) *From Geographic Discipline to Inquiring Student: Final Report on the High School Geography Project*. Washington, D.C.: Association of American Geographers.

Raiffa, H. (1968) *Decision Analysis*. Reading, Mass.: Addison-Wesley.

Ross, B. M., and Levy, N. (1958) "Patterned predictions of chance events by children and adults." *Psychological Reports* 4:87–124.

Savage, L. J. (1954) *The Foundations of Statistics*. New York: Wiley.

Schelling, T. C. (1968) "The life you save may be your own." In S. B. Chase, ed., *Problems in Public Expenditure Analysis*. Washington, D.C.: Brookings Institution.

Schiff, M. (1970) *Some Theoretical Aspects of Attitudes and Perception*. Toronto: Natural Hazards Research Working Paper No. 15.

Schlaifer, R. (1969) *Analysis of Decisions Under Uncertainty*. New York: McGraw-Hill.

Simon, H. A. (1956) "Rational choice and the structure of the environment." *Psychological Review* 63:129–38.

———. (1959) "Theories of decision making in economics and behavioral science." *American Economic Review* 49:253–83.

———. (1960) *The New Science of Management Decision*. New York: Harper & Row.

Slovic, P., and MacPhillamy, D. (in press) "Dimensional commensurability and cue utilization in comparative judgment." *Organizational Behavior and Human Performance*.

Smedslund, J. (1963) "The concept of correlation in adults." *Scandinavian Journal of Psychology* 4:165–73.

Starr, C. (1969) "Social benefit versus technological risk." *Science* 165:1232–38.

———. (1972) "Benefit-cost studies in sociotechnical systems." In *Perspectives on Benefit-Risk Decision Making*. Washington, D.C.: National Academy of Engineering, Report of the Committee on Public Engineering Policy.

Steinbrugge, K. V., McClure, F. E., and Snow, A. J. (1969) *Studies in Seismicity and Earthquake Damage Statistics*. Washington, D.C.: U.S. Department of Commerce, Report (Appendix A) COM-71-00053.

Tversky, A. (1969) "Intransitivity of preferences." *Psychological Review* 76:31–48.

———, and Kahneman, D. (1971) "Belief in the law of small numbers." *Psychological Bulletin* 76:105–10.

———, and Kahneman, D. (1973a) "Availability: a heuristic for judging frequency and probability." *Cognitive Psychology*, 5:207–232.

———, and Kahneman, D. (1973b) "Anchoring and calibration in the assessment of uncertain quantities." *Oregon Research Institute Research Bulletin*, in preparation.

von Neumann, J., and Morgenstern, O. (1947) *Theory of Games and Economic Behavior*, 3d ed., 1953. Princeton, N.J.: Princeton University Press.

Ward, W. C., and Jenkins, H. M. (1965) "The display of information and the judgment of contingency." *Canadian Journal of Psychology* 19:231–41.

Water Resources Council (1971) "Proposed Principles and Standards for Planning Water and Related Land Resources." *Federal Register* 36:245.

White, G. F. (1945) *Human Adjustment to Floods: A Geographical Approach to the Flood Problem in the United States*. Chicago: University of Chicago, Department of Geography, Research Paper No. 29.

———. (1961) "The choice of use in resource management." *Natural Resources Journal* 1:23–40.

———. (1964) *Choice of Adjustment to Floods*. Chicago: University of Chicago, Department of Geography, Research Paper No. 93.

———. (1966) "Optimal flood damage management: retrospect and prospect." In A. V. Kneese and S. C. Smith, eds., *Water Research*. Baltimore: Johns Hopkins Press.

———. (1970) "Flood-loss reduction: the integrated approach." *Journal of Soil and Water Conservation* 25:172–176.

———, Bradley, D., and White, A. (1972) *Drawers of Water: Domestic Water Use in East Africa*. Chicago: University of Chicago Press.

Wiggins, J. H., Jr. (1972) "Earthquake safety in the city of Long Beach based on the concept of balanced risk." In *Perspectives on Benefit-Risk Decision Making*. Washington, D.C.: National Academy of Engineering.

York, H. (1970) *Race to Oblivion: A Participant's View of the Arms Race*. New York: Simon & Schuster.

25. Economic analysis of natural hazards: an ordered choice approach

University of Pennsylvania

The large body of empirical evidence compiled by geographers and psychologists indicates that individuals have a difficult time dealing with hazardous events before they occur. Discussions of the rationale for this behavior as well as of specific examples from field studies appear elsewhere in this volume. This chapter provides additional empirical evidence on this point and develops a framework of decision making which should improve our understanding of why individuals are not concerned with the potential losses from hazards.

Insurance experience in the United States

One way that people can protect themselves against the consequences of natural hazards is by taking out insurance; however, the available statistics suggest that except in countries such as New Zealand where government insurance schemes are compulsory or widespread, most people are not adequately covered against damage from such natural hazards as floods and earthquakes. In the United States the private property loss in the San Fernando Valley following the 1971 earthquake was estimated at over $250 million but total losses payable by insurance were less than $45 million (U.S. Department of Housing and Urban Development, 1971). Most houses in California are of the wood-frame variety where the premiums are relatively modest (15¢ per $100 with a 5 percent deductible clause), yet in 1965 the total premiums written for earthquake insurance in the state were less than 5 percent of those for fire insurance (Underwriter's Report of California, 1973).

More conclusive evidence on the unwilling market for earthquake insurance in the United States is provided through a recent experiment by the Insurance Company of North America following the San Fernando earthquake (Syfert, 1972). The company mounted a serious campaign in October 1971 to market earthquake insurance in California, by placing newspaper ads in the major dailies, advertising on TV, and enabling all their California agents to mail special brochures and announcements to their customers. The following month only 61 policies were sold and then sales dropped off during the next seven months to an average of 17 per month. The response to the campaign was not exactly earthshaking.

A similar pattern of behavior may exist with respect to flood insurance coverage, but the national program is so new that it is too early to generalize. Even though Rapid City qualified for the federal government's subsidized National Flood Insurance in April 1971, only 29 policies were in force at the time of the June 1972 flood. Analogous behavior was evident with respect to the states hit by Tropical Storm Agnes; only 634 residential policies were sold in Pennsylvania, 2,097 in New York, and 619 in Maryland before the disaster occurred.

The only natural hazards where property owners have substantial protection today are fires, windstorms, and hail, all of which are required by banks and commercial institutions as a condition for a mortgage.

Recognizing that few individuals had protected themselves voluntarily from the consequences of natural hazards, the federal government recently developed a liberal disaster relief policy which bails out victims by giving them low-interest loans and forgiveness grants. In other words, the current federal relief policy treats disasters as if they were a public responsibility so that every taxpayer in the United States bears a fraction of the costs of disaster-induced damage to portions of the country.

The main purpose of this paper is to develop a formal model of decision making which may help explain the reluctance of individuals to concern themselves with natural hazards during the predisaster period. The approach has been stimulated by the evidence from the field and laboratory reviewed in the previous section and is couched within the bounded rationality framework developed in more detail by Slovic, Kunreuther, and White (chap. 24 above). In this volume a model of ordered choice under uncertainty henceforth designated as the lexicographic approach will be utilized to analyze individual decisions and will be illustrated through a simple one-goal model. (The word "lexicographic" indicates that there is some preference ordering by individuals in their decisions. The term is used in the same sense as that used by Georgescu-Roegen, 1954; and Robinson and Day, 1970.) The implications of this framework of analysis with respect to an individual's perception and choice of adjustment to natural hazards are initially discussed under the implicit assumption that he must bear the full loss of the disaster. The concluding section demonstrates the effect of current federal disaster relief policy on the decision maker's perception of hazards and suggests an alternative system of federal response based on comprehensive disaster insurance.

Framework for analysis

The lexicographic model of choice under uncertainty explicitly addresses the perceptual problem of extreme events such as natural hazards. This framework of analysis appears to be a more accurate way of looking at the decision-making process under uncertainty than the expected utility model and it should yield different policy prescriptions. The proposed model is based on concepts both from organization behavior and economic theory which stress the limits of man's rationality and suggest ways in which he copes with uncertainty.

Antecedents in the literature

Central to the analysis is the notion first postulated by Simon (1949) that organizations and individuals make decisions which satisfy short-run goals rather than maximize some long-run objective function. This concept has been utilized by Cyert and March (1963) in their work on the behavioral theory of the firm and was extended by Williamson (1970) in his study of the hierarchical decision-making process and the multifunction firm which characterizes the modern corporation.

As of now, the behavioral theory of the firm has had little impact on the thinking of most economists. One possible reason is its emphasis on individual decision making without great concern for market behavior which forms the basis for most work in traditional economic theory. Perhaps an even more important reason is its prevalent use of standard operating procedures as a basis for decision making and its emphasis on goals other than maximizing profits. This view is at wide variance with the model of rational economic man which is the cornerstone of microeconomic theory and consumer behavior.

The use of lexicographic programs for describing the decision-making process was first proposed by Georgescu-Roegen (1954) in his pathbreaking analysis of consumer behavior and has since been utilized by Encarnacion (1964) and Day (1970) in developing a unified theory of rational, economic behavior. The proposed model extends the work of these authors by couching the analysis in terms of the state-preference approach formulated by Arrow (1964), and expanded upon by Hirshleifer (1970) in his illuminating book on investment decisions under uncertainty and by Marshall (1969) in developing a theory of insurance. The analysis also builds on the works of Markowitz (1959) and Tobin (1958) by stressing the importance of variance as well as expected return in determining allocation decisions by firms or individuals. By using these concepts, the differences in individual behavior implied by a lexicographic model can be compared explicitly with expected utility theory.

Developing a model

To begin with, consider an individual or firm with present wealth (W_0) which can be allocated to activities x and y next period. In the case of a farmer such as a Tanzanian peasant the value of W_0 consists of a certain amount of land as well as labor and capital and his decision governs the amount of land devoted to maize (x) and cotton (y) during a given season. The two crops represent different adjustments to the availability—

excess or deficiency—of moisture. For a firm the value of W_0 is its available work force, production capacity, and inventory on hand and its allocation decision determines how much of each of two products to manufacture next year. For a consumer the decision might be related to as a budget allocation between food (x) and housing (y). The future is represented by a point in time (time 1) in which there are two alternative states of nature—either a normal year (state a) or an extreme year (state b) with respective probabilities of occurrence p_a and p_b. The extreme year may be caused by some natural hazard such as a flood or drought. If W_0 is invested solely in x, then the net return at time 1 is a random variable X which can take on one of two values, X_a or X_b. The random variable Y is defined in a similar manner.

Let m represent the proportion of W_0 allocated to activity x with the residual, $1 - m$, allocated to y. Each value of m represents a distinctive adjustment by an individual to future states of nature. The outcome of a specific allocation decision after a particular state of nature has occurred will be designated as a state of the world. This definition is identical to the one used by Arrow (1964) and Hirshleifer (1970) in their treatment of individual preferences. Since there are only two states of nature in this example, each value of m implies two possible states of the world, Z^a and Z^b, for wealth at the end of period W_1. Thus if $m = 1$, the two possible states of the world are $Z^a = X_a$ with probability p_a and $Z^b = X_b$ with probability p_b. When $m = 0$, then Y_a and Y_b are the two possible states of the world. If an individual diversifies by allocating 2/3 of W_0 to activity x (i.e., $m = 2/3$) and the remainder to y, then with constant returns to scale, $Z^a = 2/3 \, X_a + 1/3 \, Y_a$ with probability p_a and $Z^b = 2/3 \, X_b + 1/3 \, Y_b$ with probability p_b. The states of the world resulting from each of these three alternatives are depicted in Table 25-1.

The lexicographic approach will distinguish between productive and protective activities. In the case of a firm, productive activities yield positive returns to the investor in the form of profits while for a consumer such activities satisfy basic needs. The sole function of protective activities is to reduce the probability of certain states of nature occurring or to reduce the loss to the individual should such a state of nature occur. Hence, a homeowner may purchase a house in the

Table 25-1. States of the world for three alternative allocations of W_0

Alternative Allocations of W_0	State of Nature	
	Normal Year (a)	Extreme Year (b)
$m = 1$	X_a	X_b
$m = 0$	Y_a	Y_b
$m = 1/3$	$2/3 \, X_a + 1/3 \, Y_a$	$2/3 \, X_b + 1/3 \, Y_b$

floodplain (productive activity) and then decide to flood proof the structure and/or take out flood insurance (protective activities). Floodproofing reduces the probability of a state of nature (severe flooding of the house) while insurance redistributes income toward a hazardous state of the world (insurance payment if water damage occurs to the home). The individual derives utility from these activities only to the extent that they protect his original investment. Our treatment of insurance and other protective mechanisms thus differs from the recent analysis by Ehrlich and Becker (1972) and follows the spirit of Arrow (1965).

We will first treat the case where an individual must invest his entire wealth in productive activities so that his sole method of reducing risk is through diversification. The following section will then discuss when an individual will want to allocate part of his wealth to protective activities.

The individual would like to allocate his initial wealth in such a way as to maximize the expected value of W_1, i.e., $E(W_1)$, but has one-period goals which may constrain his behavior. For example, a farmer may want to maximize the expected return from his land but has certain minimum subsistence requirements which may be critical to his future survival. Firms may want to maximize the expected return from their productive activities but may also specify certain short-run profit targets as well as sales targets which may critically influence their decision on how much of W_0 should be invested in x and y. The individual is assumed to be able to rank these goals in order of importance to him with 1 being the most important goal and R the least.

The possible outcomes of each goal i are given by a random variable Z_i with predetermined target value denoted by Z_i^*. For example Z_i may be the random variable "annual dollar sales" and Z_i^* will be the sales target. The probability distribution of Z_i will be determined by the proportions of W_0 allocated to producing x and y respectively. For each goal i the decision maker is assumed to be willing to tolerate a maximum risk level α_i^* that $Z_i < Z_i^*$. The value of α_i^* may be determined by cost considerations or by some personal preference function.

To illustrate, it is common for firms to base their production and inventory policy on achieving at least a certain service level with respect to meeting demand. The firm frequently specifies this service level explicitly by saying that it wishes to have a 95 percent chance of satisfying all incoming orders within a prescribed period through stock on hand. If this objective is classified as goal i, then the corresponding value of $\alpha_i^* = 0.05$.

The question here is not how the decision makers specify the appropriate target values and maximum risk levels when confronted with the risk of natural hazards, although that is an important topic for empirical research. It is simply asserted that individuals specify values of Z_i^* and α_i^* implicitly or explicitly and that these values critically influence their decisions.

Given these assumptions, the appropriate model to describe an individual's behavior in allocating resources in the face of a natural hazard is:

$$\max E(W_1) \tag{1}$$

subject to

$$\text{probability } (Z_i < Z_i^*) \leq \alpha_i^* \; i = 1, \ldots, R \tag{2}$$

This model is of the chance-constrained programming variety, which has been treated extensively in the management science literature. (For a recent set of references on chance-constraint programming see Eisner, Kaplan, and Soden, 1971.) If there is at least one feasible solution to this problem, then the individual chooses the portfolio yielding the highest $E(W_1)$. If, on the other hand, there is no feasible alternative which satisfies all the constraints, then the individual will be forced to relax one or more of his restrictions.

One solution to this latter problem would be to assign different penalty functions for deviating from each of the goals and trying to minimize the overall costs. This approach has been labeled as goal programming and has recently received wide coverage in the literature. (For a discussion of the goal programming approach and list of references see Dyer, 1972.) The main drawback of the goal programming approach is that it forces the decision maker to specify simultaneously the costs of deviations from each of his goals. Not only may it be difficult to obtain these appropriate cost functions but it is somewhat unrealistic to assume that in practice a decision maker will modify all his constraints simultaneously. Rather, he is likely to change one constraint at a time to see if he can obtain a feasible solution.

It is argued here that an individual modifies his constraint set using a system of priorities dictated by the relative importance of each of the goals, lowered-numbered goals being more important than higher-numbered ones. In other words, he first is willing to accept a lower probability of achieving goal R than any of the others. His modified objective is thus:

maximize

$$\text{probability } (Z_R > Z_R^*)$$

subject to

$$\text{probability } (Z_i < Z_i^*) \leq \alpha_i^* \; i \leq 1, \ldots, R-1$$

If there is still no feasible alternative, then goal $R-1$ will be relaxed and the first $R-2$ constraints will be maintained, and so on down to the last possible case where the objective function is to maximize the probability $(Z_1 > Z_1^*)$. This model of ordered choice is similar to one suggested by Encarnacion for the deterministic case. In terms of the Arrow-Hirshleifer state-preference formulation, this approach implies that for *each* goal i there is a critical state of the world Z_i^* toward which the decision maker is aiming. Given uncertainty, he is assumed to tolerate some maximum risk level α_i^* that future wealth will fall below this state of the world.

A one-goal example

To illustrate the implications of this type of ordering rule for decision making we will focus on a one-goal model that has been labeled "safety-first" or "safety-fixed" in the literature. (For a detailed discussion of safety-first and safety-fixed models and a comparison of these approaches with expected utility theory see Pyle and Turnovsky, (1970.) In more formal terms, suppose the individual wants to allocate W_0 among a set of activities so as to:

$$\max E(W_1) \qquad (3)$$

but that he is concerned with the possibility of W_1 falling below a critical minimum return (Z^*) as reflected in the constraint

$$\text{probability } (W_1 < Z^*) \leqslant \alpha^* \qquad (4)$$

where α^* is the maximum risk level. The above model has been utilized by Telser (1955), Roumasset (1971), and Kunreuther (1972) to explain firm or individual decision making under uncertainty.

To see whether (3) and (4) yield a feasible solution the *minimum* possible risk level can be determined for a given value of Z^* simply by allocating W_0 in such a way as to satisfy the following objective function:

$$\min [\text{probability } (W_1 < Z^*)] \qquad (5)$$

Designate the resulting risk level as α'. If $\alpha' < \alpha^*$ then there is some portfolio of activities, or adjustment to the risk, which will satisfy both (3) and (4). If $\alpha' > \alpha^*$ the individual will have to modify his target value Z^* downward and/or α^* upward. If it is assumed that Z^* remains fixed, the decision maker will set $\alpha^* = \alpha'$ and utilize the portfolio specified by (5).

An illustration of the sequential nature of the decision-making process for a prespecified value of W_0 is depicted in Fig. 25–1 where an extreme set of points in the $Z^* - \alpha$ plan (curve EE) is plotted based on (5). Naturally as Z^* increases, the probability of achieving $W_1 > Z^*$ decreases and hence α' increases. Any point within the hatched area of Fig. 25–1 represents a feasible solution to (3) and (4) (e.g., point A). Point B, on the other hand, is outside this feasible region, so the individual must utilize (5) and raise the acceptable risk level from α^* to α'. In other words initial wealth (W_0) is then allocated in such a way as to minimize the probability of disaster. This latter criterion is identical to the one proposed by Roy (1952) in his classic article on the safety-first principle. By following a sequential approach, the constraint in the first stage becomes the objective function whenever a second state is required.

Perception of natural hazards and choice of adjustment

The above lexicographic model has implications with respect to the allocation of initial wealth between productive and protective activities. This will be most easily

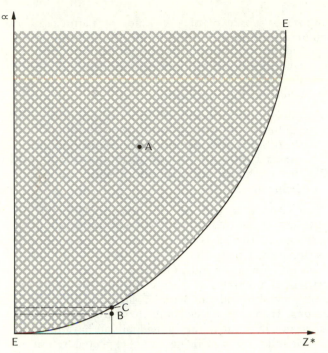

Fig. 25–1. The sequential nature of the decision-making process for a prespecified value of W_0

seen by grouping states of the world and states of nature into mutually exclusive sets determined by target values and maximum allowable risk levels. So as to keep the analysis simple, the one-goal safety-first model will be used for illustrative purposes. The same concepts apply in treating the more general multiple goal model given by equations (1) and (2).

Theoretical treatment

To begin with, suppose the individual would like to determine the distribution of returns from a *specific* portfolio of activities, or adjustment, given some initial wealth W_0 and N possible states of nature. For each state of nature j there will be a corresponding state of the world Z^j representing a possible value of W_1. Given a target level (Z^*), the individual can conveniently divide the states of the world into two distinct sets

$$Z' \quad \text{states where } Z^j < Z^*$$
$$Z'' \quad \text{states where } Z^j \geqslant Z^*$$

Turning now to the states of nature which produce the values Z^j, suppose that for a specific portfolio the states of nature are ordered so that the event which produces the lowest value of W_1 (i.e., the worst state of the world) is ranked 1 and its corresponding outcome is specified as Z^1. Continuing in this manner the state of nature producing the highest value of W_1 will be ranked N and its outcome will be Z^N. In other words, for each portfolio the states of nature are ordered on the basis of the resulting state of the world (i.e., the value of W_1).

Suppose a subset of the N states of nature, say n, produces states of the world in set Z'. These n states of nature are of prime concern to the individual because they cause W_1 to fall below the critical target Z^*. Given a maximum risk level (α^*) these n states of nature can now be divided into two distinct sets:

D' those ordered states of nature whose joint probability of occurrence is less than or equal to α^*. In other words, state i is only included in D' if states 1 though $i - 1$ are already part of this set.

D'' the remaining states of nature which result in Z^j being in Z'

The states of nature in D'' are of concern to the individual since they violate the constraint given by equation (4). Hence, the target value Z^* and the critical risk level α^* jointly determine those states of nature which may force the individual to modify his allocation decision. If no states of nature fall in D'', then the given portfolio will be feasible, since equation (4) is satisfied. If, on the other hand, there are some states of nature in D'', then the individual will have to revise his portfolio by sacrificing some expected return to achieve greater safety. Eventually, he will either determine a set of activities which maximizes $E(W_1)$ while having D'' empty, i.e., which satisfies both (3) and (4), or he will modify α^* upward by satisfying (5).

By introducing a set of protective activities into the picture the individual has an opportunity to modify the probability of certain states of nature occurring (e.g., floodproofing a structure) and/or to redistribute income to certain hazardous states of the world (e.g., through insurance payments). It is assumed here that the decision maker only considers allocating a part of W_0 to those protective activities which can modify *unsatisfactory* states of nature and/or *unsatisfactory* states of the world as defined by target levels and critical risk levels. Specifically insurance and other protective measures will be treated as a response to an initial allocation plan which consists solely of productive activities but does *not* satisfy the constraint set given by equation (2). In other words, whether or not to divert some of the W_0 to protective activities is viewed here as a second stage in the decision-making process.

In contrast, Ehrlich and Becker (1972) treat both productive and protective activities as satisfying basic needs. It follows from their analysis that both types of activities would be examined initially across all possible states of nature and states of the world. An optimal portfolio would be defined by them as one that maximizes expected utility.

How do the sets D' and D'' affect an individual's desire to protect himself against the consequences of natural hazards if his behavior is characterized by a lexicographic model? Since events such as rare floods and earthquakes tend to cause extreme damage to all activities in the affected region, it is likely that these

states of nature will be the first candidates for admission into D' for any given portfolio, in which case they will be ignored by the decision maker. To determine whether a state of nature is in D' or D'' for a particular portfolio, it is necessary to collect only a limited amount of information on damage and probabilities. Specifically, if the expected damage from a specific hazard causes Z^j to be in Z'', he can ignore the events. On the other hand, if Z^j is in Z' and the state of nature j is in D'', then the individual must determine whether or not to allocate some of W_0 to protective activities and/or to further diversify his portfolio among productive activities.

The implications of the lexicographic model with respect to loss preventation from natural hazards should thus be clear. If these extreme states of nature are in D', then individuals will not even bother to collect information on the potential damage from such events nor will they investigate the cost of protective activities such as floodproofing or insurance.

The lexicographic model also implies that high income people and large firms will tend to be the least concerned with natural hazards in their decision-making process. Although Z^* will normally increase as initial wealth increases, the ratio of Z^*/W_0 should decline, particularly if targets are considered to be minimum acceptable levels. Furthermore, individuals and firms with a large initial wealth have reserves to bail them out in case of an emergency and have relatively easy access to funds at the market rate of interest. Hence they will likely be willing to accept a larger value of α^* than their lower income counterparts. The net effect of these two changes should be to reduce the set of events in D'' and hence minimize the importance of their protecting themselves against damage from hazards.

The other extreme case is the low-income individual who is initially faced with a large number of states of nature in D'' for a wide variety of adjustments. In his case he will likely be concerned with the effects of natural disasters on his future wealth and hence will consider diverting some initial wealth to protective activities and/or diversifying W_0 by investing in productive activities with both a relatively low return and a low variance. Even with these adjustments he still may be unable to specify a feasible portfolio which satisfies both Z^* and α^* and may choose to increase his maximum risk level. This revision may force him to invest a larger proportion of his wealth in productive activities which yield a relatively high expected return but also result in a high probability of disaster.

It is thus conceivable that rich and poor individuals will allocate W_0 in a similar manner but for entirely different reasons. The high-income person can afford to overlook hazards and will tolerate a higher variance in return for greater expected return; the low-income individual is fully aware of the dangers of a disaster but requires a higher expected return in order to stand a chance of meeting his target level (e.g., earning enough to feed his family) and is thus forced to increase the risk of not having enough to meet his needs.

An illustrative example

A numerical example will illustrate the decision-making process of a Tanzanian farmer who can allocate his land (W_0) to either cotton (x) or maize (y). To keep the computations simple only two states of nature will be assumed—a normal year (a), with $p_a = 0.9$, and a drought (b), with $p_b = 0.1$. For simplicity, constant returns to scale from land are assumed so that a farmer who invested his entire land in cotton would maximize $E(W_1)$. Drought insurance is the protective activity open to the farmer. Illustrative states of the world for the three portfolios in Table 25–1 as well as the values of $E(W_1)$ are portrayed in Table 25–2.

We will consider three cases.

Case 1: $Z^* = 1$; $\alpha^* \geqslant 0.1$. Suppose that the farmer allocates all his land to cotton (i.e., m = 1) so that $Z^a = 9$ and $Z^b = 0$. In other words,

$$Z' = 0 \; ; Z'' = 9$$

The decision maker must now characterize those states of nature which are responsible for any states of the world in Z': in this case only a drought (state b). Given $\alpha^* \geqslant 0.1$, it is implied that

$$D' \text{ - state } b \; ; \; D'' \text{ is empty}$$

Hence the allocation of W_0 which maximizes $E(W_1)$ also satisfies the minimum return constraint and is, therefore, optimal. In this case the farmer would not consider purchasing any type of drought insurance.

Case 2: $Z^* = 1$; $\alpha^* < 0.1$. By allocating all of his land to cotton the farmer would have the same states of the world in Z' and Z'' as in case 1, but now he would find that

$$D' \text{ is empty; } D'' = \text{ state } b$$

Hence the initial portfolio is not a feasible one. The farmer then has the option of either purchasing some

type of drought insurance and/or diverting some of his land from cotton to maize so as to be able to satisfy the minimum return constraint. If he chooses not to buy any insurance, then he would like to set m as large as possible while still satisfying his probabilistic constraint. For this highly simplified example his optimal investment pattern is to set $m = 2/3$, which yields $E(W_1) = 7.3$, as seen from Table 25–2. From Table 25–2 he can also see that if $m = 2/3$, $Z'' = 8.1$ and Z' has no states of the world. As a result both D' and D'' will be empty so the farmer has specified a feasible portfolio. Note that if more than 2/3 of the Tanzanian's land is allocated to cotton, then $Z^b < 1$; hence $Z' = Z^b$ and $D'' = $ state b, indicating that such a portfolio is infeasible.

Another extreme course of action would be for the farmer to plant only cotton and purchase insurance guaranteeing him a return of 1 if a drought occurs. The attractiveness of this strategy over crop diversification depends on the price of insurance, the opportunity cost of the money used to pay the premium as well as the difference in the expected value of W_1 between growing only cotton on his land and allocating 1/3 of his acreage to maize (i.e., $\Delta E(W_1) = 8.1 - 7.3 = 0.8$ in the above example). In a similar manner the farmer might also compare a set of mixed strategies consisting of some crop diversification and of some insurance.

Case 3: $Z^* > 3$; $\alpha^* < 0.1$. Now the farmer cannot specify any portfolio of productive activities which satisfies both Z^* and α^* levels. He would then either purchase some drought insurance, or if the premiums were sufficiently high, increase his maximum risk level by utilizing equation (5). The second alternative illustrates the plight of the low-income farmer who finds it necessary to gamble by tolerating a higher risk level than he had originally intended. In this example if he maintains $Z^* > 3$, he must raise $\alpha^* = 0.1$ to determine a feasible portfolio. He will now find it optimal to allocate his entire land to cotton, thus following the same pattern as the case 1 farmer but for entirely different reasons.

It should be emphasized here that in practice individuals may modify both their goals and maximum risk levels simultaneously. Thus the low-income farmer of case 3 could revise Z^* downward as well as α^* upward, in which case he might allocate portions of his land to both corn and cotton.

Empirical evidence

There is some empirical evidence to support the view that individuals employ a lexicographic rule in making decisions. An example is provided in a study by White, Bradley, and White (1972) of how people drawing water for household use in rural areas where water may be subject to contamination choose among alternative

Table 25–2. States of the world and $E(W_1)$ for three alternative allocations of W_0 to cotton (x) and maize (y)

Alternative Allocations of W_0	State of Nature		$E(W_1)$
	Normal year (a)	Drought (b)	
m = 1	9	0	8.1
m = 0	6	3	5.7
m = 2/3	8	1	7.3

sources. They found that the users classified the sources as good or bad solely on the basis of perceived health effects. If more than one source was satisfactory on this primary dimension, the remaining "good" sources were then discriminated on the basis of the economic costs of transporting the water. There was little indication that the users were willing to "trade off" lower-quality water with accompanying health hazard for lower economic costs. The two dimensions were simply not compensatory.

A laboratory experiment of risk taking by Lichtenstein, Slovic, and Zink (1969) suggested that individuals process diverse dimensions of information by setting a criterion level on one or more of these dimensions. Alternatives that did not promise to meet that criterion were rejected. For those alternatives that remained, another dimension was employed as a basis for discrimination.

In a study of farmer's crop-growing decisions (Kunreuther, 1972) it was hypothesized that peasant farmers seek reasonable assurance of survival when deciding how to allocate their resources among crops varying in risk and expected yield. Only for those allocation plans in which survival needs are likely to be met does maximizing expected net return become a consideration. Empirical evidence from Bangladesh presented in that paper supports the theoretical model.

Recent experience under the National Flood Insurance Program in the United States suggests that individuals' behavior can be described more accurately by a lexicographic model than by the expected utility theory. Despite the fact that an individual only pays a premium which is approximately 10 percent of the actuarial cost of insurance, relatively few individuals moved promptly to take out coverage before a disaster. This lack of interest persisted after a disaster. In Harrisburg where damage from Tropical Storm Agnes was unusually severe only 556 new policies were sold in the following 6 months despite concerted publicity efforts on the part of the insurance industry. Liberal federal relief following Agnes undoubtedly dampened enthusiasm for coverage, but an equally important factor appears to be the cost of collecting and processing information on unlikely events. A similar point is made by Hirshleifer (1973), who suggests that the costs of attention may be responsible for the lack of interest in insurance against natural hazards. There is strength in this line of reasoning but such transaction costs provide the rationale for using the above lexicographic model to describe decision making under uncertainty rather than attempting to incorporate them into the expected utility framework.

Implications for federal policy in the United States

The foregoing analysis is that if a disaster occurs, the individual will be forced to bear the entire cost himself. Thus a farmer who suffers severe losses from a flood will

be forced to utilize his reserves or borrow funds or food to sustain his family during the remaining part of the season. If Z^* represents the minimum return from his land, he will undoubtedly be forced to raise this value in an attempt to return to his predisaster wealth position. If α^* remains the same, an increase in Z^* implies that the number of states of the world in Z' will increase and hence there will be more states of nature in D''. As a result, an individual will choose a portfolio of productive activities which has a lower variance or he may now decide to divert part of W_0 to protective activities such as insurance. Such a decision will be reinforced if the recent catastrophe has caused the person or firm to increase the subjective probability of the occurrence of the extreme event.

History of disaster relief

This sequential pattern does not depict the current situation in the United States. During recent years federal policy has treated disasters as if they were a public responsibility with every taxpayer bearing a part of the cost of disaster-induced damage to a few. The program in the United States is somewhat similar to those of other countries lacking national schemes (Vaughn, 1971) and has interesting contrasts and similarities to the national program in New Zealand described by O'Riordan (chap. 26 below).

With the passage of the Small Business Act in 1952, the SBA began to offer low-interest loans to homeowners and businesses suffering injury from natural disasters. The general purpose of SBA disaster loans is "to restore a victim's home or business property as nearly as possible to its pre-disaster conditions." Before the Alaskan earthquake, the agency provided 3 percent loans with a maximum repayment period of 20 years to cover the exact amount of physical damage. It was understood that the borrower would use the entire loan strictly for the purpose of rebuilding or repair. The severity of the damage in Alaska caused concern that unless the SBA liberalized its policy many individuals would not qualify for a disaster loan because of their inability to pay off their old mortgages and other debts and still make monthly payments to the SBA.

Perhaps the most significant revision of SBA policy was the authorization of loans for substantial debt retirement for any homeowner or business suffering losses from the earthquake. He was given funds not only to repair his damaged structure, but also to retire old debts (for example, outstanding mortgages, accounts payable) which might have nothing to do with the disaster. Instead of continuing to pay conventional 7 or 8 percent rates on these outstanding claims, the borrower could now retire them at a subsidized 3 percent rate. A further reduction in the size of a victim's monthly payment was achieved by permitting a 30-year amortization period instead of the normal 20-year maturity of SBA loans. If the property owner requested it, the agency would waive any repayments of principal and interest during the first year of the loan and repayments of principal up to an additional four years. The victim's burden was minimized and in a number of cases, particularly for businesses, the borrower was financially sounder after than before the "catastrophe."

Although the SBA made it clear that its action in Alaska was taken to meet a special situation, the agency did not retreat to its more stringent policy. Using the Alaskan case as a precedent, a congressional bill was passed at the end of June 1965, authorizing the SBA to permanently extend its maximum loan period from 20 to 30 years. The Southeast Hurricane Disaster Relief Act of 1965 authorized the S.B.A. to "forgive" a part of each loan up to a maximum of $1,800, a provision which had not been permitted even in Alaska. (For more details of the equity and efficiency of the SBA disaster loan policy, see Kunreuther, 1968; Dacy and Kunreuther, 1969, chaps. 9–10 and Kunreuther, 1973).

The Disaster Relief Act of 1970 increased the forgiveness amount to $2,500 and permitted the annual interest rate on loans to be set up to 2 percent below the rate for 10–12-year government securities but never higher than 6 percent per year. A congressional act (PL 92-385) inspired by the Rapid City floods and Tropical Storm Agnes liberalized SBA policy even further. An owner of a home or business which received damage from a disaster occurring between January 1, 1972, and July 1, 1973, could obtain a forgiveness grant of up to $5,000 and a loan at 1 percent per year to cover the remaining losses.

The liberalized federal disaster relief policy reinforced individuals' decisions not to protect themselves against future potential losses. The states of nature causing natural disasters often trigger liberal relief from the federal government so that an individual's future wealth is only slightly decreased; in a few cases the "disaster" may be a financial blessing. An extreme case from the Fairbanks flood of August 1967, reported by the New York *Times,* illustrates the latter situation. A motel owner suffered damage estimated at slightly over $140,000 but received a 3 percent loan for $894,000 to be paid over a 30-year period; $704,000 of this loan was used to pay off outstanding debts and $50,000 for working capital. Assuming the conventional loan rate was 7 percent per year, the present value of the debt retirement portion of the loan was only $566,000. Thus despite the loss of $140,000 the motel owner was $188,000 better off after the disaster than before the event. It is no surprise to find people in Anchorage and Fairbanks saying that the earthquake and flood were the two best things that happened to Alaskans in recent years.

Earlier it was hypothesized that extreme events do not affect individuals' actions prior to occurrence because they belong to set D'. Current federal policy after a disaster reinforces this behavior because of its effect on W_1. By providing liberal relief to an individual

suffering losses, the state of nature "flood" or "earthquake" will not substantially decrease his wealth and hence may result in a state of the world belonging to set Z''.

If victims of a disaster were forced to bear the costs themselves, it has been shown above that they would have a larger incentive to protect themselves against future catastrophes. Current federal policy encourages individuals not only to continue to ignore these events in future but actually to take steps to profit from the next earthquake or flood. It was not unusual to hear that residents of the San Fernando Valley and Alaska were attempting to obtain the smallest down payment on their home mortgage and longest possible maturity so as to be in the best possible position to take advantage of the SBA policy should another earthquake hit the area.

A suggested program

A lexicographic model of choice implies that individuals have not voluntarily protected themselves against the potential losses from natural hazards because these states of nature are in the set D'. It is precisely because hazards have been ignored by most individuals that the government has liberalized federal relief following recent major disasters. As a result, the individual has less incentive today than in the past to invest in protective activities.

If one wants to shift the cost burden from the public to the private sector, the implications for policy arising from this model of ordered choice should be clear: *require* the individual to divert some of his initial wealth into protective activities. The expected utility model, on the other hand, would not necessarily imply such a radical solution to the problem. For example, the treatment of insurance by Ehrlich and Becker (1972) implies that if premiums are sufficiently subsidized, the consumer will want to purchase coverage voluntarily as a way of maximizing his expected utility. The recent experience with the National Flood Insurance program is not encouraging in this regard but there is insufficient empirical data to analyze precisely why most people in flood-prone areas have failed to take out coverage. For a detailed description of the provisions in the National Flood Insurance Act of 1968, see U.S. Office of Emergency Preparedness (1972).

A modified version of the National Flood Insurance program may still serve as a prototype for developing a realistic plan to mitigate future disaster losses. Specifically, some form of subsidized disaster insurance can be made available to homes and businesses in a hazard-prone area but only after the community has taken positive steps toward reducing potential losses by enforcing adequate land-use measures and building code regulations. The initiative can thus lie with the communities rather than the federal government. In essence, the federal government would help pay the costs of protecting individuals now residing in hazard-prone areas from future disaster losses while requiring that the communities make these areas safer places in which to live. New residents would be charged the actuarial rate since they have the option to choose their location.

For such a plan to have any chance of success, the federal government would have to withdraw its liberal disaster assistance programs such as providing low-interest loans and forgiveness grants. There would otherwise be no financial incentive for communities to develop land-use measures or building codes. The most effective way of achieving a favorable reaction by communities toward such a program would be for federal agencies, such as the Veterans Administration and Federal Housing Administration as well as private lending institutions, to require some form of comprehensive disaster insurance as a condition for mortgage. Insurance would thus be compulsory for the large majority of property owners and hence the political pressure for reinstating liberal relief following the next major disaster would be greatly reduced.

Acknowledgments

I would like to express my appreciation to Anthony Fisher, Jack Hirshleifer, David McNicol, Paul Slovic, and Gilbert White and members of the Industrial Organization Workshop at the University of Pennsylvania for their comments on a preliminary draft of this paper.

References

Arrow, K. (1964) "The role of securities in the optimal allocation of risk bearing." *Review of Economic Studies* 21:91–96.

——. (1965) *Aspects of the Theory of Risk Bearing.* Helsinki: Yrgo Jahnssonin Saatio.

Cyert, R., and March, J. (1963) *A Behavioral Theory of the Firm.* Englewood Cliffs, N.J.: Prentice-Hall.

Dacy, D., and Kunreuther, H. (1969) *The Economics of Natural Disasters.* New York: Free Press.

Day, R. (1970) "Rational choice and economic behavior." *Theory and Decision* 1 (3):229–51.

Dyer, J. (1972) "Interactive goal programming." *Management Science* 19 (1):62–70.

Ehrlich and Becker. (1972) "Market insurance, self-insurance and self-protection." *Journal of Political Economy* 80:623–48.

Eisner, M., Kaplan, R., and Soden, J. (1971) "The E-model of chance constrained programming." *Management Science* 17 (5):337–53.

Encarnacion, J. (1964) "Constraints and the firm's utility function." *Review of Economic Studies* 31 (86):113–20.

Georgescu-Roegen. (1954) "Choice, expectations and measurability." *Quarterly Journal of Economics* 68:503–32.

Hirshleifer, J. (1970) *Investment Interest and Capital.* Englewood Cliffs, N.J.: Prentice-Hall.

——. (1973) Personal communication Jan. 15.

Kunreuther, H. (1968) "The case for comprehensive disaster insurance." *Journal of Law and Economics* 11:133–163.

——. (1972) "Risk-taking and farmer's crop growing decisions." Chicago: University of Chicago, Center for Mathematical Studies in Business and Economics, Report No. 7219.

—— (1973) *Recovery from Natural Disasters: Insurance or Federal Aid?* Washington, D.C.: American Enterprise Institute.

Lichtenstein, S. C., Slovic, P., and Zink, D. (1969) "Effect of instruction in expected value on optimality of gambling decisions." *Journal of Experimental Psychology* 79:236–40.

Markowitz, H. (1957) *Portfolio Selection.* New York: John Wiley, p. 2.

Marshall, J. (1969) "Theory of insurance and applications to insurance regulation." Cambridge, Mass.: MIT, Ph.D. dissertation.

New York Times (1968) January 19, p. 18, col. 7–8.

Pyle, D. H., and Turnovsky, S. J. (1970) "Safety-first and expected utility maximization in mean-standard deviation portfolio analysis." *Review of Economics and Statistics* 52:75–81.

Robinson, S., and Day, R. (1970) "Economic decisions with lexicographic utility." Madison: University of Wisconsin, SSRI Report No. 7047.

Roumasset, J. (1971) "Risk and choice of technique for peasant agriculture: safety first and rice production in the Philippines." Madison: University of Wisconsin, SSRI Report No. 7118.

Roy, A. (1952) "Safety first and the holding of assets." *Econometrica* 20:431–49.

Simon, H. (1949) *Administrative Behavior.* New York: Free Press.

—— (1959) "Theories of decision making in Economics and Behavioral Sciences," *American Economic Review* 49:253–283.

Syfert, Robert K. (1972) "The unwilling market for earthquake insurance." *Best's Review* 73:14–18.

Telser, L. (1955) "Safety first and hedging." *Review of Economic Studies.* 23 (1):1–16.

Tobin, J. (1958) "Liquidity preference as behavior towards risk." *Review of Economic Studies* 25:65–86.

Underwriter's Report of California (1973) Roy Pasini, editor (San Francisco, Cal.: May 31, 1973) p. 14.

U.S. Department of Commerce. (1969) Coast and Geodetic Survey, *Studies in Seismicity and Earthquake Damage Statistics.* Report prepared for the Department of Housing and Urban Development, Office of Economic and Market Analysis.

U.S. Department of Housing and Urban Development. (1971) Federal Insurance Administration, *Report on Earthquake Insurance to the Congress of the United States.* Washington, D.C.: Government Printing Office.

U.S. Office of Emergency Preparedness. (1972) *Disaster Preparedness,* vols. 1–3. Washington, D.C.

Vaughn, J. (1971) "Notes on insurance against loss from natural hazards." Toronto: University of Toronto, Natural Hazards Working Paper No. 21.

White, G. F., Bradley, D., and White, A. (1972) *Drawers of Water: Domestic Water Use in East Africa.* Chicago: University of Chicago Press.

Williamson, O. (1970) *Corporate Control and Business Behavior.* Englewood Cliffs, N.J.: Prentice-Hall.

IV National reviews

26. The New Zealand natural hazard insurance scheme: application to North America

TIMOTHY O'RIORDAN
Simon Fraser University

New Zealand offers a unique case study of a national natural hazard insurance policy which covers all classes of natural hazard and which in theory at least comes very close to the optimum in hazard management. In practice, however, the scheme is running into numerous difficulties which tend increasingly to divert implementative policy from legislative policy to the ultimate detriment of the program. This paper will review the New Zealand scheme, account for this unfortunate divergence of policy, and offer suggestions for incorporating the lessons learned from this analysis into hazard management practice in North America.

The New Zealand policy as conceived in the legislation

The New Zealand policy is essentially a national insurance scheme to cover damage from all classes of natural hazard. Any building covered by a fire-insurance policy is automatically subject to an additional levy of 5¢ per $100 of cover. This revenue (which amounts to $13 million annually) is paid to an administrative body called the Earthquake and War Damage Commission, and is diverted into two accounts. Nine-tenths goes into the Earthquake and War Damage Fund. This fund now stands at about $120 million and from it the commission pays out on any damages (up to the indemnity value of the property involved) caused by earthquake or, in the event of war, enemy action. Ten percent of the annual revenue is diverted into the Extraordinary Disaster Fund, which now stands at about $600,000. This fund is used to repair or replace (to its indemnity value) property damaged by events considered to be "abnormal and unforeseen and of extraordinary effect." Such events include volcanic eruption, flood, storm, tempest, landslide, or tsunami—in fact any natural disaster is covered.

In theory the New Zealand scheme offers a valuable example of near-optimum hazard management in a natural context.

The scheme is based upon the philosophy that the "true" hazard is that event which exceeds the normal adjusting and buffering capabilities of the human system to absorb it (Kates, 1970). In New Zealand such events may occur anywhere, may affect any property owner, and are essentially unpredictable. Potential damage (particularly catastrophic earthquake) may be considerable (a severe earthquake in Wellington could result in over $600 million in damages) (Power, 1968), and since private insurance companies are unable to provide the

necessary cover (due to lack of statistical knowledge of the event, unwillingness to cover only the risky clients who are the only ones to apply, and inability to build up large reserves or collect the necessary reinsurance cover) (Steinbrugge, 1968), a communal fund is the only feasible means of providing disaster relief. On the assumption that the "true" hazard is random in incidence and unpredictable in occurrence, the communal fund is theoretically as well as administratively appropriate.

The "normal" or "reasonably expected" event is supposed to be safeguarded against by the property owner. The commission can enforce this by ensuring that all properties meet building code requrements designed to protect each class of building from foreseeable damage. Should a property not meet these requirements, the commission can classify it into a category of higher risk and set deductible clauses and premiums accordingly. Where a property is in an area of very high risk, (e.g., a floodplain or unstable slope) the commission can refuse to provide cover. In addition, all property owners must seek private insurance cover as protection against the "normal" event. The commission is supposed to cover damage only in the case of extraordinary and unforeseen disasters.

The commission can initiate updated building codes and can in theory advise against any new development in perilous areas. Failure on the part of local authorities to meet either requirement, should the commission wish it, can result in reclassification of any property involved into higher hazard risk or even rejection of its eligibility for cover.

Claim files provide an invaluable source of information on the extent and nature of damage for every natural disaster in the country. This is a vital record to enable managers to determine more precisely the characteristics (location, frequency, extent, patterning) of the natural events system. Armed with this information, the commission is in a better position to assess hazard risk and set appropriate rates and conditions.

Problems in implementing the New Zealand policy

In practice, however, the New Zealand scheme is running into difficulties. This is due largely to public misinterpretation of the nature of the commission's work and reflects how the public evaluates the incidence of natural hazard in their lives. The commission is seen as a backroom, claim-processing administrative body sitting on millions of dollars of public money which should be used; not saved. The concept of hedging against a long-term catastrophic damage is not widely understood, yet a severe earthquake in Wellington could set back the economy by ten years (Powers, 1968). As a result the commission has been forced to expand its coverage to events which are not in a sense unforeseen, such as landslide damage on slopes of known instability or flood damage in areas with a previous record of flood. Because its role is seen at best as advisory and at worst as merely a paper pusher, the commission has no powers to enforce building codes nor to restrict development in areas of forseeable hazard. To reclassify property would require knowledge of the hazard risk, which the commission has on file but does not have the technical personnel to enforce. In any case such an action would affect local property values and would be strongly resisted locally. As a result local authorities fail to enforce earthquake building codes which are designed more to protect loss of life than to control structural damage, and ignore potential hazards in development proposals (Steinbrugge, 1968).

Another problem is political. When the commission is faced with a number of damage claims from a particular area, no matter how poorly maintained each or all of the buildings and no matter how foreseen the disaster, it must pay up. To discriminate between claims would be political suicide in view of the fact that the average New Zealander feels he has paid his share, the funds are vast, and he feels he suffered from a "natural" disaster. Thus spurious claims are frequent—cracking of walls due to clay shrinkage attributed to earthquake, poorly constructed sheds blown over by a strong breeze, slump-damaged houses sitting on clay-lubricated subsoils. Consequently claim payments are rising annually though the incidence of hazard does not appear to be increasing.

Private insurance companies are loath to cover high-risk, high-value property and whenever possible leave such coverage to the commission. Many properties do not seek private insurance coverage, as the rates may be 10 to 100 times higher than the subsidized scheme provided by the commission. As a result, though about 7 percent of all claims on the Extraordinary Disaster Fund are from commercial properties, these claims account for about 70 percent of the annual payments (Table 26–1). Damage to poorly constructed barns make up much of the remaining claims.

Table 26–1. Cost of claims on the extraordinary disaster fund, 1966

	Domestic	Commercial/ industrial	Farm outbuildings
% of total claims	7.7	7.7	84.6
Amount paid	$2,850	$76,178	$52,204

Source: Earthquake and War Damage Commission.

Thus unless public clarification of the theory of the scheme is made the commission appears locked into a system from which it has little escape. The larger the funds grow the greater the pressures after every disaster to expand the coverage: the wider the coverage the greater the amount of claims; increasing claims deplete the funds; depleted funds encourage the commission to raise the premium; and raising the premium enlarges the funds.

The new New Zealand government has recently proposed legislation to widen the concept of communal insurance against unpredictable events—events which could afflict anybody beyond what might be regarded as a reasonably expected accident—to include accidents occurring during work, in the home, on the street, or in an automobile. A fund similar to the Earthquake and War Damage Fund is to be established, paid for communally from vehicle registration fees and a 1 percent levy on income tax. Claims can then be made by persons who suffer disability from accidents, wherever and whenever they occur, as long as it can be shown that such accidents cannot reasonably be safeguarded against. This clearly is an extension of the natural hazards insurance scheme with an interesting application into the area of social and personal hazard adjustment. Thus, it is not simply workmen's compensation, but compensation against domestic accidents and acts of personal violence.

Applicability to North America

The New Zealand scheme would appear to be applicable on a state- or province-wide basis in those areas subject to random and unpredictable natural disasters such as earthquakes in California and Alaska, tornadoes in the Midwest, hurricanes in the South and Southeast, hailstorms in the prairies. It could be designed to provide cover where private insurance interests cannot protect or are unwilling to protect property at reasonable rates. To be successful the appropriate hazard management agency would have to provide constant public information on the extent of the policy to offset the inevitable selfish desire to make use of a communal fund. The agency would have to be armed with the necessary teeth, with the necessary technical personnel, and be in the appropriate agency of government to ensure that building and zoning regulations are being met. Part of the revenue could be diverted to research aimed at improved hazard forecasting and upgrading techniques of building construction and risk estimation. The agency would then be in a position to devise a suitable combination of hazard avoidance and hazard modification behavior depending upon the characteristics of both the human use and natural events systems. This would be the "post industrial multiple means" stage of Kates's model (Kates, 1970). The agency could initiate this policy with all new development, then ensure that high-value property was suitably protected, and finally upgrade existing lower value property. As a result, damage from "normal" events should be drastically reduced while damage from truly catastrophic events would be covered by legitimate relief.

Acknowledgments

This study was supported by a grant from the National Committee for Water Resources Research of the Canadian Department of the Environment, under whose direction the author conducted the field work and preliminary interpretations.

References

Kates, R. W. (1970) *Natural Hazard in Human Ecological Perspective: Hypotheses and Models.* Natural Hazards Research Working Paper No. 14. Toronto: Univ. of Toronto, Dept. of Geography.

Power, C. A. (1968) "Earthquake insurance in New Zealand and the problems of reconstruction." *New Zealand Engineering* 34:23–27.

Steinbrugge, K. V. (1968) *Earthquake Hazard in the San Francisco Bay Area: A Continuing Problem in Public Policy.* Berkeley: University of California Institute of Governmental Studies.

27. Natural hazards and hazard policy in Canada and the United States

HAZEL VISVADER
University of Colorado

IAN BURTON
University of Toronto

The choice of a strategy that a nation will adopt in dealing with extreme geophysical events usually involves an assessment of the character of the events themselves, and a canvass of the range of actions available. The choice is also consistent with attributes of the nation itself, including its size, level of wealth, and social institutional traditions.

Beyond these broad dimensions, however, there is little to guide a nation toward a strategy that will be maximally effective in its own circumstances. Certainly there is no blueprint or formula that can be universally applied (Burton, Kates, and White, forthcoming). Under these circumstances it is instructive to compare the hazard policies of Canada and the United States. This paper illustrates some of the ways which each country has chosen to respond to a varied set of natural hazards. The data do not permit a confident overall appraisal of the success achieved, but a number of the consequences of each nation's strategy can be qualitatively described.

The social context

In Canada, natural disasters are much less frequent than in the United States, and in consequence the national level of awareness is lower. The lower rate of occurrence of disasters results in part from the geographical distribution of the major types of hazard over the North American continent, and from the much smaller population in Canada. Although larger than the United States in area, Canada has a population of only one-tenth the size (U.S. 196,920,000; Canada 20,050,000—1966 estimates). Density of population in Canada is approximately 2 per square kilometer (5 per square mile) compared with 21 per square kilometer (54 per square mile) in the United States. The Canadian population, however, is heavily concentrated in the contiguous border areas. Considering the approximately equal level of economic development in the heavily populated areas of both countries, it might be expected that losses from natural hazards would be roughly proportional on a per capita basis. What limited evidence exists, however, suggests that losses in Canada are probably less than 10 percent of United States losses.

Although both countries share an inheritance of the English common law and both have written federal constitutions, the distribution of political power is quite different. The framers of the Canadian Constitution (British North American Act of 1867) saw need for a strong central authority to hold a geographically dispersed bilingual and bicultural nation together. In the United States a strong reaction against remote central authority was uppermost in the minds of the revolutionary colonists in 1776. Contrary to these intentions, however, the power of the federal authority is much less in Canada than in the United States. Partly as a result of this different evolution of governmental powers and structure, it has been possible and perhaps necessary for the United States federal government to become involved in natural hazard problems to a much greater extent than the Canadian federal government.

The 40 years after 1933 witnessed rapid growth in the assumption of the need for federal action in dealing with some major natural hazards in the United States. The trend was in the same direction in Canada, but to a much smaller degree. There is evidence that in the case of flood problems, heavy United States federal involvement since the Flood Control Act of 1936 has served to encourage reliance on control technology (e.g., dams, levees, floodwalls) and has discouraged or not given equal weight to other possible adjustments such as land-use regulation, floodproofing, and flood warnings and emergency action (White et al., 1958). The extent to which a relatively smaller involvement by the Canadian federal government may have encouraged use of a broader range of adjustments is not easy to assess. The experience suggests that some provincial governments have followed a flood policy not unlike that of the U.S. federal government (Sewell, 1965). On the other hand the 1959 plan for flood control in Toronto was one of the first anywhere to provide for the public purchase of floodplain land (Metropolitan Toronto and Region Conservation Authority, 1959).

In both countries the knowledge that national governments can be relied upon to provide post disaster relief may well have decreased the incentives for individual action, and discouraged the search for alternative adjustments. Generally speaking, the higher the level of government involved, the greater its command of highly skilled scientists, engineers, and economists, and the greater its financial resources. This is reflected in a United States flood policy which has become highly sophisticated and now encompasses a broader approach, especially in the creation of adjustments that would be unavailable to lower-order governments or communities (U.S. House of Representatives, 1966). These include expensive multipurpose dams, subsidized flood insurance, and large-scale floodplain information programs. In Canada it appears that less federal involvement has meant that expensive control technology has been used less and that communities have been made to rely more heavily on their own resources. In the economic analysis of projects, it has been possible to depend much more on financial criteria in Canada, and the use of benefit-cost analysis has been less well developed. In the United States benefit-cost analysis has been used as a screen to filter out less justified proposals. When responsibility for the cost of projects is kept in the community or at a lower level of government, the need for refined evaluation devices is less (Burton, 1964).

The more comprehensive hazard policies now beginning to emerge in the United States offer an opportunity to achieve a high degree of equity in protection from extreme natural events, and at the same time avoid the creation of new disaster potential. For many years the decentralized Canadian approach has meant that the people in parts of the country have greater security against some losses from natural hazards than others. While this remains true also in the United States, a comprehensive national policy may well emerge that would minimize such discrepancies. The inequity between regions seems likely to be more persistent in Canada, and in both countries the achievement of a consistent multihazard policy is likely to remain unrealized.

Hazard characteristics and common adjustments

We have attempted to summarize the major characteristics for nine major hazards in terms of magnitude, extent, and frequency, and to describe the adjustments most commonly practiced. This serves to illustrate the diversity of situations and responses found in two contiguous countries. It does not constitute a model approach but is suggestive of some of the variables that need consideration in national assessment of hazard policy.

Avalanches

Avalanches in Canada occur in the mountainous areas of British Columbia, Alberta, and the Yukon, and to a lesser extent in the highlands of southeast Quebec. The major hazard is at avalanche sites in the vicinity of Rogers Pass in the Selkirk Range of British Columbia. In the United States, avalanche-prone areas are located in the 11 westernmost states and Alaska. The major hazard area is in the Cascade Range of Washington.

At Rogers Pass, large accumulations of snowfall, strong winds, and steep slopes cause numerous avalanches which seriously affect transportation on the Trans-Canada Highway in spite of a route chosen carefully in 1956. Along a 35-mile stretch, avalanches from 74 sites reach the valley floor from both sides, and the road was closed due to avalanche hazard for a total of 62 days in the winter seasons beginning late 1953 and ending early 1959.

In the Cascades, the popularity of winter activity, heavily traveled roads, and unique snow conditions create a formidable hazard. In the United States, the most destructive single incident involved a passenger train stranded on the tracks by a snowstorm in March 1910. Ninety-six people died from an avalanche meeting the train in the vicinity of Stevens Pass, Washington. In the last two years, only 17 people in the United States have died from avalanche falls. The improvement to a large extent came because of heeding avalanche warnings.

Precise forecasting of avalanche occurrence for a given area in North America is impractical because of the microscale nature of the hazard and the vast area to be covered. In Canada, the use of the "testing" method is common. It involves intermittent inspection of snow layers (pit method) and subsequent testing of snow stability (e.g., with explosives) when necessary. An avalanche rating scale from 0 (none) to 5 (very high) determines what specific restrictions on travel are placed by authorities in avalanche areas (Schaerer, 1962). The

United States relies more on the "analytic method." It involves several years of data collection and experience at individual sites. Some of the factors in avalanche movement are the presence of an underlying layer of "depth hoar" (fragile, recrystallized snow), heavy snowfalls within 24 hours, heavy rainfalls, and wind-created "slabs."

Adjustments used in the two countries do not yet include avalanche zoning ordinances, as in certain regions of Switzerland. In addition to temporary measures (explosives for stabilization of snow cover and triggering an avalanche of minor proportions, and road closures), permanent measures are adopted to some degree, but their costs generally limit their use to populated or well-traveled sites. Various types of snow barriers are constructed to hold in the snow. In addition, reforestation of denuded avalanche paths, especially in the region of known breakaways, is infrequently practiced. Because of the use of temporary retaining barriers to protect the young trees, these projects also involve high costs.

Droughts

Large-scale droughts have been largely confined to the Great Plains and the southwestern states. The drought-prone area of the plains extends north into the Canadian prairie provinces and southeastern British Columbia. An exception to this pattern was the drought of 1961–66 in the northeastern United States, when acute water shortages in municipal supplies were experienced by late 1964.

The main effect of drought is upon agriculture. With repeated crop failures, the small and marginal farmers can be forced out of business. This happened in a dramatic way during the drought of the 1933–37 period in the Great Plains and is evocatively portrayed in John Steinbeck's novel *The Grapes of Wrath*. Urban residents and industries may suffer some restrictions on water use, but because of the higher level of community resources the economic consequences for an individual or firm are usually quite small.

A wide range of adjustments to drought are available and many of them are now practiced. These include water conservation practices such as summer fallowing, stubble mulching, strip cropping and contour plowing, and the use of drought-resistant crops and livestock. Various steps may also be taken to protect water supplies and to increase water availability such as concrete lining of irrigation canals and accelerated well-digging operations. There is much that an individual farmer can do to reduce his vulnerability to drought and he is helped in both Canada and the United States by the availability of government crop insurance schemes. When a severe drought does occur, assistance to the stricken area can be provided from governmental sources. In consequence, the misery and deprivation described by Steinbeck is largely a thing of the past. The

Great Plains droughts of 1952–56 was climatically more severe than that of the 1930s, but the economic and social effects were much less. The availability of government subsidies, insurance, and the wide range of agricultural adjustments adopted by farmers have greatly mitigated the effects of drought in North America. Economic hardship can still occur, but drought is not the threat to survival that it can be in other parts of the world.

Earthquakes

The danger of earthquakes has been a matter of great public concern in the United States in the last decade. By contrast, it has very low salience in Canada. The difference may be attributed in large part to the absence of major earthquake disasters in Canada and to two recent disasters in the United States which in each of two years (1964 and 1971) caused estimated property damages exceeding $500 million and many fatalities. By contrast, the last earthquake causing significant damage in Canada occurred in Cornwall, Ontario, in September 1944. Damages were estimated at only $1 million and no lives were lost.

National maps of earthquake risk have been prepared by both countries and these reveal interesting discrepancies (Figs. 27–1 and 27–2). The differences reflect varying assumptions and approaches. In Canada, areas of little or no seismic activity (zone 0) incorporate large sections including the prairie provinces and western Ontario. Areas of major activity (zone 3) center in relatively uninhabited areas of the Arctic and Yukon, but also in the populated British Columbia coastal areas and the St. Lawrence region. Although the great majority of shocks occur in uninhabited regions, Vancouver and Montreal are located in zone 3 and 2 respectively, and other eastern Canadian cities may be considered to have some small degree of risk.

In the United States the Pacific region (including Nevada) is an area of major seismic activity associated with fault lines in the coastal ranges of California and southern Alaska. Moderate to severe shocks have occurred in eastern and southern states, notably near Charleston, South Carolina in 1886, Massena, New York in 1944, and Wilkes-Barre, Pennsylvania in 1954. The Mississippi Valley experiences small but frequent earthquakes of limited destruction (with the exception of a major disaster at New Madrid, Missouri in 1811–12).

If we assume that the United States and Canadian definitions of zone 3 are comparable, the greater vulnerability of the United States population to earthquake hazard can be estimated. The population of zone 3 in Canada is approximately 760,000 (1960 estimate) and amounts to about 4 percent of the national population. The figure for zone 3 in the United States is approximately 31 million (1970 estimate), or 15 percent of the national population. Applying the rule that the United States population and hazard exposure is ten times that

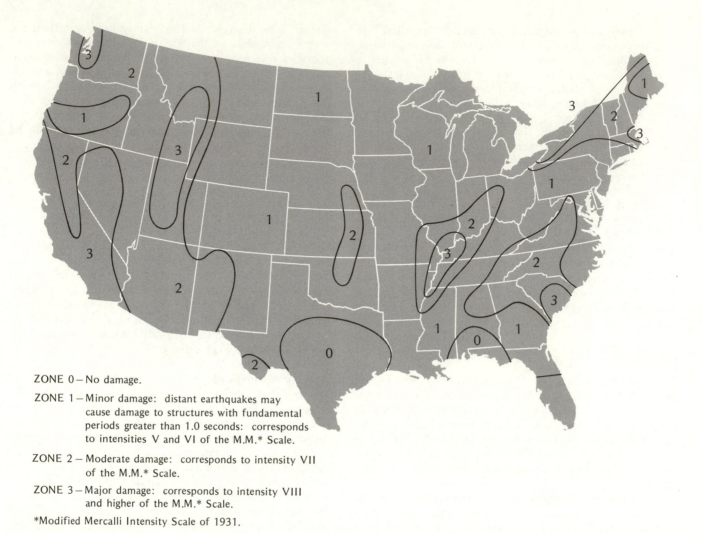

ZONE 0 – No damage.

ZONE 1 – Minor damage: distant earthquakes may
cause damage to structures with fundamental
periods greater than 1.0 seconds: corresponds
to intensities V and VI of the M.M.* Scale.

ZONE 2 – Moderate damage: corresponds to intensity VII
of the M.M.* Scale.

ZONE 3 – Major damage: corresponds to intensity VIII
and higher of the M.M.* Scale.

*Modified Mercalli Intensity Scale of 1931.

Fig. 27–1. Seismic risk in the United States. This map is based on the known distribution of damaging earthquakes
and the M.M. intensities associated with these earthquakes; evidence of strain release; and consideration of major
geologic structures and provinces believed to be associated with earthquake activity. The probable frequency of
occurrence of damaging earthquakes in each zone was not considered in assigning ratings to the various zones. Source:
U.S. Office of Emergency Preparedness (1972).

of Canada, we would expect far more population in
earthquake zone 3 areas in Canada or less in the United
States.

Table 27–1 lists some of the more destructive earthquakes in the United States. In spite of these dramatic
disasters, the average annual loss on a per capita basis
amounts to less than 55¢.

Adjustments to earthquake hazard include seismic
recording stations and research directed at prediction
and forecasting, the adoption of building codes and
land-use restrictions in connection with known fault
localities, relief and rehabilitation, disaster emergency
preparedness, and earthquake insurance. A much greater
level of activity is found in the United States than in
Canada. Although building codes and seismic risk map-

ping have been undertaken in California, this does not
appear to have been highly effective in guiding development away from areas of high risk.

Earthslides

The danger of sudden earth movement, landslides,
and rockfalls is greatest in the areas of steep slope and
heavy rains in the mountainous western states and provinces. Due to the generally sparse population density,
damages and loss of life remain relatively small. One of
the most dramatic events to occur was the Frank slide
of 1903. The top of Turtle Mountain in Alberta (a
limestone overthrust on softer sandstones and shales)
collapsed, hurtling 30–34 million cubic meters 354–401

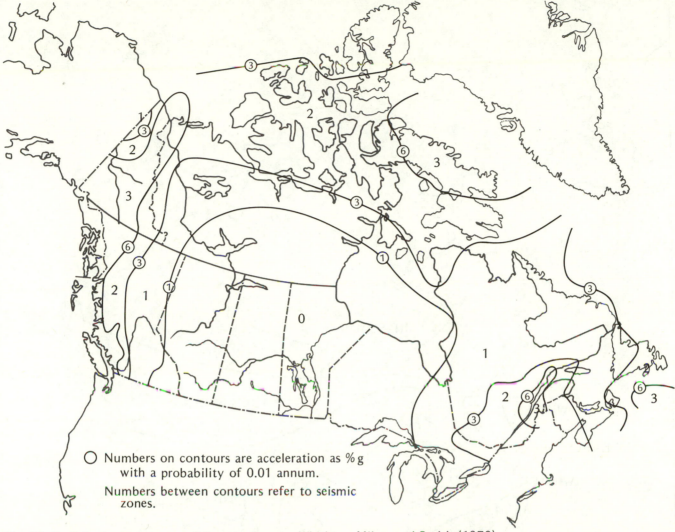

Fig. 27–2. Seismic zoning map of Canada. Source: Whitham, Milne, and Smith (1970)

Table 27–1. Major U.S. earthquakes

Year	Place	Lives lost	Damage ($ million[a])
1906	San Francisco	700	85.0[b]
1933	Long Beach, Calif.	115	133.0
1952	Kern County, Calif.	14	77.0
1964	Alaska	131	597.5[c]
1971	San Fernando, Calif.	65	553.0

[a]1971 prices.

[b]Plus $1,780 million fire damage.

[c]Including tsunami damage.

million cubic yards) of rock onto the mining town of Frank. Deaths totaled 70 and the Trans-Canada Railroad line was blocked. In 1959, 28 people were killed by an earthquake-generated slide in Montana and a new lake (Earthquake Lake) was formed. Considerable damage

was caused by slides and the resulting wave action during the severe Alaska earthquake of 1964.

An unusual landslide problem occurs also in the St. Lawrence and Saguenay lowlands of Quebec. Here the presence of an unusual clay formation deposited by the postglacial Champlain Sea in an area of relatively dense agricultural settlement constitutes a major hazard. When oversaturated, the clay liquefies and flows out at a fast rate. Areas of potential landslide occur principally in areas of steep slopes or along the banks of incised rivers where slides occur frequently and occasionally involve areas of human settlement. The most recent disaster occurred at Saint-Jean-Vianney, on May 4, 1971, in which 40 houses were destroyed and 31 persons were killed.

Adjustments include mapping of potential slide areas, land-use regulations to keep settlements away from vulnerable places, and engineering measures such as removal of unstable materials, improvements in slope

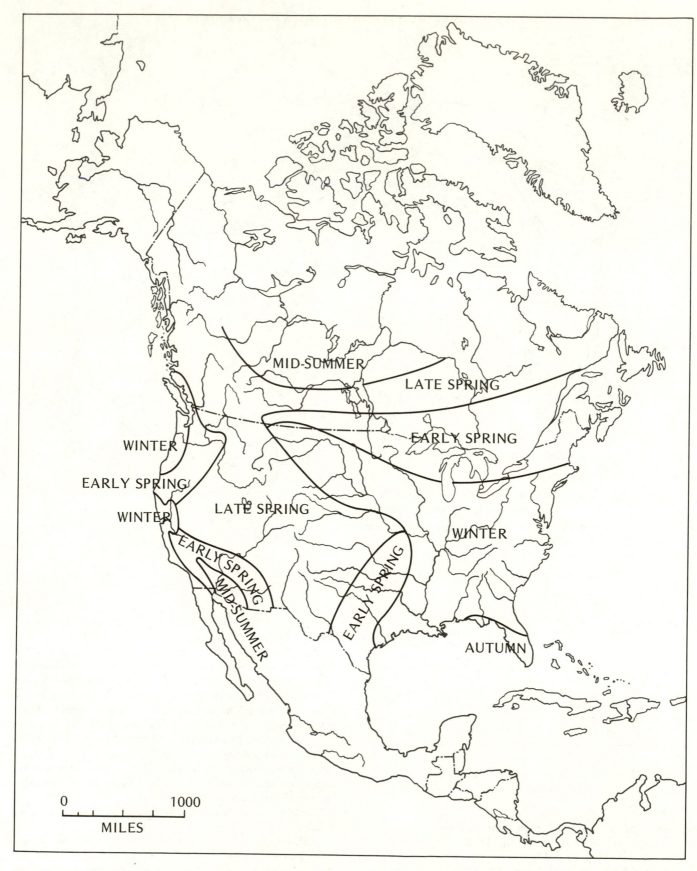

Fig. 27–3. Flood seasons in North America. Adapted from Hoyt and Langbein (1955)

drainage and the construction of retaining walls. Precise forecasts are not yet possible but research to improve forecasting ability is under way in both countries.

Floods

Flood hazard in North America has been characterized by few deaths and a high and growing damage level. Unlike earthquake hazard, the risk from floods appears to be relatively at the same level in the United States and Canada. It is estimated that between 1 and 2 million people live on flood plains in Canada (Sewell, 1965), compared with 10 million in the United States (U.S. Office of Emergency Preparedness, 1972). The ratio is thus approximately in accord with relative population size.

Principal flood seasons in North America are shown in Fig. 27-3. In Canada, major flood disasters have occurred on the lower Fraser River (British Columbia in 1948, on the Red River (Winnipeg, Manitoba) in 1950, and on a number of small rivers in the Toronto area in 1954, associated with the passage of Hurricane Hazel. Major flood disasters in the United States are more frequent and have resulted in damages estimated at $300 million annually (Dacy and Kunreuther, 1969). Average annual losses for 5-year periods are shown in Fig. 27-4. With the inclusion of losses from the recent June 1972 catastrophic floods at Rapid City, South Dakota (236 deaths, over $100 million in damages) and in the eastern states from Hurricane Agnes (118 deaths and $3.8 billion in damages as estimated in U.S. Depart-

ment of Commerce, 1972b, it appears that damages in the period of 1970–74 will be by far the highest ever recorded for a 5-year period. Heavy losses in the Mississippi Valley in April–May, 1973 will add substantially to the total.

There is evidently a slow decline in the death rate from floods, but per capita damages have certainly not fallen and may well be rising sharply in spite of high levels of expenditure on flood protection and control at a current rate of expenditure estimated at $500 million a year (U.S. House of Representatives, 1966). Quite a few problems are involved in estimating an average annual loss figure. Data for Canadian national losses are not readily available, as Quebec is the only province to conduct a survey of property damage. This amounts to about $2 million annually. When this is extrapolated for the country as a whole (based on 30.1 percent of the national population), some $6.6 million may be estimated for Canada. This figure is nowhere near the loss estimated by National Weather Service (Department of Commerce) for the United States as a whole, which is on the order of $322 million, averaged annually for the years 1965–70 (U.S. Department of Commerce, 1972a). Due to inadequate coverage by surveys of real-property damages, this is considered to be rather low, and an estimate of at least $1 billion per year is more accurate for the United States (U.S. Office of Emergency Preparedness, 1972). The resulting per capita comparisions (using conservative figures) are $0.33 for Canada and $1.58 for the United States, still a substantial difference.

Adjustments to floods include control works, land-use regulations, floodproofing of buildings, flood insurance, forecast and warning systems, and emergency relief and rehabilitation. Of these, control works have dominated, but a distinct trend toward adoption of a broader range of adjustments is underway in both countries.

A distinct feature of United States policy is the National Flood Insurance Act of 1968. Under this act, 175,000 flood insurance policies were sold by April, 1973. Henceforth, restrictions will be imposed upon the use of federal relief funds where insurance has not been purchased. The availability of flood insurance in a community is also dependent on the adoption of floodplain management policies. In Canada, insurance is not generally available except at high rates given by private companies. Another difference has been a great reluctance on the part of the Canadian federal government to become involved in relief operations through the declaration of a national disaster. Only when it is convinced that the magnitude of a disaster is beyond the capacity of the local and provincial governments to handle does the federal government interfere.

Hail

Hail is a significant danger to crops and has also been known to kill livestock and damage airplanes and cars.

Fig. 27-4. Average annual losses in United States floods, 1925–69. Dollar losses standardized to 1972 levels. Source: U.S. Department of Commerce (1972a)

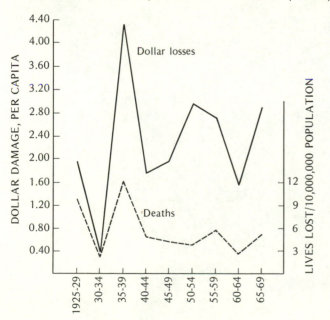

Glass, especially where exposed as in greenhouses, may be broken.

Areas of greatest hail frequency and intensity occur on the Great Plains in the lee of the Rocky Mountains and in western Alberta in Canada. The crops most susceptible to hail damage are fruits, tobacco, and certain vegetables. Soybeans, grains, corn, and cotton may also be damaged or rendered inaccessible to harvesting machines (Changnon, 1972).

Estimates of hail damages put the Canadian figure as high as $75–$100 million annually. One United States estimate is $284 million (U.S. Department of Agriculture, 1968), although it is thought possible that average annual crop losses may exceed $500 million (Boone, 1973). Of these losses, private insurance companies reimburse losses for $65 million annually in the United States.

In Alberta alone, for the period 1961–68, average annual crop losses are estimated at $22.8 million plus a further $11.5 million in indirect losses to business activity and $1.5 million in property damage (Summers and Wojtiw, 1971). Storms lasting only a few minutes may take a surprising toll. A storm in the Edmonton area of Alberta in August 1969 caused an estimated $3–$5 million crop damage. One in southeastern Missouri in April 1971 caused over $27 million damages to 250,000 acres.

The average number of days with hail is shown for the United States in Fig. 27–5, and a point-frequency map for western Canada (or approximate number of hail-days a year) is shown in Fig. 27–6.

Adjustments to hail include forecasting, which can be effective in reducing property damage, but not crop damage. Efforts at hail suppression are underway in several experiments. These currently include the Alberta Hail Studies Project and the National Hail Research Experiment in Colorado. The theory of these experiments is that cloud seeding with silver iodide will decrease the average size of hailstones.

Individual farmers can avoid growing hail-sensitive crops, or scatter landholdings so as to spread the risk. The extent of these practices is not known. A common adjustment is the purchase of special hail insurance or the inclusion of hail in an all-risk policy. In Alberta, it is estimated that only 20 percent of the farmers carry hail insurance due to high premiums. The figure is probably higher for hail areas in the United States.

Hurricanes

The hazard from hurricanes is described in detail elsewhere (chaps. 3, 4, and 30) and need not be repeated in this summary.

The hurricane season is from June to November with

Fig. 27–5. Hail occurrence in the United States, based on two hundred first-order Weather Bureau stations, 1899–1938. Source: U.S. Department of Agriculture (1941)

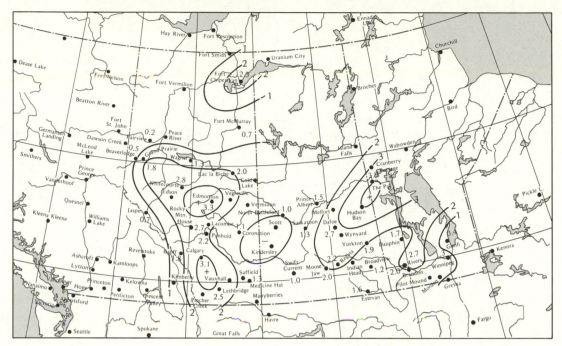

Fig. 27—6. Point frequency of hail per year in the Canadian prairies (based on ten years 1950—59). Source: Summers (1970)

maximum frequency of occurrence in September. The areas affected are the Gulf of Mexico and east coast in the United States. The tracks of some major hurricanes are shown in Fig. 27—7. Canada is relatively unaffected, but some hurricanes do retain considerable force as they pass northeastward over the Maritime Provinces. Hurricane Hazel (October 1954) followed an unusual track and moved inland over Pennsylvania, crossing Lake Ontario and passing over Toronto. In that city, 81 deaths resulted from the subsequent flooding.

In the United States, hurricanes average about five a year. Total fatalities in the period 1915—70 have exceeded 5,900 and property damage exceeds $7.9 billion for the same years. Deaths have averaged 107 per year and damages $147 million (U.S. Office of Emergency Preparedness, 1972).

The trend in fatalities is downward, due to more effective warning systems and provisions for emergency evacuation. Damages are rising due to continued invasion of the outer coastal areas largely by recreational development. Much current debate centers on the flexibility and desirability of hurricane modification through seeding.

Snow

Snowfall in both countries is generally considered to have substantial benefits. Water is accumulated and stored in the form of snow for later use in the spring growing season. Also the skiing industry can suffer economic loss if there is insufficient snow.

The adverse effects of snow are felt largely in the cities, where traffic disruption occurs, and offices, schools, and factories may be forced to close. The major cities of the northern United States and Canada spend hundreds of millions of dollars annually in snow clearing and removal operations.

The rate of snow accumulation is an important factor in determining the disruptive effects in urban areas (Rooney, 1966). Occasionally snow loads become heavy enough to cause roof collapse, as when a theater roof fell in and killed 96 people in Washington, D.C., in 1922. In this storm, 710 millimeters (28 inches) accumulated in about 32 hours.

The amount of disruption caused in traffic flow is dependent upon the scale and efficiency of snow removal operations as well as on wind velocity, temperature, and time of day or week. When people are unable to go to work, loss of income results. A study in Toronto showed that 13 percent of the work force were so affected and that in one winter, losses totaled about $3.5 million for a work force of 900,000 (Archer, 1970). A single storm in Boston (February 1969) has been credited with causing 44 deaths and creating damages of $150 million. This includes the costly delays caused to transportation, especially airlines and railways. Power supplies and telephone services may also be cut off due to collapse of transmission lines.

Annual snow and ice removal costs are estimated at about $500 million in the United States (*Public Works*, 1968), and $125 million in Canada (Leggett and Williams, 1964). Much of this is spent on snowplowing, salting, and sanding of roads. The city of Montreal alone spent $5.5 million on snow removal in 1961.

Other adjustments to snow include the designation of snow emergency routes, forecasting, and highway design

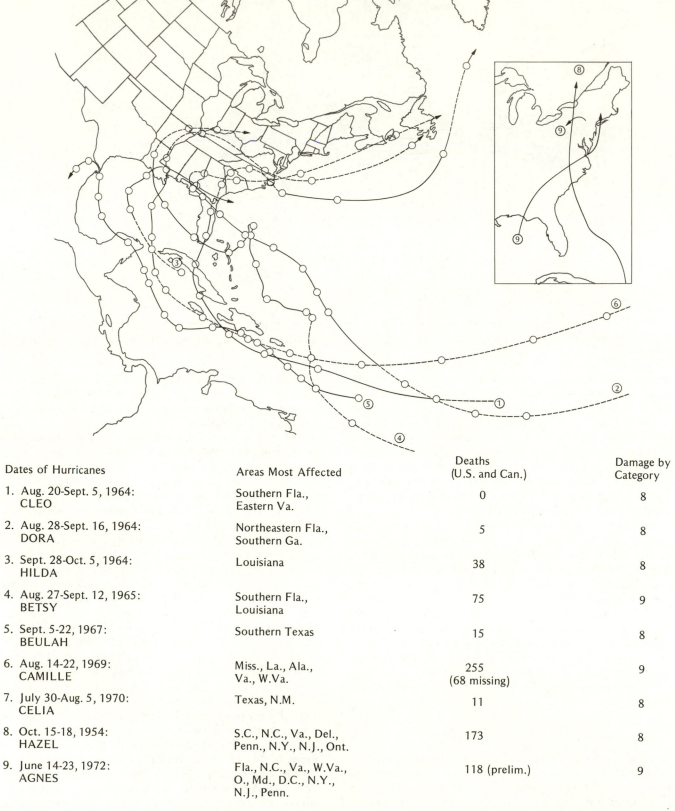

Dates of Hurricanes	Areas Most Affected	Deaths (U.S. and Can.)	Damage by Category
1. Aug. 20-Sept. 5, 1964: CLEO	Southern Fla., Eastern Va.	0	8
2. Aug. 28-Sept. 16, 1964: DORA	Northeastern Fla., Southern Ga.	5	8
3. Sept. 28-Oct. 5, 1964: HILDA	Louisiana	38	8
4. Aug. 27-Sept. 12, 1965: BETSY	Southern Fla., Louisiana	75	9
5. Sept. 5-22, 1967: BEULAH	Southern Texas	15	8
6. Aug. 14-22, 1969: CAMILLE	Miss., La., Ala., Va., W.Va.	255 (68 missing)	9
7. July 30-Aug. 5, 1970: CELIA	Texas, N.M.	11	8
8. Oct. 15-18, 1954: HAZEL	S.C., N.C., Va., Del., Penn., N.Y., N.J., Ont.	173	8
9. June 14-23, 1972: AGNES	Fla., N.C., Va., W.Va., O., Md., D.C., N.Y., N.J., Penn.	118 (prelim.)	9

Fig. 27–7. Tracks of recent major hurricanes. Not shown are the track for Celia, the origin and development of Hazel and Agnes. Category 8: damage ranges from $50 million to $500 million; category 9: from $500 million to $5 billion. Updated from U.S. Office of Emergency Preparedness (1972), vol. 3

and the use of snow fences to control drifting. For individuals, adjustments include use of snow tires and chains, use of alternate means of transport, and rescheduling of activities.

If the impact of a natural hazard upon a society is a sum of the efforts expanded on adjustment plus the residual damages, then snow is an example of a hazard where most of the expenditure is on adjustments and residual damages remain small.

Tornadoes

Tornadoes are concentrated chiefly in a belt running north from Texas through Oklahoma into Kansas. They are comparatively infrequent in Canada, although severe damage has occurred in southwestern Ontario (Fig. 27–8).

In the past 50 years, approximately 9,000 people have died in tornadoes in the United States, with property damage over the last two decades exceeding $50 million annually (U.S. Office of Emergency Preparedness, 1972). Death trends are shown in Table 27–2. As in the case of other hazards, the trend in deaths has been downward while damages have been rising. The increased number of tornado sightings resulted from more complete reporting after 1953 rather than an increase in the frequency of tornadoes.

Table 27–2. Tornado losses, 1920–69

Years	Average annual deaths	Average annual tornado sightings
1920–29	316	133
1930–39	194	169
1940–49	179	155
1950–59	141	484
1960–69	94	678

Source: U.S. Office of Emergency Preparedness (1972).

Adjustments to tornado hazard include preparations to take emergency evasive action upon receipt of warnings. Tornado forecasting has been improved and information is made available to individuals about what action to take. In the areas of greatest frequency, the construction of tornado shelters in the basement of houses is a common adjustment. Insurance for wind damage is available but the extent of coverage is believed low.

Toward a comprehensive approach to natural hazards

The wide diversity of hazard characteristics and of adjustments as illustrated in the United States and Canada suggests that the formulation of coherent national pol-

Fig. 27–8. U.S. tornadoes, 1916–55. (Isolines based on total number by 2° squares, counting first point of contact with ground of 7,206 tornadoes.) Source: Wolford (1960)

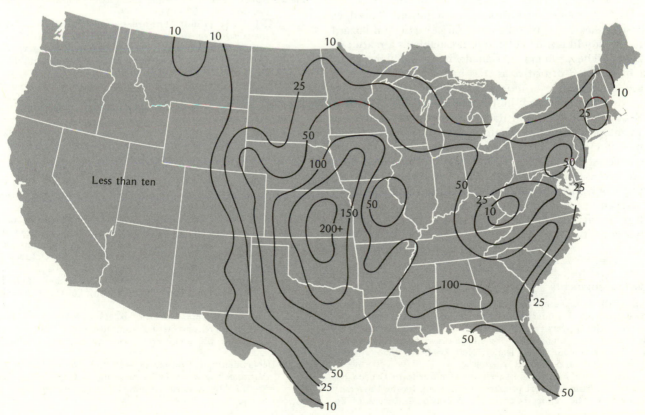

icies which deal consistently with all hazards and can provide a high degree of equity of treatment between hazards, regions, and individuals will not be achieved easily.

Where a hazard is of little significance or salience, it may be better to leave responsibility for dealing with it at the local or regional level. The assumption of responsibility by a national government does not necessarily lead to improved strategies, and it may serve to increase the magnitude of the problem.

Carefully designed and comprehensive national policies for single hazards seem indicated where the damage potential is high and where the national government has the resources at its disposal to deal effectively with the problem. Societies, however, can find ways of coping with natural hazards without high-level government involvement being regarded as necessary.

The task of selecting an appropriate strategy is complicated by the absence of valuable data for evaluation. The success of a policy may be gauged in terms of the degree to which it has succeeded in reducing the total damages, including residual damages, plus the cost of adjustment. Where damage data are available for the United States, these almost invariably reveal a rising tendency, or at least a failure to reduce damages in spite of considerable expenditures on adjustments. This strongly suggests that the policies have not been effective.

Similar data is almost totally lacking for Canada, and where individual damage estimates are made, these are insufficient to permit inferences to be drawn about trends without some additional investigations. This does not necessarily imply that national statistics on hazard damage should not be collected in Canada as a matter of urgency. The problem in Canada is not as high on the list of national priorities as it is in the United States. In these circumstances a more ad hoc approach to natural hazard policy may well be the best approach. Certainly, collection of data would itself be an expense and while it might permit a more accurate assessment of the situation, experience in the United States shows that the mere collection of data would not by itself lead to a more successful or effective policy. A strategy of benefiting from the experience of the United States without large-scale expenditures on research or data collection offers an attraction for Canada and perhaps also for other countries.

Acknowledgments

Much of the information on hazards in the United States has been drawn from work underway in the Natural Hazards Assessment Project at the University of Colorado, Boulder. Those working on this project and on whose work we have drawn are Elwood Beck, Waltroud Brinkmann, Fred Dauer, Neil Ericksen, Alan Murphy, Patricia Trainer, and Richard Warrick.

Much of the information on natural hazards in Canada has been drawn from a report prepared by a group at the University of Toronto. The report is entitled "A National Review of Selected Natural and Man-made Hazards in the Canadian Environment" and was prepared under the general direction of Ian Burton with the editorial assistance of Neil Ericksen. Those who worked on the report and on whose work we have drawn are Neil Ericksen, Christopher de Freites, David Harper, Edgar Jackson, Lynne May, and Paul Wilkinsen.

References

Archer, P. E. (1970) *The Urban Snow Hazard: A Case Study of the Perception of Adjustments to, and Wage and Salary Losses Suffered from, Snowfall in the City of Toronto during the Winter of 1967–8.* Toronto: University of Toronto: Department of Geography, unpublished M.S. thesis.

Boone, Larry M. (1973) Personal communication.

Burton, I. (1964) "Investment choices in the public sector." In A. Rotstein, ed., *The Prospect of Change: Proposals for Canada's Future.* Toronto: Mc-Graw Hill, pp. 149-73.

———, Kates, R., and White, G., (forthcoming) *The Environment as Hazard.* New York: Oxford University Press.

Changnon, S. A., Jr. (1972) "Examples of economic losses from hail in the U.S." *Journal of Applied Meteorology,* 11 (7):1128–37.

Dacy, D. C., and Kunreuther, H. (1969) *The Economics of Natural Disaster.* New York: Free Press.

Hoyt, W. G., and Langbein, W. B. (1955) *Floods.* Princeton, N.J.: Princeton University Press.

Leggett, R. F., and Williams, G. P., (1964) "Snow removal and ice control in Canada with a note on snow and ice research." In *Snow Removal and Ice Control.* Ottawa, Ont.: National Research Council, Technical Memorandum No. 83.

Metropolitan Toronto and Region Conservation Authority. (1959) *Plan for Flood Control and Water Conservation.* Woodbridge, Ont.

Public Works. (1968) "Plowing and loading snow—one machine can do both." 99 (11).

Rooney, J. F., Jr. (1966) "The urban snow hazard: an analysis of the disruptive impact of snowfall at ten cities in the central and western United States." Worcester, Mass.: Clark University, Department of Geography, unpublished Ph.D. thesis.

Schaerer, P.A. (1962) "The avalanche evaluation and prediction at Rogers Pass." Ottawa, Ont.: National Research Council of Canada, Division of Building Research, Technical Paper No. 142.

Sewell, W. R. D. (1965) *Water Management and Floods in the Fraser River Basin.* Chicago: University of Chicago, Department of Geography, Research Paper No. 100.

Summers, P. W. (1970) "Present status of hail suppression." In J. Maybank and W. Baier, eds., *Weather Modification.* Ottawa, Ont.: Canada Department of Agriculture, Research Branch.

———, and Wojtiw, L. (1971) "The economic impact of hail rainfall parameters." Reprint of a paper presented at the American Meteorological Society's 7th Conference on Severe Local Storms, Kansas City, Mo., October 5–7, 1971.

U.S. Department of Agriculture. (1941) *Climate and Man: Yearbook of Agriculture,* Washington, D. C.: Government Printing Office.

———, (1968) *Economic Aspects of Weather Modification.* Report to the Interdepartmental Committee on Atmospheric Sciences by the Economic Research Service, Washington, D.C.

U.S. Department of Commerce. (1972a) *Climatological Data, National Summary 1971.* Asheville, N.C.: National Oceanic and Atmospheric Administration, Environmental Data Service.

_____, (1972b) *Storm Data,* monthly publication. Washington: NOAA, Environmental Science Services Administration, Environmental Data Service.

U.S. House of Representatives. (1966) "A unified policy for flood loss reduction." 89th Congress, 2d Session, House Document No. 465. Washington, D.C.: Government Printing Office.

U.S. Office of Emergency Preparedness. (1972) *Disaster Preparedness,* vol. 3. Washington, D.C.: Executive Office of the President.

White, G., et al. (1958) *Changes in Urban Occupance of Flood Plains in the U.S.* Chicago: University of Chicago, Department of Geography, Research Paper No. 57.

Whitham, K., Milne, W. G., and Smith, W. E. T., (1970) "The new seismic zoning map for Canada, 1970 edition." *Canadian Underwriter,* June 15.

Wolford, V. L. (1960) *Tornado Occurrence in the United States.* Washington, D.C.: U.S. Department of Commerce, Weather Bureau.

28. Natural hazards: report from Japan

Compiled by T. NAKANO
Tokyo Metropolitan University

With the cooperation of:
H. KADOMURA
T. MIZUTANI
M. OKUDA
T. SEKIGUCHI

Natural and social characteristics

Japan is an island arc which belongs to the monsoon region, and is under the influence of warm and moist air masses in summer and cool air masses in winter. The moisture which is taken in the lower layers of the air masses over the sea is poured on the country by typhoon in summer, by snowfall in winter, by the Bai-u Front in June and July, and by depressions and fronts in all seasons. The average amount of precipitation is 1,800 millimeters (70 inches) a year. This is two or three times the amount received in other areas of the same latitude. In the southern Pacific coast areas it amounts to 4,000 millimeters (160 inches). Owing to Japan's slender shape and complicated landforms, areal differences of climate are great.

Three-quarters of the land is mountainous with generally high relief. Chains of spinal mountains, some reaching 3,000 meters (10,000 feet) above sea level, run through the center of the narrow and long country. Consequently, the rivers are generally short with steep gradient. Erosion and devastation in the mountain areas are very rapid. Rivers are flooded soon after a heavy rain.

Japan is also situated in the circum-Pacific seismic zone, and suffers from severe seismic and volcanic activities. Active volcanoes in Japan make up one-tenth of the world's total.

The greater part of the population of more than 100

million live in narrow plains with exceedingly high densities. As a result of rapid and disordered urbanization, inappropriate land use prevails. Artificial changes of natural environments are rapid and large, accompanying the great increase in economic activity and exploitation.

Kind and nature of natural hazards

Owing to the above-mentioned characteristics, Japan suffers from various kinds of severe and frequent natural hazards. Their effects on social and economic life are significant.

The damage listed in Table 28–1 covers 80–90 percent of the total damage due to natural hazards during 1946–70. The greater part is caused by heavy rain and extreme wind due to typhoons or tropical cyclones, frontal cyclones, and extratropical cyclones. Over time the damage due to severe earthquake, though infrequent, may be considerable. Although typhoons are less frequent than frontal and extratropical cyclones, typhoon damage amounts to about one-half of the total because the destructive force of a typhoon is generally stronger and the afflicted area is wider. Frequency of hazard from the Bai-u Front amounts to three-quarters of that from frontal and extratropical cyclones. The activity of the Bai-u Front is often strengthened by a typhoon. Thunderstorm hazard is local. Hazards of typhoons and frontal and extratropical cyclones occur in

Table 28–1. Damages caused by main natural hazards in Japan, 1946–70

	No. of events	Deaths	Houses destroyed	Houses flooded
Typhoon	59	13,745	576,378	4,479,665
Heavy rain due to front and ETC[a]	77	7,372	60,877	3,681,042
Heavy wind due to ETC[a]	12	784	4,941	16,487
Earthquake	11	5,490	113,339	81,654
Landslide	5	86	143	–
Hail and thunderstorm	4	28	847	23,482
Heavy snow	2	242	1,734	7,062
Volcanic eruption	1	12	12	–

[a]ETC = extratropical cyclone.

the form of flood, inundation, storm surge, extreme wind, surging wave, land collapse, mudflow, and so on.

Recently, floods from large rivers have become more rare, and damages by land collapse, mudflow, and floods of small rivers have become significant. Owing to land subsidence caused by overpumping of groundwater and the urbanization of lowland areas, inundation results from much lighter rain than before. When a typhoon is accompanied by a high storm surge, severe damage is caused in coastal cities. Damage by heavy wind due to typhoon is generally small compared to that by heavy rain, except on small islands. Hazards due to heavy wind and high wave are often caused along the coastal areas in northern Japan by winter cyclones, and bring serious damage to fishing boats.

Frequencies of hazards due to climatic anomalies which cause severe damages to agricultural products are: cold summer, 2: drought, 1: frost, 3: and unusually long rainy spell, 2.

The number of earthquakes felt in Japan exceeds 1,000 per year; few earthquakes cause damage, but when a strong earthquake does occur near a large city, disastrous damage results. Tsunamis occur on the coast facing the Pacific Ocean. Damage from landslide is small and local. Thunder occurs in all parts of Japan, mainly in summer, but the occurrence of heavy thunderstorm with hail is regionally limited. When heavy snowfall occurs in the plains areas, there is a large disturbance to economic activities. Although there are many active volcanoes, damage from their eruption is rare; current extensions in habitation near volcanoes increase damage potential. Climatic anomalies such as cold summer,

drought, and unusual long rains cause decreases in yields of agricultural products.

Seasonality of occurrence

In Japan extreme natural events frequently occur in the period from June to September, mostly from typhoons and the Bai-u Front. The occurrence of typhoons is virtually limited to June to October; the greatest frequency is in August. Almost all typhoons causing disastrous damages have attacked Japan in September. In 1955–63 the average number of typhoons was 26.2 per year; the average number attacking the coast was 3.9, with 2.6 of them causing considerable damage.

Heavy frontal rain occurs in June, July, and August. In June and July, the Bai-u Front stagnates above Japan and induces flood and land collapse. The Bai-u Front is most active in July.

In August, storms occur when extratropical cyclones with active fronts pass over the Japan Sea and when the period of the Bai-u is prolonged in northern Japan. Extra tropical depressions pass over Japan frequently in spring and autumn, and floods due to extratropical depressions often occur in autumn. In winter and in spring a cyclone as strong as a typhoon often passes over the shore of Hokkaido and brings heavy winds and surging waves which cause serious damages in coastal areas.

Landslide tends to occur in the period of snow melt. Hail, together with heavy thunderstorm caused by the cold fronts, occurs mainly from April to July. Heavy snow hazard occurs mainly in January and February. Summer cold damage is caused by low temperatures and deficiency of insolation in spring and summer. Damage by frost and unusual cold weather occurs mainly in spring.

Social and economic effects

In Japan the numbers of dead and missing, damaged houses, flooded houses, and sunken ships are recorded by the police. Amounts of damage to public equipment and agricultural products caused by natural hazards are evaluated by the government. The mean annual damage to public equipment and agricultural products during the period 1965–69 is 332 billion yen (U.S. $122.84 billion). Adding damages to non-public equipment and properties, mean annual direct damage caused by natural hazards is estimated at 350–400 billion yen (U.S. $129.5–$141 billion) in recent years (Fig. 28–1). This is equivalent to 0.8–0.9 percent of the annual mean of the gross national product during the period. The rate of increase of Japanese GNP is very high in recent years; in 1970 Japanese GNP was 72.7 trillion yen ($26.899 trillion).

At the same time there has been no significant increase in the amount of damage from natural hazards. Consequently, the ratio of damage to GNP is decreasing.

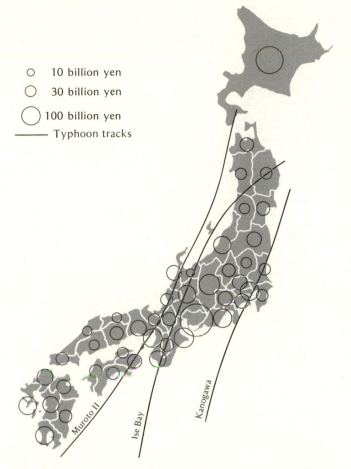

○ 10 billion yen
○ 30 billion yen
◯ 100 billion yen
— Typhoon tracks

Fig. 28–1. Amount of damage from natural hazards, by prefectures (1955–63)

Table 28–2. Mean annual damage caused by natural hazards in Japan, 1955–63

	Deaths	Houses destroyed	Houses flooded	Damage (in million yen)
Typhoon	864	19,513	258,020	168,647
Heavy rain due to front and ETC	280	1,097	186,881	83,200
Heavy wind due to ETC	203	351	994	2,990
Earthquake	19	527	3,964	3,004
Heavy snow	91	184	1,920	7,466
Thunderstorm	33	53	167	115
Climatic anomaly	—	—	—	48,566
Total	1,491	21,724	451,945	313,988

The ratio in 1970 was 0.6 percent. Taking account of annual fluctuations in amount of damage, 0.5–1.0 percent may be an adequate value for the ratio in recent years. In the period of 1955–63 when large disasters occurred, the ratio was 2 percent and the ratio of damage to national wealth was 1 percent (Table 28–2).

The mean annual number of deaths due to natural events in 1966–70 was 382, which is equivalent to 2.5 percent of the annual mean of deaths caused by traffic accidents. In 1955–63 the number of deaths was 1,490 and the ratio was 15 percent, showing that the number of deaths due to natural hazards has decreased remarkably.

The mean annual number of damaged houses in 1966–70 was 7,468. This is about three-quarters the number of houses that burned during the same period. However, when transportation and communication facilities are damaged, social and economic activities are influenced over wide areas, and this disruption tends to increase with high development of social functions.

Large disasters

Catastrophic disasters have caused the loss of more than 30,000 lives in the last 100 years. On the average, four catastrophic disasters out of seven were caused by earthquakes. Damage was enormous. Because most Japanese houses are lightly constructed and made of wood, fires accompanied by earthquakes in densely built-up areas are especially significant. In the case of the Kanto earthquake of 1923, 447,000 houses burned, mainly in Tokyo and Yokohama. Storm surge due to typhoon is also highly hazardous; most of the damages from the Muroto and Ise Bay typhoons were caused by extraordinary high storm surges in coastal lowland areas of Osaka and Nagoya.

Fifteen large disasters occurred in 1945–70 as shown in Table 28–3 (disasters due to climatic anomalies are excluded): they include typhoon disasters, 10 (8 of them in September); heavy rain due to the Bai-u Front, 4; and earthquake, 2. Typhoons account for 15 percent of all disasters, but for 70–80 percent of total damage. Storms and floods by frontal and extratropical cyclones are 5 percent of all disasters, but account for up to 50 percent of damage.

Temporal change of natural hazards

Records have been kept of the number of meteorological disasters, by prefecture, during 5-year periods since 1868. The numbers, however, do not reflect the magnitude or severity of the disasters. A gradual increase in number before about 1940 and rapid increase after then may be identified. But, this secular change does not always coincide with secular variation of meteorological conditions.

Until the nineteenth century, cold summer, drought, and large fires enlarged by meteorological conditions

Table 28–3. Recent natural disasters in Japan

	Year	Deaths	Houses destroyed	Remarks
Makurazaki typhoon	1945	3,756	89,340	Minimum pressure at Makurazaki = 916.6 mb
Nankaido earthquake	1946	1,443	32,356	M (magnitude) = 8.1
Kathleen typhoon	1947	1,540	12,751	Maximum daily precipitation = 438 mm
Fukui earthquake	1948	3,769	51,851	M = 7.2
Ion typhoon	1948	838	18,017	Minimum pressure at landing = 940 mb
Jane typhoon	1950	508	56,089	Storm surge in Osaka Bay = 2.37 m
Ruth typhoon	1951	943	72,362	Maximum wind speed, 51.4 m/s
Nishi Nippon flood	1953	1,013	17,370	Maximum daily precipitation = 500 mm
Minami Kinki flood	1953	1,124	9,829	Maximum daily precipitation = 565 mm
Typhoon No. 13	1953	478	26,071	Minimum pressure at landing = 930 mb
Toyamaru typhoon	1954	1,708	30,951	Maximum wind speed = 55.0 m/s
Isahaya flood	1957	992	4,372	Maximum daily precipitation = 1,109 mm; maximum hourly = 144 mm
Kanogawa typhoon	1958	1,269	4,294	Minimum pressure = 877 mb; maximum daily precipitation = 694 mm
Ise Bay typhoon	1959	5,098	153,930	Storm surge in Ise Bay = 3.89 m; maximum wind speed = 55.3 m/s
Heavy rain by the Bai-u front	1961	372	3,653	Maximum daily precipitation = 644 mm
Muroto typhoon II	1961	202	61,932	Maximum wind speed = 84.5 m/s; storm surge in Osaka Bay = 2.7 m
Niigata earthquake	1964	26	2,250	M = 7.5
Matsushiro earthquake	1965	—	—	M = 5.4
Ebino earthquake	1968	3	368	M = 6.1
Tokachi-oki earthquake	1968	52	689	M = 7.9

were most feared. However, these are less important, local matters today. Earthquakes and typhoons have been feared since ancient times. There is a possibility that seismic damage will increase, owing to the progress of urbanization and the storage of combustibles in urban areas.

Damage due to floods has increased as a result of increased exploitation of lowland areas. For about 10 years after the Second World War, large rivers flooded frequently. Since 1961 detailed statistics have not been available, but storm and flood damages are estimated to be rapidly decreasing because of the progress of adjustments against the disasters. Recently, however, floods of small rivers and storm inundations or urban areas have become more significant.

Storm surges due to typhoons have become more threatening to coastal cities because of progress of land subsidence. Following urbanization of hilly land, land collapse and mudflow caused by heavy rain become significant causes of disaster. Heavy snowfall and accumulated snow is now recognized as a threat; once it was regarded as a normal part of the natural environment. Damages due to climatic anomalies (cold summer and drought) have been much decreased as agricultural techniques have developed. Air pollution became a serious problem in cities and factory districts in the 1960s.

Protective measures and warning systems

Inasmuch as flooding is a main natural hazard in Japan, flood prevention works have claimed great interest from early times, and rivers are well embanked. The invest-

ment in the prevention of flood disaster is currently more than 200 billion yen (U.S. $74 billion) a year. The greater part of it is for flood prevention and erosion control works, where annual expenditure is 100–150 billion yen (U.S. $37–55 billion). The Ministry of Construction deals with the problem of dike or embankment construction to prevent floods. The necessity of systematic measures for flood prevention was recognized acutely after the Ise Bay typhoon in 1959, and the government established the Standard Law for Measures of Disasters. Areal plans for disaster prevention are now prepared in every prefecture.

The Japan Meteorological Agency under the Ministry of Transportation issues forecasts and warnings of typhoons, heavy rain, and other meteorological phenomena, but forecasting local heavy rain, which often causes serious damage with short lead time, is still difficult. To predict earthquakes would be very effective in minimizing seismic damage, and many investigators are studying earthquake prediction systematically; thus far, however, the results are not fruitful.

Regionality of natural hazards in Japan

Hazard Areas

Typhoons are formed mainly in the Carolines area and approach Japan from the south (Fig. 28–2). Consequently, storms and floods due to typhoon are frequent in the Pacific coast areas from southern Kyushu to southern Kanto. Storm surge caused by ty-

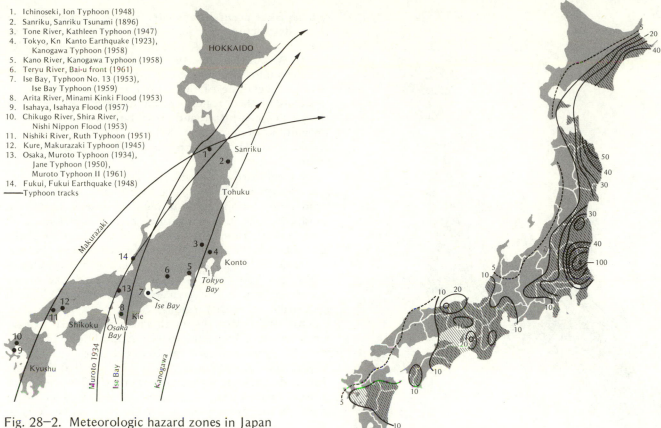

1. Ichinoseki, Ion Typhoon (1948)
2. Sanriku, Sanriku Tsunami (1896)
3. Tone River, Kathleen Typhoon (1947)
4. Tokyo, Kn Kanto Earthquake (1923),
      Kanogawa Typhoon (1958)
5. Kano River, Kanogawa Typhoon (1958)
6. Teryu River, Bai-u front (1961)
7. Ise Bay, Typhoon No. 13 (1953),
      Ise Bay Typhoon (1959)
8. Arita River, Minami Kinki Flood (1953)
9. Isahaya, Isahaya Flood (1957)
10. Chikugo River, Shira River,
      Nishi Nippon Flood (1953)
11. Nishiki River, Ruth Typhoon (1951)
12. Kure, Makurazaki Typhoon (1945)
13. Osaka, Muroto Typhoon (1934),
      Jane Typhoon (1950),
      Muroto Typhoon II (1961)
14. Fukui, Fukui Earthquake (1948)
——Typhoon tracks

Fig. 28–2. Meteorologic hazard zones in Japan

Fig. 28–3. Annual frequency of felt earthquakes (average, 1921–58)

phoon occurs in the bays which open to the south, and is most frequent in Tokyo, Ise, and Osaka bays. Storms and floods due to frontal and extratropical cyclones occur frequently in the Japan Sea coast areas from Hokkaido to northern Kyushu. Heavy rainfall caused by fronts is most frequent in northern Kyushu and Hokuriku. Heavy wind due to winter cyclone attacks coastal areas of northern Japan. Since snow is brought by the seasonal winds from Siberia, snow hazards occur in the coastal regions of the Japan Sea.

Earthquakes occur frequently in the eastern part of Japan (Fig. 28–3). High tsunamis are limited to the coast of Sanriku, Kii Peninsula, and southern Shikoku, which have rias coasts open to the Pacific Ocean. Hokkaido and northern Tohoku suffer from cold summmers. Damages by hail and frost occur mainly in the inland areas of central Japan.

In Fig. 28–1 the amount of damage is shown by prefectures for 1955-63. Damage is great in the south coast regions, due to typhoon damage. Japan can be roughly divided into two regions according to type of hazard: the southern Pacific coast region with typhoons, and the northern Japan and Japan Sea coast regions with frontal and extratropical cyclones and winter hazards. In the inland area an intermediate region may be noted.

Although the frequency and degree of danger of typhoons is relatively high in the western part of Japan (Fig. 28–4), an intense typhoon may cause damage to

parts of Japan; no region is entirely free from severe typhoons. The frequency of heavy rain is high in the southern Pacific coast areas (Fig. 28–5). Even in areas where the average amount of rainfall is small, localized rainfall may cause disasters, with flood and land collapse. In most of the mountainous regions land collapses and mudflow due to heavy rain are threats. Most of the alluvial lowlands are threatened by flood. Dangerous areas are delimited for each lowland by microrelief. Lowland areas of Tokyo, Nagoya, and Osaka are in danger of storm surge, and in these cities the risk grows because of the expansion of land subsidence. The area below sea level in Japan amounts to 390 square kilometers (150 square miles). The danger of land collapse caused by heavy rain is great in cities situated on hilly land such as Yokohama, Kobe, and Kure, and in areas of pyroclastic flow deposits such as southern Kyushu. Landslides occur frequently in Hokuriku and Shikoku.

Most severe earthquakes in the past have occurred under the sea bottom of the Pacific Ocean along the east and south coasts of Japan. Expectation of maximum seismic intensity is greatest in southern Kanto. In cities situated on areas of soft ground and land subsidence, earthquakes are particularly hazardous; here, the Koto

Fig. 28–4. Distribution of frequency of typhoons (1926–55)

lowland in Tokyo is the most dangerous area.

Although the evidence is inconclusive, it is maintained by some people that there exists a cycle of 69 years in the occurrence of strong earthquakes in Tokyo and that the beginning of the next dangerous period is 1978. If an earthquake like the Kanto earthquake of 1923 were to attack Tokyo now, millions might be killed.

The most active volcanoes, classed "A" by the Japan Meteorological Agency, are Aso and Sakurajima in Kyushu, Asama in central Japan, and Mihara in Izu Oshima. Damage from cold summer weather is high in Hokkaido and the east coast of Tohoku, and damage from drought is high on the hilly land of western Japan.

Natural hazards resulting from lithogenic and geomorphic events

Earthquakes

Earthquakes are the most serious natural hazard in Japan, in terms of damage. About 10 percent of world earthquakes occur in and around the Japanese islands. Most destructive earthquakes have occurred in the outer zone of the island arcs, i.e., the sea areas off the Pacific coast. Within this zone the Hokkaido-Sanriku coast and the Tokaido-Nankaido coast are most active; tremors with a magnitude of more than 8.0 have occurred repeatedly in these regions. Earthquakes also occur on the Japan Sea coast and land areas, but movements with a magnitude of more than 8.0 are rare. The Nobi earthquake of 1891, with a magnitude of 8.4, is an exception.

A great earthquake usually results in an integrated hazard over broad areas. It causes disasters not only by land deformation, such as faulting, cracks, landslides, land collapses, uplift, subsidence, and liquefaction of loose sandy ground, but also by direct vibration of the ground. Fire disasters frequently accompany the destruction of houses and result in serious damages, especially in urban areas. When a great earthquake occurs under the sea, violent tsunamis (great sea waves) result and strike the coastal regions.

Over 400 destructive earthquakes are known to have occurred since the sixth century. According to historical records, great damage was frequent in the coastal regions from Kanto to Kyushu, i.e., the Tokaido and Nankaido regions, where approximately 70 percent of Japan's population is presently located. The major metropolitan areas, namely Tokyo, Nagoya, and Osaka, and the main industrial areas are within this vulnerable zone.

Earthquakes which occurred before the Meiji era (1868) are too many to mention in detail. Beginning in 1096, nine earthquakes of a magnitude of 6.9–8.6 caused serious damage. During the earthquake of 1855 at Edo, the area occupied by present-day Tokyo suffered great damage, with 10,000 persons killed and 14,346 houses destroyed. In December 1854 two great earthquakes occurred at an interval of 32 hours, one off Tokaido and the other off Nankaido; over 70,000 houses were destroyed and 4,000 people killed by strong vibrations of the ground, fire, and tsunami. These two earthquakes, together with the Edo earthquake, caused tremendous damage to the economy and society under the Shogunate regime and contributed to its decay.

The Nobi earthquake of 1891, occurring some 50 kilometers (30 miles) west of Nagoya, was characterized by the formation of a lateral fault 80 kilometers (50 miles) long—the longest on land in historical times. During the Kita-Izu earthquake of 1930, faults were formed at Tanna Basin in the northern part of Izu peninsula. The return period of earthquakes producing faults in this area has been approximately 1,000 years. The Tokaido and New Tokaido railway lines now tunnel through these active faults.

The Kanto earthquake of 1923 is representative of earthquake hazard in urban areas. The main cause of loss of human life and damage to houses was fire. In Tokyo, fire broke out at more than 130 places immedi-

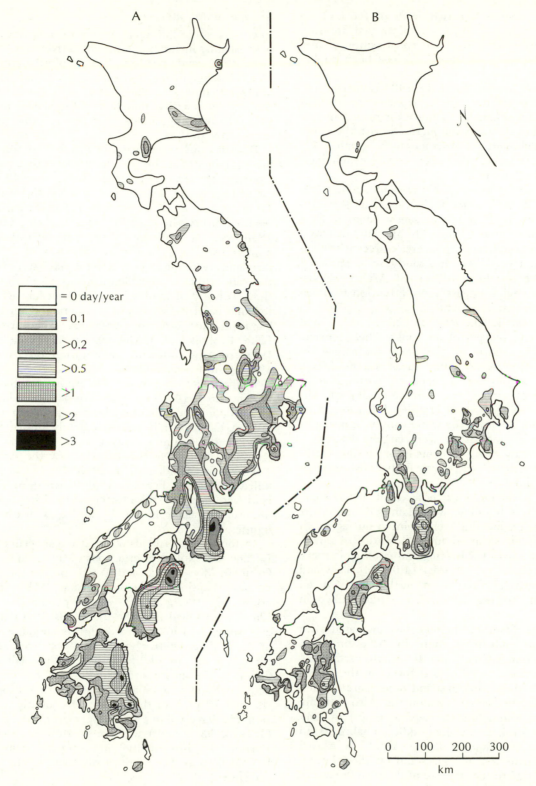

Fig. 28–5. Frequency distribution of heavy rainfall (1956–65). A: daily precipitation > 200 mm. B: daily precipitation > 300 mm

ately after the earthquake; two-thirds of the built-up area of the city was completely destroyed by fire. Thereafter, prevention of fire and construction of fire-proof and earthquake-proof buildings have been emphasized.

Another lesson learned from the 1923 disaster is the relationship between damage to wooden houses and ground conditions. Damage to houses is principally related to ground conditions, and increases with the thickness of soft sediments. Construction of earthquake-proof buildings, such as reinforced concrete buildings, was accelerated after 1923.

The Niigata earthquake of 1964 was characterized by damage to reinforced concrete buildings in the city, caused mainly by liquefaction of sandy layers or by a quicksand phenomenon (Table 28–3). The damaged buildings were concentrated in an area of recently abandoned river courses where fine sandy soils saturated with groundwater are thickly deposited. After this disaster, foundation engineers gave more attention to possible liquefaction of sandy ground.

The Niigata earthquake also caused tsunamis, and fires originating in big petroleum tanks. These impacts came sequentially; most of the areas submerged by tsunami inundation were below mean sea level as a result of land subsidence induced by withdrawal of groundwater in obtaining water-soluble natural gas. Urban activities of Niigata were crippled for a long time because the main facilities for water supply, transportation, and telecommunication were largely destroyed. Niigata raised many new problems in preventing earthquake losses in urban areas, particularly those situated on coastal lowlands.

During the Tokachi-oki earthquake of 1968, some reinforced concrete buildings were damaged, and errors in seismological engineering were pointed out again. In cities in Tohoku, fires started due to upsetting of stoves. With petroleum stoves for heating and cooking in every house, it is estimated that at least 14,000 stoves would overturn and general fires if an earthquake with an intensity similar to Kanto were to strike Tokyo in winter.

Potential loss from earthquake has increased with urbanization and industrialization. In the event of an earthquake with an intensity like the Kanto earthquake in 1923 the damage may be most severe in the Tokyo deltaic lowland where buildings and houses are agglomerated on thickly bedded soft sediments. Also there are over 65 square kilometers (25 square miles) of land below mean sea level, with some 500,000 inhabitants. In these conditions, earthquake damage would be caused not only by destruction of houses and buildings but by fire, and flood due to the fracture of dikes and seawalls. Urban redevelopment based on disaster prevention projects is urgently needed to reduce the complex hazard. Recently, precise survey observations which it is believed may be useful in predicting earthquakes are being carried out in southern Kanto.

Volcanic eruptions

There are some 200 Quaternary volcanoes in the Japanese islands, including about 80 active ones. The main destructive agents resulting from explosive eruption are lava flows, mudflows, pyroclastic flows and falls, gases, blasts, earthquakes, landslides, and tsunamis. It is through these agents that human life is lost and houses, villages, towns, farmlands, forests, and other properties are destroyed.

Disasters due to eruption in the past are too many to mention in detail. Before the nineteenth century two violent eruptions occurred, one killing over 15,000 people at Unzendake, Kyushu in 1792, and the other killing 1,151 people at Asamayama in 1783. During the nineteenth century the eruption of Bandaisan, Tonoku, in 1888 was the greatest. There were 461 deaths, and numerous villages, and many hectares of forests and farmlands were destroyed. Several lakes were formed in the valleys due to the damming up of rivers by mudflow. This is typical of how violent volcanic activities on the one hand cause serious disasters and on the other hand produce beautiful landscape. The formation of Lake Taishoike at Kamikochi in the middle of the Japanese Alps following an eruption of Yakedake in 1915 is another example.

In the twentieth century the eruption of Sakurajima, Kyushu, in 1914 caused eight villages to be buried and 140 people to be killed by immense lava streams and ash falls. A narrow strait was completely buried with lava flow, and Sakurajima was connected with the main Kyushu Island. On the occasion of the eruption of Tokachidake, Hokkaido, in 1925–26, 144 people were killed and 1,200 hectares (3,000 acres) of cultivated land were damaged by a torrential mudflow. This volcano erupted again in 1962, causing 5 deaths and 12 injuries.

In the volcanic islands of Izu damage occurs frequently due to explosive eruptions. Miharayama (in Izu-Oshima), Miyakejima, and Torishima are the main volcanoes erupting in recent years. When Torishima erupted in 1902, all 125 of the people living on the island were killed and the island has been uninhabited ever since. In 1962 a great eruption suddenly occurred at Miyakejima, and five houses were destroyed by fire, but there was no loss of life because of immediate escape action by the inhabitants. Miharayama, one of the more active volcanoes in Japan, has had ten eruptions during the first 70 years of this century, frequently with extrusion of a large volume of lava; the eruption of 1950–51 (1 dead, 53 injured) was the greatest in recent years. Fortunately, loss of life and property damage were limited because there are no dwellers in the area near the crater.

A new volcano 406 meters (1,300 feet) high was born near Usudake, Hokkaido, during 1943–45. This lava dome was named "Showa-Shinzan", which means "born in the Showa era," and attracted the attention of world geologists.

Many villages, towns, and cultivated fields are located on the lower slopes of volcanoes which can be expected to erupt in future. Active volcanoes constitute the main scenery of national and quasi-national parks, and numerous hot spas are near their slopes. Volcanoes attract many people, and visitors for sightseeing, mountain climbing, and skiing are numerous throughout the year. Dwellers in hazardous areas are rapidly increasing with the development of the country. Even at the most active volcanoes, the number of visitors within a day exceeds 10,000. About 300,000 people live in the hazardous zone of Sakurajima, within a distance of 10 kilometers (6 miles) of the active crater. These social conditions accelerate the potential hazardousness of explosive eruption.

Because the energy is too enormous to curb technologically, escape from the danger area is the only effective way to cope with volcanic violence. Therefore, prediction based on precise observation is most important. Active volcanoes of Japan are classified as A, B, and C, according to type of activity and degree of hazardousness. Precise and continuous observations are now made for class A and B volcanoes, 16 in all, by the Meteorological Agency of Japan. Four volcanoes—Asamayama, Miharayama, Asozan, and Sakurajima—are noted as most active (class A). Warning systems are in operation for class A and B volcanoes.

Tsunamis

Although a tsunami may be caused by volcanic eruption, the most common cause is an earthquake occurring under the sea, i.e., a seismic sea wave. Most tsunamis originate near the Pacific coast, and resulting sea waves attack the coastal regions from Hokkaido to the Ryukyu Islands, causing great damage. Tsunamis originating off the Kamchatka Peninsula, the Kuril Islands coast, and the Chilean coast of South America also cause serious damages to the Pacific coast of Japan. The Chilean tsunami of 1960 is a recent example.

The distribution of wave height on the coastal areas varies according to the type of tsunami. V-shaped narrow bays are believed to provide conditions favoring the increase of wave height toward the bay bottom. Convergence of the energy of the tsunami and the seiche of the bay are the main factors increasing wave height. Lowland areas at the head of embayments are usually occupied by agglomerated settlements, mostly fishing villages. Thus, natural and social conditions are interrelated to favor tsunami disasters. This is the main reason why the rias-type coast frequently suffers serious disasters. The Sanriku coast and the eastern coast of Kii Peninsula, where rias is developed typically, are most hazardous (Fig. 28–2). Also these coastal regions face the sea under which great earthquakes are liable to occur. Both areas are, therefore, the region of the greatest hazard from tsunami disasters within the Pacific coast; they have been attacked by great tsunamis repeatedly throughout historical times.

There are many records of tsunami disasters in Japan. About 150 tsunamis resulting in damage are known to have occurred since the seventh century. Among historical tsunami disasters the damage caused by the Meiji Sanriju tsunami of 1896 is the greatest. This originated with an earthquake (magnitude 7.6) on the offshore area of the Sanriku coast, and struck the coastal regions from Hokkaido to the Oga Peninsula. Maximum wave height exceeded 24 meters (80 feet), with 27,122 people killed and 9,316 injured, and 10,617 houses and 7,032 ships destroyed.

In the coastal regions of Tokaido-Nankaido since 1096 nine tsunamis caused by earthquakes have resulted in notable disasters. Each of these killed thousands of people and destroyed houses and ships. Greater damages usually occur along the coastlines of Izu and Kii peninsulas, and the southern coast of Shikoku Island. Some of those tsunamis entered Tokyo and Osaka bays and inundated towns situated on the lowlands, causing slight damage.

The Chilean tsunami originating off southern Chile on March 22, 1960, struck the coast of Japan from Hokkaido to the Ryukyu Islands. Although it arrived on the Japanese coast 22–24 hours after the earthquake, no warning was given before the first arrival. Maximum wave height exceeded 6 meters (20 feet) in some bays of the Sanriku coast. Total Japanese loss from this tsunami was 139 killed and 872 injured, with 2,830 houses and 1,500 ships destroyed. The total number seriously affected exceeded 160,000. At Shima, situated on the eastern coast of Kii Peninsula, where pearl culture is carried out in narrow and deep embayments, many pearl beds were destroyed by violent currents. Fish farming suffered serious damage elsewhere.

Even in the case of tsunamis generated near Japan, the first arrival of the sea wave is generally 10–30 minutes after an earthquake. It is, therefore, possible to predict its occurrence. In the coastal regions where tsunami disasters strike repeatedly people know from experience how to reduce the risk in tsunami disasters. Escape to high places is the usual way to prevent loss of life when people receive warning. Though essential, this is only an emergency measure. In order to make permanent adjustments other countermeasures are necessary: relocation of houses to higher places, construction of seawalls, tidewater control, afforestation, and shelter harbors. These have been undertaken, but not sufficiently, in most tsunami-striken areas. Prediction of earthquakes is essential for preventing tsunami damages, and consolidation of warning systems with earthquake detection and prediction is of basic importance. Recently, the pan-Pacific tsunami warning system has been organized through the cooperation of countries concerned. If it had been completed at the time of the Chilean tsunami, much damage might have been prevented.

Landslides and related events

Though great landslides and related events are sometimes caused by volcanic eruptions and earthquakes, these geomorphic processes are more frequent due to climatic events. In Japan, landslides, land collapses, mudflows, and rocky mudflows take place in mountainous and hilly regions every year due to heavy rainfall caused by typhoons and the Bai-u Front as well as snow melting. These events result in loss of human life and damage to houses, roads, railways and other properties. The annual loss has increased rapidly with the development of Japan. The geographical distribution of landslides and related events is principally related to geology and geomorphology. Landslides in Japan can be classified into three types in relation to geology: those occurring (1) in fractured zones, (2) in Tertiary sediments, and (3) in hot-spa areas.

The fractured zone type of landslide is found mainly in the outer zone of southwestern Japan, where crystalline rocks are strongly faulted and fractured. Among crystalline rocks, black schists provide most favorable conditions, notably on hillslopes in the Shikoku Mountains. This type generally moves slowly over a long time; rapid movement due to heavy rainfall rarely occurs.

In Tertiary sediments, especially those consisting of black mudstones or shales of the Miocene age, landslides are found throughout Japan. This type is concentrated in the young folded zone of the Japan Sea side of northeastern Japan. The movement of this type is rather rapid and sometimes causes torrential mudflows. Rapid movement is caused by heavy rainfall, a spell of long rains, and melting snow. In Hokuriku, rapid-movement mudflow is often caused by melting of snow in early spring after a heavy snowfall. In March 1963 a large landslide occurred on the Japan sea coast due to sudden melting of snow, and a train on the Hokuriku line was hit by a torrential mudflow. In some areas, for instance northwestern Kyushu, Tertiary mudstones are capped by lava saturated with groundwater in its pores, contributing to the occurrence of landslides.

In volcanic areas landslides frequently occur on a large scale. The occurrence of some is related to chemical decomposition of rocks due to sulfurization in the hot-spring areas. The landslide at Sounzan, Hakone Volcano, in 1953 is typical of this type.

Landslides are closely related to land use in mountainous regions. Gentle slopes as well as replenishment of soils resulting from repeated landslides provide suitable conditions for agriculture and settlements in mountainous regions where steep hillslopes are dominant. In hilly country of Tertiary sediment, most gentle slopes caused by landslides have been used as paddy fields for rice cultivation since long ago, and in mountain lands in the outer zone of southwestern Japan, gentle slopes resulting from repeated slow landslides are the main places for agriculture. The mountain villages on landslide areas have been supported by repeated

landslides. Evacuation of houses from hazardous sites to adjoining safe sites, and rehabilitation and reallotment of cultivated land have often been necessary in these landslide-stricken areas. In present-day Japan most active landslide areas are designated by the Law of Preventing Landslides and Related Events, and various kinds of technological countermeasures are carried out by national and local governments.

Several landslides resulting in mudflows and/or rocky avalanches of more than 100 million cubic meters (3.5 billion cubic feet) volume have occurred in historical times. In the high mountains of central Japan many great landslides have occurred since the sixteenth century. On the western slope of Mount Fuji a great valley, named Ohsawakuzure Slide, heading at a point close to the summit, still vigorously erodes the volcanic body with frequent rockslides. The prevention of such rockslides is difficult because of high altitude and steep slope. Recently, discussion has centered on preventive measures to preserve the beautiful shape of the mountain as well as to protect farmlands and settlements on the foot slopes.

Railways and roads passing through mountainous regions are hit frequently by landslides and related events, and interruption of traffic for over a month is not rare. Damage to roads and vehicles increases with the construction of new networks to mountainous regions and the increase of traffic. Even in Hokkaido, the most underdeveloped region in Japan, landslides and resulting disasters have accelerated due to construction of roads and dams. In March 1961 the Tokaido line and National Route No. 1, the most important traffic routes in Japan, were in danger from a massive landslide at Yui, Shizuoka Prefecture, central Japan. In order to prevent slides from occurring on hillslopes of Tertiary sediment close to the two traffic routes, intensive countermeasures were taken with the cooperation of Japanese National Railways and the ministries concerned at a cost of 20 billion yen (U.S. $56 million).

When urban areas are stricken by rapid landslides and related events, extensive damage may be expected to occur, as in Kobe, Kure, Moji, and Isahaya. In July 1938, Kobe, which is situated on a narrow lowland at the foot of granitic mountains, was struck by rocky mudflows generated by heavy rainfall at the end of the Bai-u season, with 461 deaths and over 100,000 destroyed houses recorded. Kobe was hit by landslides and rocky mudflows again in 1961 and 1967. In September 1945, Kure, also situated on narrow valleys and lower slopes of granitic mountains, was stricken by violent rocky mudflows caused by heavy rainfall accompanying the Makurazaki typhoon, resulting in 1,154 deaths (Fig. 28–2). The city was hit by landslides and rocky mudflows again in 1967. Disasters at Moji in 1953 and Isahaya in 1957 were also severe.

The Kanogawa typhoon attacked Tokyo and Yokohama, on September 26, 1958, and resulted in the heaviest rainfall ever experienced in Tokyo, 392.5 millimeters (150 inches) in 24 hours. One thousand twenty-nine landslides and land collapses, including collapses of retaining walls, occurred within the hill lands of Yokohama, and 61 were killed. At the end of June 1966 both cities were hit again by landslides, owing to a heavy rainfall caused by Typhoon No. 6,604.

As a result of recent urbanization, risk from landslides and related events has increased in these and many other urban areas of Japan. Construction of new towns on hill lands and slopes usually causes strong modification of the natural system of the land, resulting in modification of hydrologic and geomorphic cycles. It is man-induced changes of the land system that accelerate rapid geomorphic processes. Prevention of such effects is considered one of the major tasks in newly built urban areas.

Areal distribution of storm and flood damages in Japan

Figure 28–6 indicates the areal distribution of the percentage of houses of each prefecture destroyed or

badly damaged. The regions where the percentages are large are Kyushu, Kinki, and Chubu districts. Figure 28–7 shows the distribution of percentages of farmland damages for each prefecture. The highest percentages are found in northern Kyushu, southern Shikoku and Kinki district.

Review of the relation between daily amounts of precipitation and flood damage of farmland and of the public construction works in Miyazaki Prefecture in the Kyushu District reveals different magnitudes of damages for critical values of daily precipitation (Table 28–4).

Estimation of the areal distribution of the rate of flooded houses

Based on studies by Takahashi and Arai, the number of flooded houses is roughly proportional to a cube root of daily precipitation. By this means, we can estimate an

Fig. 28–6. Geographical distribution of the areal percentage of damaged farmland against the total areas of each prefecture in Japan (1955–63)

Fig. 28–7. Geographical distribution of the percentage of destroyed houses (completely and/or badly destroyed) against the total number of houses of each prefecture in Japan (1955–63)

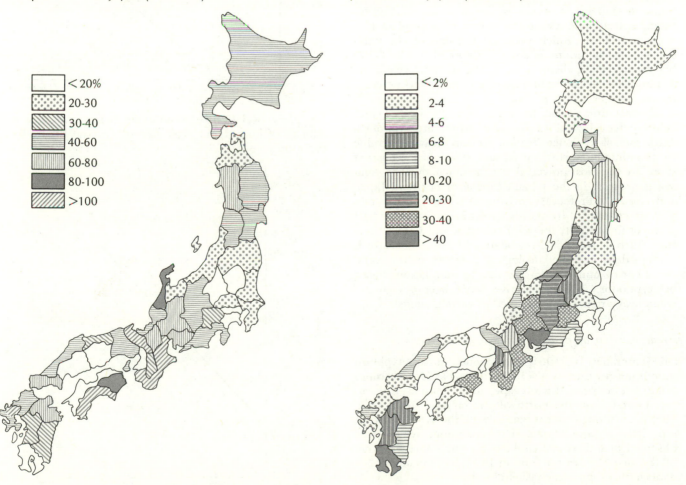

Table 28–4. Daily precipitation: critical values in causing serious damage

Damage	Precipitation (millimeters)
Rice field (10 ha)	40
Farm land (10 ha)	80
Breaks in road (10 places)	80
Breaks in embankment (10 places)	120
Bridges (10)	120

areal distribution of houses endangered by floods from the map of frequency of daily precipitation amounts.

Natural hazards resulting from meteorological events

Typhoons

Immense damage is caused by storms accompanied by extratropical cyclones, and by heavy rains produced by stagnated fronts which come at the end of the Bai-u season. The typhoons in August, September, and October create severe disruptions in the national economy. Typhoons cause more damage than any other meteorological disaster in Japan.

The magnitude varies by region (Fig. 28–4). The chief chief meteorological causes of disasters are typhoons, extratropical cyclones, frontal storms, heavy rains due to the thunderstorms, gusts, and monsoons. The first three are the most powerful. Extratropical cyclone damage is greatest in the Japan Sea side of northern Japan and Nagasaki Prefecture in Kyushu District; heavy rain damage caused by fronts is greatest in the northern part of Kyushu District, the Kii Peninsula, and Gifu Prefecture. Such geographical regionality of natural hazards is explained by the meteorological diversity of the country. Loss of houses is largely due to inundation caused by typhoons. The destructive force becomes greater when heavy rain is accompanied by strong winds.

Areal distribution of wind damage by typhoons

On September 16, 1961, the second Muroto typhoon (minimum pressure was 930.9 millibars at Muroto-cape) came ashore near Muroto-cape, moved northeastward, and passed near the north of the city of Osaka (Fig. 28–2). It brought great damage in the Kinki District especially around Osaka. The maximum wind speed observed in and around Osaka was comparable with that of the first Muroto typhoon in 1934, but the amount of damage this time was considerably less.

The areal distribution of the percentage of destroyed

and badly damaged houses can be compared with the number of houses in various administrative units. The areas with damage rate of more than 10 percent almost exactly correspond to the areal distribution of peak gust of more than 50 meters per second (110 mph). A similar relationship was found for the Ise Bay typhoon in 1959, the surrounding area being Nagoya.

Relationship between flooded houses and destroyed or badly damaged houses

When we consider simple flood damage caused by heavy rainfall, there is a close relation between number of

Fig. 28–8. Relation between flooded houses and the number of houses completely or badly damaged by heavy rainfall for Japan, excluding Kyushu District

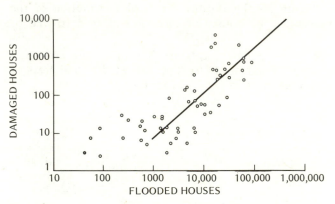

Fig. 28–9. Relation between estimated amount of wind damage of houses and peak gust by cities at the Ise Bay typhoon

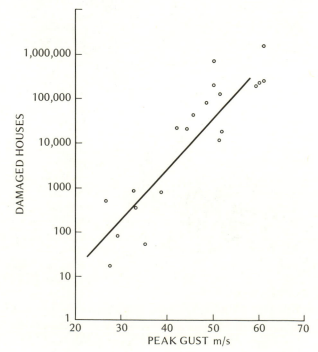

flooded houses and number of destroyed and badly damaged houses. Figure 28–8 shows the relationship, first found in Kyushu District and then extended to the whole of the country. The abscissa indicates the rate of flooded houses against population for each prefecture; the ordinate indicates numbers of destroyed and badly damaged houses, and the solid line is the regression curve of their relationship. The regression equation is:

$$D = e^{\log d^{10}} \ (1.175 \log_{10} F - 2.663)$$

D = number of destroyed houses and F = number of flooded houses.

This relationship is found in other districts in Japan, as well.

Relation Between Wind Damage of Wooden Houses and Wind Speed

It is difficult to separate out the component of wind damage in multicomplex disasters such as typhoons by observation. However, by using the regression equation we can numerically separate it out.

The equation is to estimating wind damage to wooden houses caused by the Ise Bay Typhoon. Figure 28–9 indicates the relation between estimated wind damage of houses and peak gust.

Disaster at sea

In Japan, many shipwrecks result from storms, dense fog, and snowfall. The distribution of frequency of shipwrecks due to unfavorable meteorological conditions according to distance from the coast was calculated. The highest frequency of shipwrecks (of both the fishing boat and the general ships) occurs within 5.5 kilometers (3 nautical miles) of the coast.

29. Natural hazards in the territory of the USSR: study, control, and warning

I. P. GERASIMOV
Academy of Sciences of the USSR

T. B. ZVONKOVA
Moscow University

Natural hazards derive from highly dynamic processes, whose elemental essence consists in their indefinite and equivocal manifestation. Natural hazards that cause human losses and great material damage to the economy are called disastrous and catastrophic. Two aspects of natural hazards are recognized: their potential disaster and their real catastrophic results. Not all potentially dangerous hazards are disastrous. The intensity of disasters in contrast to their potential danger which remains the same is determined by five major factors.

1. Historical and social conditions, as well as the level of economic development of a country or a region, affect its vulnerability. In underdeveloped countries even potentially slightly dangerous hazards may prove disastrous. It is known, for instance, that in middle Asia even at the beginning of the twentieth century, barkhan sands caused great damage by covering some towns and irrigation systems. At present, eolian processes are no longer considered disastrous, though they continue to be destructive for the Republics of Soviet Central Asia. Thus, in 1966 the clearing away of sands from 500 meters (1,600 feet) of the railway line in the western part of the Kara-Kum desert cost about 100,000 rubles (U.S. $110,000). After the railroad had been removed from the zone of a wind-sandy stream, it no longer required clearing. Now, neither towns nor agricultural fields in Soviet Central Asia are covered with sand.

2. The scope of damage that may be caused by natural hazards is determined by the conditions of land use. For whom and what are the natural hazards dangerous? For instance, a snow cover 400 millimeters (16 inches) deep disturbs operating conditions and requires a large amount of work connected with clearing railways and highways. However, the same depth of snow cover proves beneficial to winter crops.

3. The degree of disastrous results of natural hazards is affected by the geographical position of a region as a sequence of processes. For instance, in the European

part of the USSR winter crops are destroyed by frost when the snow cover is less than 300 millimeters (12 inches) thick, whereas in Siberia damage occurs with 400–500 millimeters (16–19 inches). A complicated and economically detrimental situation can be observed in west Siberia when in water logged areas the usual sequence of processes—frosts on soils followed by snow—is disturbed. In cases when snow covers thawed peaty soil, transportation of wood, equipment for oil mines, and other goods becomes hardly possible. Not long ago this was disastrous for the population of the Tyumen region. Nowadays "Antei" planes and helicopters are used in such cases, minimizing the effects of an otherwise disastrous situation.

4. Many natural hazards are not dangerous when taken separately, but if they are combined with other processes, the circumstances become difficult. In the Magadan region of our country, when temperatures of −35° C. (−31° F.) are combined with strong winds, outdoor work ceases. Landslides on the Black Sea coast would not be disastrous if they were not combined with intense erosion of the shoreline.

5. The degree of disaster of most natural hazards is determined not by a single characteristic, but by the mass character of spread, duration, and intensity of the phenomena. Thus, while natural hazards are potentially disastrous always and everywhere, the damage caused by them may vary depending on social, economic, and geographical conditions.

The problem of natural hazards should be studied with allowance for present-day scientific-technical advances and for the continuously growing influence of man's activity on the natural environment. This activity disturbs and modifies the natural dynamics of the processes concerned. Hence, self-regulation of natural processes either disappears, or is weakened. Man, who has disturbed the natural state of the environment, should be a regulator of these processes.

Diversity of natural conditions in the territory of the USSR is responsible for the polygenetic character of natural hazards. At the same time, potentially disastrous natural phenomena in the USSR territory are regularly associated with certain regions and geographical zones.

These phenomena frequently occur on the interface of physically different natural environments: sea and land, atmosphere and ionosphere, and so on. In the Soviet Union, potentially dangerous regional processes are attributed to volcanism, earthquakes, tsunamis, mudflows, avalanches, floods, and hurricanes. In some regions these phenomena cause considerable material danger to the economy as well as human losses.

Volcanism

In the USSR, incidents of recent volcanic activity have been recorded in the Far East only—in Kamchatka and the Kuril Islands (38 volcanoes). Most of these volcanoes have acted periodically in the form of lava emis-

sions, expulsion of rock fragments ("bombs"), sand, ashes, and gas. In this case the majority of large earthquakes, occurring every 7 to 10 years, were reported in the group of Klyuchevskoi volcanoes, as well as the Karymsky and Avachinsky volcanoes.

Though the eruptions in Kamchatka and the Kuril Islands were very strong, there was no great damage, owing to the fact that these areas have never been densely populated. There are various well-known methods of study and forecasting of volcanic activity. The main research work on this subject is carried out at the Institute of Volcanology in Kamchatka and the Institute of Physics of the Earth of the USSR Academy of Sciences.

Earthquakes

Up to 20 percent of the USSR area is subject to earthquakes with a force shock greater than 8 on the intensity scale of the Soviet Academy of Sciences. These are predominantly mountain areas of the south and the Far East of the country: Kopetdag, Tien Shan, Pamirs, south Siberia, Kamchatka, the Kuril Islands. In these areas live roughly 20 percent of the population of the USSR, and such cities as Tashkent, Alma-Ata, Dushanbe, and Irkutsk are situated there.

In the Carpathians and the mountain Crimea, earthquakes occur with a shock force from 6 to 8. The main area of earthquakes in the Carpathians is a sharp bend of the mountain ridge. In the Crimea the principal focuses of earthquakes are confined to the south shore, i.e., to the zone of a great fault.

Predominant in the Caucasus are earthquakes with shock force 6 to 7, although more intensive ones (up to shock force 8 to 11) take place in the volcanic uplands of Transcaucasia and on the periphery of the Kuro-Araksin hollow. Eleven catastrophic earthquakes with shock force over 9 to 10 have caused damage to cities in Soviet Central Asia: Alma-Ata (1887, 1911), Krasnovodsk (1895), Andizhan (1902), Dushanbe (1903), Fergana (1907, 1946), Ashkhabad (1929, 1948), Kazandzhik (1946), and Tashkent (1966).

The mountains of south Siberia are characterized by earthquakes with shock force up to 6 to 9. The majority of them are confined to the basin of Lake Baikal and the rift zone surrounding it. Destructive earthquakes in the area took place in 1814, 1902, 1908, 1931, 1946, and 1959. Kamchatka and the Kuril Islands are also characterized by high seismicity, the earthquakes recorded there having shock force 8 to 9.

It is well known that the destructive consequences of earthquakes are diverse. When strong earthquakes take place in large cities, thousands of people perish and a great number of houses are destroyed. Major damages occurred in the city of Vernyi (Alma-Ata) in 1911; in Ashkhabad in 1948; and in Tashkent in 1966, though human losses were not numerous. However, no less dangerous are other damages caused by earthquakes: the destruction of roads, bridges, canals, dams, as well as intensive mountain avalanches, stone and landslides, and cracks in the ground. In the mountains of Soviet Central Asia and the Caucasus many dammed lakes are formed during strong earthquakes.

Presently the USSR has a well-developed state seismological service which registers accurately all the movements of the earth's crust caused by earthquakes. This service has accumulated vast amounts of information on earthquakes. Combined with different geological and geophysical data, this information serves as a basis for both general and detailed seismic zoning of the USSR territory. Primarily the zoning aids in the designing of antiseismic construction and other protective measures.

The principal efforts of present-day research in the USSR are directed toward the elaboration of methods for forecasting the exact place and time of earthquakes by detecting their forerunners. With this aim, studies are undertaken of recent tectonic movements of the earth's surface in seismic regions with the help of repeated levelings of high precision and of tiltmeter measurements. A detailed study is also carried out of all the symptoms of changes in the earth's crust which take place before earthquakes (the relation of linear and transverse waves, electric and magnetic fields, geochemical phenomena, etc.). Although all these investigations have not given final reliable results so far, they are undoubtedly highly promising.

Tsunamis

Tsunamis are long waves caused primarily by seaquakes and volcanic activity. They prove very dangerous for the population of the shoreline of Kamchatka, the Kuril Islands, and Sakhalin. Tsunamis appear suddenly and move with a velocity of about 400 to 500 kilometers per hour (250-300 mph). When approaching the coastline, the waves form a series of rollers (3–7) with an average height of 5–10 meters (15–30 feet). Waves reaching 20 meters (65 feet) are formed in small areas of the coastline, predominantly in bays of the fjord type.

Table 29–1. Material transported by selected seli in the USSR

River	Volume in m^3	Wash down in m^3 from 1 km^2 of area of active seli formation
Chkheri River (Caucasus)	1,440,000	180,000
Kishgei River (southern slope of the Greater Caucasus)	3,000,000	120,000
South Kazakhstan	3,500,000	30,000

Especially strong tsunamis in the USSR were recorded in 1737, 1780, 1898, 1918, 1923, 1952, and 1963. The degree of danger of tsunamis has been quite varied and was determined not so much by strength of seaquakes, but the elements of relief and depth of the ocean on the routes of moving tsunamis.

In the Soviet Union, there is a warning service which uses two principal methods for detecting the location of tsunamis: seismic and hydroacoustic. As the velocity of spread of seismic and hydroacoustic waves is much greater than that of tsunamis, a warning service is able in most cases to issue an alarm signal 30–40 minutes in advance. Worthy of attention is that the areas of rocky bays in Kamchatka that can contribute to tsunamis are not populated.

Mudflows

As to activity and damage to the economy caused by mudflows, or seli, in mountain regions, the USSR occupies first place. The main areas subject to seli are the north Caucasus, Transcaucasia, the mountains of Soviet Central Asia and east Kazakhstan, the mountain Crimea, Carpathians, and the near-Baikal region.

Seli happen suddenly and carry very rapidly great amounts of loose material, mud, and water along the mountainous rivers. As indicated for selected mudflows reported in Table 29–1, the volume of solids transported may reach as much as 180,000 cubic meters (6.4 million cubic feet) from 1 square kilometer (0.4 square mile).

During the last 80 years hundreds of destructive seli have been recorded, some of them catastrophic. Such, for instance, was the glacial seli that took place in July 1963 in the central part of the Zailiysky Ala-Tau. Its runoff (its volumes being almost a million cubic meters or 35 million cubic feet) filled the basin of Issyk Lake entirely (thickness of the loose strata was 25 meters or 80 feet) and produced seli that took place in March 1965 in the Valley of Fergana, and in August 1967 in the north Caucasus leading to abundant floods. Finally, destruction by the seli of 1971 occurred in the region of the Krugo-Baikal railroad. Large cities of the USSR such as Alma-Ata, Erevan, Dushanbe, and Frunze are located in dangerous seli areas.

Several generalizations may be made about the distribution of seli within USSR territory: (1) Seli are associated with areas of the latest and most recent mountain folding. (2) There is a close relation between seli and heavy snowfall or sudden snow melt. This is present chiefly in medium- and low-mountainous areas having no glacier alimentation. This type of torrent is most widespread in the USSR. (3) The disastrous character of glacial seli originates in areas of old and recent moraines in periods of maximum ice melting (July-August). (4) Seli are related to earthquakes, as well as avalanches and landslides, increasing the danger of mud- and stone-laden torrents. (5) Intensification of seli phenomena results from various misuse of mountain slopes, such as elimination of forests.

Seli processes are studied in the USSR through establishing a potential danger of seli in an area, and through direct forecasting of seli phenomena (synoptical situation). In addition, measures are studied for protection of areas against disastrous manifestations, involving estimation of seli danger, design and construction of protective works.

Three principal protective measures against mud- and stone-laden torrents are used: (1) organization-economical and technical measures such as prohibition of tree-felling on mountain slopes, organization of forest reserves, prohibition or regulation of pasture use, and artificial snow melting; (2) forest amelioration and agrotechnical measures such as forest planting on slopes to prevent erosion, regulation of surface runoff and terracing; (3) hydrotechnical protective construction such as jump dams for softening slopes, trap filters for retaining fragmented material, and directing dams.

In regions where seli are especially dangerous forecasts are made by hydrometeorological officers. For instance, in the region of Alma-Ata there is a special radio detection of seli, i.e., a radiotelemetric automatic design. Here, in the mountains, there also has been built a unique 100-meter (330-foot) earth dam for retaining disastrous seli.

Snow avalanches and disastrous surges of glaciers

Sliding of huge snow masses is a widespread process acting continuously not only in high mountain areas of the Caucasus, Tien Shan, but also in medium and low mountain regions of the Khibin, Urals, Sykhote-Alin, and Kamchatka.

However, the most frequent snow avalanches take place in areas of recent glaciation with heavily snow-encrusted slopes. Here, avalanches move down along the same route, ranging from one to several avalanches per year. There are three known types of avalanches: sluff, channeled, and jumping, the latter being especially disastrous. They are characterized by great velocity and shock action.

Despite the mass character of development of avalanche processes, not all of them are hazardous. Disastrous are those caused by a combination of very heavy continuous snowfalls with extremely strong winds. Under these conditions snow accumulates in unusual places and then slides down the slopes, usually devoid of avalanches. Disastrous avalanches happen seldom, but are characterized by great destructive force. For example, the snow avalanche in the region of the town of Kirovsk (the Khibin Mountains) moved down the slopes of the Aikuaiventchorr mountain on December 22, 1936, with a volume of snow mass reaching 285,000 cubic meters (10 million cubic feet). In this case persistent southward winds were blowing at a rate from 3 to 12 and even 20 meters per second (6 to 25 and even 40 mph). The daily

increment to the snow cover thickness reached 400 millimeters (16 inches). Vast avalanches also slid down from the southern slopes of the Kuril Islands on December 25, 1959, and from the west Caucasian mountains in the winter of 1962–63. The formation of snow avalanches on the southeastern and southern slopes of the Kuril Islands was favored by stormy wind (40 meters per second or 80 mph) that shifted snow masses toward the slope. Disastrous snow avalanches in the Caucasus were caused by heavy snowfalls 2 meters (6.5 feet) thick and winds.

In the first stage of study of snow avalanches, as well as some other natural hazards, a questionnaire method was used in the USSR. Questionnaires were sent to institutions dealing with the subject in question, and to the regions that had been very poorly studied in regard to avalanche danger.

The questionnaire posed the following questions:

1. Do you know the places where snow avalanches (downfalls) took place?

2. Describe the area of avalanche sliding: steep or gentle slope, its height, orientation (to the south, west, etc.), presence of forests on the slope.

3. Describe the avalanche: dimensions, snow on the avalanche, results of destruction.

4. How often did you happen to observe avalanches in the region described by you? Write down the data of sliding avalanches (date, month, year).

5. What weather preceded the avalanche?

6. What was the height of the snow cover in the places of breakaway of the avalanche?

During recent years considerable advances have been achieved in the USSR in studying snow avalanches. Use is made of (1) forecasts of danger of avalanches according to geomorphological, geobotanical, soil, and hydrogeological properties of an area; (2) sounding and study of stratigraphy of a snow mass; (3) analysis of synoptic situations; (4) estimated data on the course of meteorological elements and changes occurring in the snow cover; and (5) photogrammetry.

There has been substantial increase in the effectiveness of protective measures against snow avalanches since it became possible to carry them out on vast areas. Preventive measures include warning services which operate in especially dangerous areas, and prophylactic measures to draw avalanches off in places by bombarding them with mortars, or using chemical reagents. Engineering measures of protection provide for retaining snow on slopes, e.g., by forest planting and by direct protection of objects against avalanches by closed galleries, snow cutters, etc.

Disastrous surges of glaciers, characteristic of so-called pulsating glaciers, are known in the Pamirs, Tien Shan, and Caucasus. These surges are very dangerous owing to their suddenness and destructive action. Thorough studies of development of surges of glaciers have been carried out at Medvezhii in the Pamirs since 1963 and at Kolka in the Caucasus since 1970. The 7-kilometer (4-mile) tongue of glacier Medvezhii moves rapidly every 12–14 years. During the last surge of 1963 the glacier advanced 1,600 meters (5,300 feet) down the valley, and brought 140 million cubic meters (5 billion cubic feet) of ice into the lower part of the tongue. The next disastrous surge of ice is expected roughly in 1975. Surges of the Kolka glacier take place every 65–70 years.

Although surges of glaciers are not dangerous to life owing to their time rhythm (1–10 kilometers or 0.6–6 miles per year), they and resulting seli cause economic damage in their paths. Currently, there are no technical means of controlling glacier surges. It is hoped that comprehensive studies will enable forecasts of future disastrous surges.

Floods

In the USSR, snow and rain high waters of large rivers are recorded every year. Yet as a rule they cause no damage. To a considerable extent, this depends on an adequate regulation of river runoff on large rivers and on the use of a reliable system of protection (damming, blasting of ice blocks, etc.).

Nevertheless, there are some areas in the country with an underdeveloped economy where, under certain hydrometeorological conditions, high river floods remain disastrous. Most often they occur in areas with mutual overflows of the rivers having high waters of different times. For example, the rivers of west Siberia, particularly tributaries of the Irtysh and Ob rivers, being inundated by later floods, acquire temporary backward courses (from lower to the upper reaches), and ice jams and dams in the channels. In the mouth of the Yany River, for instance, southern floodwaters overflow an ice bed of the oceanic shelf, spread over the surface, and inundate the floodplain terrace. Such summer and autumn high waters in the basins of the Amur Zeya, Bureya, and other rivers of the Far East result from abundant and continuous monsoon rains. Catastrophic floods happen there approximately once in 7 years.

To a certain extent the floods are influenced by abundant detrital deposits in the lower reaches of a river. The Amu-Darya, in particular, is characterized by a very steep gradient of its water surface and by its load of suspended detrital material. The latter, exceeding the solid load of the Nile and that of all the rivers in Soviet Central Asia, is responsible for heavy deposition of loose material. The river course is higher than the adjacent area and submerges a flat floodplain in times of high water discharge. When this happens, new channels form and banks are intensely destroyed (degish). On some rivers wind surges take place in their mouths and cause floods (on the Neva River, for example).

Thus, disastrous floods are caused by different factors, and measures against such floods require a whole complex of meliorative and hydrotechnical means. Very important among them is long-term regulation of the rivers' runoff. On the Zeya, Amur, Bureya, Selemdzhe, and many other rivers of the country dams and water reservoirs are constructed for that purpose.

Hurricanes

Hurricanes of a catastrophic character (wind velocity over 29 meters per second or 60 mph) happen almost every year in various parts of the USSR territory. Usually they originate in cyclones (when there is an advance of cold air) on cold fronts or on the periphery of anticyclones. Such winds arise unexpectedly and manifest themselves differently. On the sea they result in storms; on the land—as storms (often thunderstorms) and downpours. In steppe and arid zones they usually take place in the form of dust storms, during which deflation processes of plowed soils become very intense. In the south of Soviet Central Asia strong and hot winds have a special name, "afganets."

Hurricanes cause vast destruction to property and sometimes, human life. However, weather forecasts of hurricanes become more and more reliable, as losses decrease. The government pays the people insurance compensation for damage caused by hurricanes if the wind velocity in the given region is over 15.3 meters per second or 34 mph: wind force 8).

The natural phenomena related genetically to certain geographical zones include droughts, dry winds, heavy snowfalls, and low temperatures. The above processes, of regional character, are usually confined to relatively small areas and may prove disastrous even when they happen only once. The zonal processes are catastrophic with a considerably extensive area of distribution, and frequent recurrence or duration. These processes are most clearly manifested when external values of meteorological factors are not compatible with conditions of a given zone. These natural phenomena are not disastrous as a rule, nor do they bring human loss of life, though they frequently cause considerable economic damage.

Droughts and dry winds

Droughts and dry winds are rather frequent (Fig. 29–1) in forest-steppe and steppe zones, with a considerable gap by cultivated plants (during the vegetation period) between a great energetic potential for transpiration and small reserves of soil moisture. On the average, in forest-steppes, droughts have been calculated to happen once or twice in a 10-year period. In the steppes they recur five to six times in 10 years (50 percent cases); they are severe and may occur 2 to 3 years running (1906–8, 1938–39, 1950–51, 1954–55). In forest-steppe and steppe zones of the USSR where the main cereal crops are grown, droughts are studied from an agroclimatic point of view, comprising biological and physiological aspects of the phenomena. The indexes for drought intensity are values of harvest decrease: up to 20 percent—insignificant drought; from 20 to 50 percent—drought of medium severity; and over 50 percent—

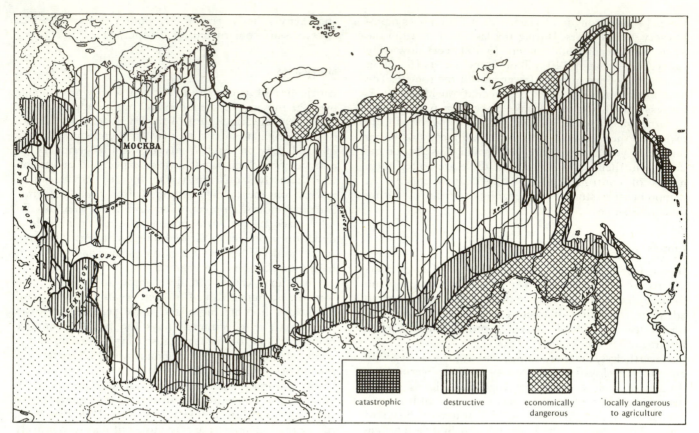

Fig. 29–1. Probability of drought in percent, according to V. A. Smirnov (1958)

severe drought. Droughts cause great damage to agriculture. However, the area of the Soviet Union being very extensive, failures of crops in one region may be compensated by favorable conditions in another region.

Droughts are frequently aggravated by severe dry winds during which an exceptionally great deficiency of moisture in air and high velocity of wind are observed, as well as insufficient reserves of productive soil moisture. Dry winds are very intense when the deficit of air moisture has reached 40 millimeters (1.6 inches), wind velocity is at least 10 meters per second (20 mph), and reserves of productive moisture are down to 0–30 milliliters (0–1 fluid ounce).

Droughts and dry winds have been studied for a long time. At present, on the basis of extensive statistical material, maps of recurrence of droughts and dry winds have been compiled in the USSR. The maps help to determine the probability of such phenomena for the coming year. An example of analysis of recurrence of dry winds is given in Table 29–2.

Irrigating fields is a basic means for protecting agricultural lands against drought. Large-scale hydrotechnical projects supply an interbasin transfer of water from the areas with excessive moisture to deficit areas. In addition, in order to decrease the influence of droughts and dry winds, the following measures are used: protective

forest belts, spring snow detention, fallow lands, contour plowing (for regulation of surface runoff), strip farming, microterracing of slopes, and purposeful selection of drought-resistant crops.

Abnormally low temperatures in autumn-spring and winter periods can be viewed as natural hazards. However, the criteria of disaster for all these processes are considerably different for various zones and economic objects. Thus, on the pastures of Soviet Central Asia the grazing stops in winter if temperature is $-16°$ C. with wind velocity 10 meters per second ($3°$ F., 20 mph) or $-27°$ C. with wind at 5 meters per second ($-17°$ F., 10 mph). In the north of the country at temperatures of $-35°$ C. to $-50°$ C. or $-31°$ F. to $-58°$ F. (wind calm) outdoor work is prohibited.

An index of abnormally severe winters is the probability of seasons with minimum air temperature less than $-30°$ C. to $-40°$ C. ($-22°$ F. to $-40°$ F.). Even on almost windless days this limits cattle grazing in winter. This temperature is dangerous for crops and offers the possibility of mass destruction of machines and other equipment.

Abnormally low temperatures, although seldom disastrous, produce additional expenses for organization of winter enclosures for cattle, for changing working conditions, and for machines.

Table 29–2. Characteristics of dry wind in the USSR (after E. A. Tsuberbiller)

Zones	Average no. of days for a non-productive period with dry winds		Average reserves of productive moisture in 1 m layer of soil (ml)	Probability of damage in crops (% yrs.)
	Intense	Medium intense		
Forest	0.1–0.3	0– 1.0	60–100	0–10
Forest-steppe	0.3–0.6	1.0– 2.0	50– 60	10–20
Moderately arid	0.6–1.5	2.0– 4.0	30– 50	20–30
Arid	1.5–3.0	4.0– 6.0	20– 30	30–50
Semidesert	3.0–5.0	6.0–10.0	10– 20	50–70
Desert (northern part)	5.0	10.0	10	70

Heavy snowfalls

In some regions of the USSR, heavy snowfalls accompanied by continuous transfer and redeposition of vast snow masses are regarded as natural hazards. The most frequent snowfalls increasing the height of snow cover by more than 100 millimeters (4 inches) have been recorded in Kamchatka (286 cases). There are well-known areas with extensive snow transfer (over 1,500 cubic meters per meter of length, or 53,000 cubic feet per foot of length) (Fig. 29–2). A considerable height (900 millimeters or 35 inches) and prolonged occurrence of snow cover have also been recorded. The most susceptible areas are the northern part of west Siberia (particularly the Taimir and the Red Sea coast), as well as the southern part of west Siberia and Kazakhstan (1,000 cubic meters per meter, or 35,000 cubic feet per foot). Extremely intense snowstorms happen in Anadyr (3.10 cubic meters per meter per hour, or 110 cubic feet per foot per hour) and in Petropavlovsk-Kamchatsky

Fig. 29–2. Snow cover and precipitation (average of maximum ten-day depths of snow cover during winter, in centimeters)

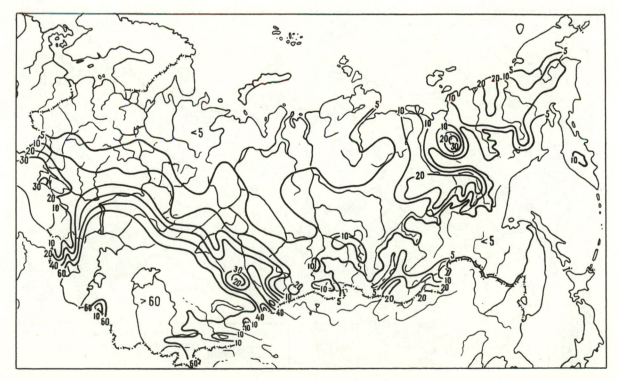

(2.55 cubic meters per meter per hour, or 90 cubic feet per foot per hour).

Snowfalls seldom cause loss of human life. Nevertheless, heavy snow results in expensive snow-clearing works on railroads and highways. Great volumes and densities of snow increase loads on the roofs of buildings. If snow falls earlier than usual in forests with remaining foliage, winds may cause limbs or trees to fall (e.g., autumn of 1971 in the vicinity of Moscow). The greatest thickness of snow cover (over 0.30 gram per cubic centimeter, or 0.17 ounce per cubic inch) has been recorded on coastlines of the northern seas and on the eastern coastline of Kamchatka (0.36 gram per cubic centimeter, or 0.21 ounce per cubic inch). Roughly 20 percent of rail lines of the USSR subject to heavy snowfalls are fenced by wooden shields; the rest of the railroad is effectively protected by forests.

Other natural hazards occur on the vast and diverse territory of the USSR, but they are either of a narrowly local character or cause insignificant damage, judged by the scale of the country.

Almost all natural hazards develop beyond any direct dependence on man's activity. Some natural hazards speed up or slow down to a certain extent in response to economic activity of man (seli, for instance). However, contamination of air and water, and so-called anthropogenic erosion of soils, are not attributed to the processes in question. These are not natural hazards, but anthropogenic processes. They are a result of an erroneous attitude of man toward the environment. This is a most important, independent problem that requires special study.

An analysis of natural hazards in the USSR territory testifies to prevalence of typical combinations. The plains of the country are characterized by droughts with the highest probability of 10 to 30 percent, spring or autumn frosts from 2 to 6 years out of 10, snowstorms, heavy snowfalls, and many other phenomena. In the mountains more common hazards are seismic processes, snow avalanches, and seli. Prevalent on seashores are hurricanes (40–50 meters per second or 80–100 mph) and tsunamis.

Twenty-nine regions have been distinguished in the USSR territory on the basis of prevailing combinations of various kinds of natural hazards. These regions are classified into four groups according to intensity of manifestation and results of natural hazards influencing man's life: region 1, catastrophic natural hazards that may cause loss of life and great damage to the economy (volcanism, earthquakes, tsunami); region 2, destructive natural phenomena that seldom cause loss of life, but result in significant damages to the economy, mainly to industry (earthquakes, hurricanes, seli, avalanches, floods combined with other natural processes); region 3, economically dangerous natural processes (droughts, floods, hurricanes, seli); region 4, local development of natural hazards that cause damage mainly to agriculture (late and early frosts, heavy rains, winds, etc.) (Fig. 29–3).

As mentioned above, a great amount of work has

Fig. 29–3. Areas of varying natural hazard intensity in the USSR

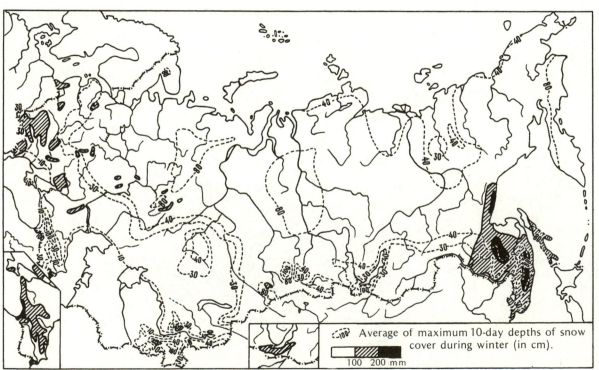

Average of maximum 10-day depths of snow cover during winter (in cm).

100   200 mm

been carried out in the Soviet Union dealing with warning and protection against natural hazards. But these phenomena have not been eliminated yet. Therefore the population and agricultural and industrial enterprises, which suffer from natural disasters, are paid insurance compensation by the government. The institutions of "Gosstrakh" have no right to obtain any profit from insurance operations and insure against the following disasters: earthquakes, floods, lightning shocks, hurricanes, seli, avalanches, downfalls, landslides, droughts, mud-laden torrents, heavy rains, hail, early autumn and late spring frosts. Agricultural land is insured not only against these processes, but against silting of soils, hoarfrost, windless weather during pollination periods of plants, etc. Animals of the extreme south and north of the country are also insured against damage caused by ice-crusted ground, thick snow, frozen snow crust, etc. According to the insurance conditions introduced in 1967, the government compensates agricultural institutions for all damages associated with losses of animals or crop failure, destruction of buildings, caused by natural processes not peculiar to a given region. By January 1, 1970, the total insurance paid, including all types of personal and domestic property among the population was 78.5 billion rubles, and that of all kinds of collective farm property also totaled 78.5 billion rubles.

References

Goldberg, I. A. (1961a) "Agroclimatic characteristic of frosts in the USSR and methods of control." *Gidrometeoizdat.*
——. (1961b) "Droughts in the USSR, their origin, recurrence and influence on crops." *Gidrometeoizdat.*
——. (1961–66) "Earthquakes in the USSR." M. Izd. *AN SSR.*
——. (1971) "Engineering glaciology." Izd. *MGU.*
Karyakin, V. A. (1970) "Avalanche prone areas of the Soviet Union." Izd. *MGU.*
——. (1971) "Determination of place of tsunami occurrence." *Gidrometeoizdat.*
Linovskaya, V. I. (1967) "Weight distribution of snow cover." *Trudy GGO,* vol. 210.
Losev, K. S. (1966) "Avalanches of the USSR." *Gidrometeoizdat.*
Mikhel, V. M., Rudneva, A. V., and Linkovskaya, V. I. (1969) "Snow transfer during snowstorms and snowfalls on the territory of the USSR." *Gidrometeoizdat.*
——, Rudneva, A. V., and Linkovskaya, V. I. (1971) "State Insurance in the USSR." A group of authors under the guidance of L. A. Motylev. Izd. *Finansy.*
Milevsky, V. Yu. (1961) "Probability of winds of different velocity on the USSR area." *Trudy LGMI,* vol. 12.
Nalivkin, D. V. (1964a) "Recommendations on control against wind and water erosion of soils." Izd. *VASKHNIL.*
——. (1964b) "Mud-laden torrents in USSR and measures against them." Izd. *Nauka.*
——. (1966) "Recent volcanism." Izd. *Nauka,* vols. 1–2.
——. (1967) "Dry winds, their origin and control." Izd. *AN SSSR.*
——. (1969) "Hurricanes, storms and whirlwinds." Izd. *Nauka,* L.

V Global summaries

30. Global summary of human response to natural hazards: tropical cyclones

ANNE U. WHITE
Boulder, Colorado

Introduction

As population grows, technology expands, and society becomes more complex, man becomes more vulnerable to damage from the occurrence and uncertainty of extreme events in nature. Social losses from avalanches, earthquakes, tropical cyclones, and many other natural hazards are increasing. This is the case even though scientific investigation of the causes of the extreme events deepens and even though new techniqes for dealing with hazards multiply and are curbing losses in some areas. Man is exposing new property to loss and also is increasing the severity of some natural events. Sophisticated means of giving relief in time of disaster are better developed than means of preventing disaster.

It is clear that continuation of present public policies in many parts of the world will increase rather than decrease the social costs of natural hazards. If the trends are to be reversed new policies will be required. Suitable policies must be based upon careful examination of the hazards and of possible and actual human adjustments to them.

This chapter deals with the hazard of tropical cyclones and the measures that people take in different parts of the globe in coping with their effects. There are many gaps in the information available; nevertheless, important insights for policy makers and scientists can be gained from an examination of experience on a global basis.

Definition

Tropical cyclone hazard arises when the violent action of one or all of the elements of the storm: wind, rain, sea, and waves, make an impact on man and his environment.

A cyclone, otherwise known as a depression or low, is a weather system in which the atmospheric pressure decreases to a minimum value at the center, with the winds blowing in a spiral toward this center. (The winds blow clockwise in the southern hemisphere and counterclockwise in the northern hemisphere.) In its intense form, it becomes a tropical cyclone, defined by the World Meteorological Organization (1966) as a "cyclone of tropical origin of small diameter (some hundreds of kilometers) with minimum surface pressure in some cases less than 900 mb., very violent winds, and torrential rain; sometimes accompanied by thunderstorms. It usually contains a central region known as the 'eye' of the storm, with diameter of the order of some tens of kilometers, and with light winds and more or less lightly clouded sky." Cyclones and the kinetic energy they produce are an integral part of the general circulation of the atmosphere. They are a necessary factor in producing what man knows as weather, but in their more extreme forms become a hazard to him.

Such storms were first called hurricanes in the Caribbean Sea, and this name has been extended in the south Pacific, southern Indian, and north Atlantic oceans to any tropical cyclone in which the wind attains great violence. In the China Sea and generally in the western part of the north Pacific the term "typhoon" is used.

Damage potential

The aspects of a tropical cyclone most likely to cause damages to people and property are the associated storm surges and sea waves, the wind, and rain.

Sea

A storm surge is the term used for the rise of the mean level of the ocean which takes place as a tropical cyclone approaches a coastline. Over the open sea, the low atmospheric pressure at the center of the storm causes the water below it to rise above the level of the surrounding surfaces. As the storm nears the coast, winds may pile the already high waters against the shore, depending upon its configuration. If this action coincides with the lunar high tide, the rise in the sea level may reach 7 meters (23 feet) or more above normal, swiftly inundating the low-lying areas of the coast, and causing the greatest threat of the storm to human life and property.

The surge itself is perhaps 15–30 kilometers (9–18 miles) wide. The general effect of waves and swell may extend as far as 350 kilometers (220 miles) from a storm center. The combination of wind, waves, and current may erode the coastline so as to destroy beaches and agricultural land as well as houses and cities. Following the inundation there may be further damage from cave-ins of land or buildings as the water subsides.

The period of high water is likely to last from about 6 hours to several days in areas of poor drainage, and may leave the soil saline and unfit for crops.

Rain

As much as 2,500 millimeters (100 inches) or as little as a trace of rain may fall in a tropical cyclone, and 500

Fig. 30–1. Types of potential damage from tropical cyclones

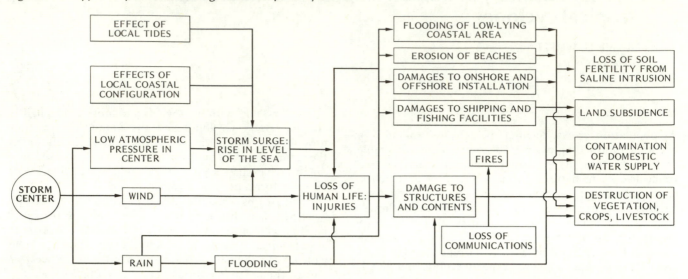

millimeters (20 inches) is fairly common. The effect of the larger amounts is likely to be flooding, especially when a storm moves across mountainous terrain, as during the Hurricane Camille in 1969 when 790 millimeters (31 inches) fell in a period of 5 hours on already saturated soil (U.S. Department of Agriculture, 1970).

Wind

The winds of a tropical cyclone may reach velocities of over 250 kilometers per hour (155 mph), causing deaths and injuries from structural collapse or flying objects, with devastating effects on buildings, vegetation, and communications systems. Tornadoes are sometimes associated with the storm. Direct wind impact is responsible for breakage of glass in windows and cars, for the lifting of roofs, and for the total destruction of some structures. Sudden high gusts may be particularly damaging as they result in unequal pressures on different sides of a building which may cause it to collapse, or the gusts may set a tall structure like a radio tower oscillating until structural failure occurs.

These three mechanisms together cause enormous disruption of human life and property (Fig. 30–1).

Extent of damages

Damages are a function of the size and intensity of the storm and the population and character of the economy of the area affected. Some of the very densest human settlement has taken place in coastal areas which lie in the general paths of tropical cyclones and the storms rank high among the natural forces in the amount of destruction caused to human activities and to ecosystems. Complete statistics are not available for all areas of the earth, but in the period 1960–70, 17 major storms in various parts of the world took the lives of

about 350,000 people, and caused very large damages (Table 30–1). In many countries there is little attempt to estimate the monetary value of damages, especially in heavily agricultural areas, possibly because of lack of statistical facilities and trained personnel.

There is extreme variation in the damage caused by tropical cyclones from year to year, but the general trend is one of increasing damages all over the world as pressures of population force people into low-lying areas

Table 30–1. Deaths and estimated damages associated with some noteworthy tropical cyclones, 1960–70

Year	Name of storm	Country	Deaths	Damage ($ million)
1970	Ada	Australia	13	12
1970		Bangladesh	300,000	a
1970	Celia	U.S.	11	453
1969	Camille	U.S.	256	1,421
1967	Beulah	U.S.	15	200
1966	Alma	Cuba	a	100
		U.S.	6	10
1966	Inez	Mexico	65	100
		Haiti	750	20
		U.S.	48	5
1965		Bangladesh	19,279	a
1965	Betsy	U.S.	75	1,421
1964	Dora	U.S.	5	250
1964	Hilda	U.S.	38	125
1964	Cleo	U.S.	3	129
1963		Bangladesh	11,468	a
1963		Cuba-Haiti	7,196	a
1961	Carla	U.S.	46	408
1960	(two storms)	Bangladesh	5,149	a
1960		Japan	5,000	a
1960	Donna	U.S.	50	426
Total			349,473	

Sources: Sugg (1971), Frank and Husain (1971), UNESCO (1970).
[a]Estimate not available.

for agricultural use, as in Bangladesh, or where denser populations in increasingly expensive structures move into vulnerable areas, as in the United States and Japan. On a smaller scale, economic development and population increase as in northeast Queensland, Australia, increases actual and potential damages. In the United States, the trend has been one of decreasing deaths but sharply increasing property damages (Fig. 30–2) which exceeded $2.4 billion in the 5-year period 1965–69 (U.S. Office of Emergency Preparedness, 1972, p. 42).

The effect of the sea is by far the most destructive mechanism of a tropical cyclone. On November 12, 1970, a tropical cyclone in the northern Bay of Bengal produced a 6-meter (20-feet) rise in the sea combined with a high tide. The storm and resulting flooding killed approximately 300,000 people and crop losses alone were $63 million U.S., but these figures do not reflect the total impact of the storm. It is estimated that 60 percent of the inland fishermen in the area were killed and 65 percent of the total fishing capacity of the coastal region was destroyed, seriously affecting the protein supply of the whole region for some time to come (Frank and Husain, 1971).

Camille, in 1969, was one of the most intense and costly hurricanes to hit the United States. It was respon-

Fig. 30–2. Deaths and damages from hurricanes in the United States. Source: U.S. Office of Emergency Preparedness (1972) p. 42

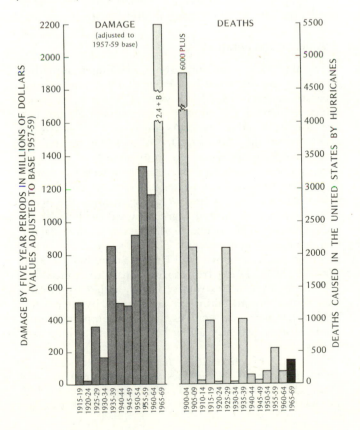

sible for 248 deaths, over 8,000 injuries, and property damage of $1.4 billion. About one-half the deaths and 10 percent of the property damages were due to inland flooding resulting from the intense rains of the storm as it proceeded northward (Dikkers, Marshall, and Thom, 1971 and U.S. Dept. of Commerce, 1969.)

Deaths may be decreasing in the parts of the world where warning systems are effective and where cars can be used for evacuation. More than 125,000 people were evacuated from low-lying coastal areas in Cuba during Hurricane Inez in 1966 (UNESCO, 1970). However, over half the population in the United States is located on or near its coastlines, and the proportion is increasing. It is estimated that the death toll might move sharply upward in the future if a severe hurricane were to make a direct hit on a crowded coastal area and evacuation routes became clogged (Simpson, 1971a). Some 63 square kilometers (25 square miles) of the urban area of Tokyo are below sea level, putting about 70,000 people at considerable risk from sea flooding (Nakano, 1970).

Extent of benefits

The effect of these storms is not all bad; benefits center principally on the precipitation associated with them which may have considerable value to agriculture and are especially significant in areas such as north Australia. Sugg (1968) estimates that nine major hurricanes in the United States since 1932 terminated dry conditions over an area of about 622,000 square kilometers (240,000 square miles). Hartman, Holland, and Giddings (1969) estimated the change in total crop value brought on by these storms occurring in different months. The losses in crops for two of the storms were $54 million and $1 million; for the third storm there was an increase in total crop value of $8 million.

Spatial distribution and areal extent

Tropical cyclones form over the warm tropical oceans where the water temperatures reach over 27° C. (81° F.) except for a zone on either side of the equator. Once formed, they have in all latitudes a general tendency to recurve from their westward initial movement poleward and then eastward (Fig. 30–3). Their path may extend to several thousand kilometers. The radius of the winds of hurricane force in a mature storm may exceed 300 kilometers (190 miles), making a belt of nearly 300,000 square kilometers (120,000 square miles) exposed to the destructive forces of the storm as it moves along. Because of the circular motion of the winds of a storm, the intensity of the wind varies in the different quarters at any one time.

Magnitude and intensity

There are immense forces involved in a tropical cyclone as a feature in the general circulation of the earth's

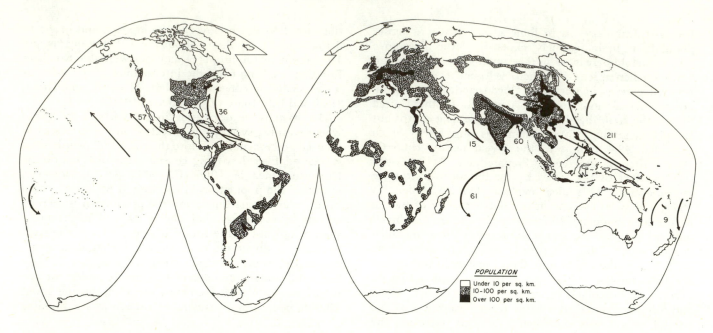

Fig. 30–3. Arrows indicate principal world regions of tropical cyclones and give roughly their directions of movement, after Tannehill (1956), p. 3. Numbers indicate frequencies of major tropical cyclones per ten years, Riehl (1954), p. 324, drawing on various studies

atmosphere. It is estimated that the energy released in one day by a storm, if converted to electricity, could supply Europe's electrical needs in 1970 for more than 6 months.

The lower the atmospheric pressure in the center of the storm, the more violent the action of wind, storm surge, and waves is likely to be. The storms can be classified by intensity, as in Table 30–2, with estimates included to show the computed damage potential in a simulated urban area for the different levels of intensity. The damage potential was computed by Friedman (1970) by assigning a location and a value for dwelling structures in a grid of six square-mile areas representing the state of Louisiana, using U.S. Census data. A hurri-

cane with a path directly over the New Orleans Metropolitan Area was then mathematically applied to the properties for each of five different storm intensities, producing "damage statistics" which have here been condensed into three intensity levels of storm.

Frequency and duration

Tropical cyclones are a seasonal phenomenon with a frequency varying from an average of 1 per year in some areas to as many as 20 per year (Fig. 30–3), a figure which is increasing as more storms are recorded through the use of satellite data. As many as 110 hurricane seedlings in the Atlantic Ocean may be spotted by satellite during the year, but only about 10 or 11 of these grow to be named hurricanes or tropical storms (Simpson, 1971b). In most regions they are confined to a less-than-6-month period of the year when the ocean waters reach the required degree of warmth (Fig. 30–4), although they may occur in the western north Pacific Ocean at any time of year. The life-span of the major recorded hurricanes over the United States has averaged 10 days (Sugg, 1971) from formation to the time when the storm loses its destructive energy either by moving over rough terrain, taking on drier air, or by encountering cooler air as it recurves into the temperate zone. Some storms last only a few hours, others as long as 3 weeks.

The lack of observation networks has in the past made estimates of frequencies rather inaccurate in some areas such as the western Pacific where there are few permanent observation stations. •The use of weather

Table 30–2. Three intensity levels of hurricanes

Description	Maximum winds (peak gust, kph)	Minimum central pressure (mb)	Aggregate computed damage, New Orleans ($ million)
Minimal hurricane	120–160	983–996	1–7
Major hurricane	161–220	949–982	8–89
Extreme hurricane	221 or more	948 or less	90–205

Sources: Dunn and Miller (1964), p. 308; Friedman (1970), p. 22 (for New Orleans metropolitan area).

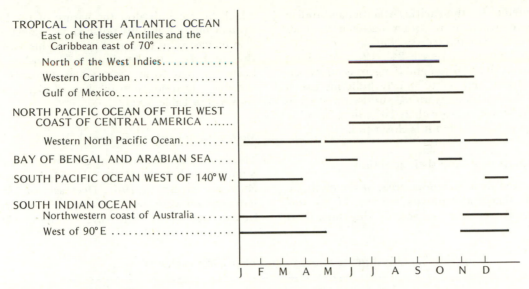

Fig. 30–4. Seasonal formation of tropical storms. Source: Riehl (1954), p. 323

satellites and radar has much improved this situation, and will make possible much more precise yearly estimates of frequencies in the oceans of the world.

Forecasting ability

There have been attempts to forecast tropical cyclones well back into the nineteenth century. The criteria developed which indicate that a new cyclone is approaching are subnormal pressure in low latitudes and above-normal pressure in high latitudes, existing disturbance of some sort, and changeable winds. In addition, sea swell, erratic tides, and microseisms may warn of the approaching storm.

International efforts under the leadership of the World Meteorological Organization have led to a system known as the World Weather Watch. The system includes some 8,500 land stations, 5,500 merchant ships, aircraft, several special ocean weather ships, automatic weather stations, and meteorological satellites. At present a tropical cyclone is usually first identified and then followed from satellite pictures. The detail in these pictures shows up anything more than 3 kilometers (2 miles) across, so seedling hurricanes show quite clearly. Observations from the network of stations on land and sea are exchanged by a complex Global Telecommunications System, and analyzed under the Global Data Processing System, as part of the World Weather Watch. There are 3 World Meteorological Centers (Melbourne, Moscow, and Washington), some 20 regional centers, and more than 130 national centers where the data are analyzed and the results made available for use by forecasters. In addition the Global Atmospheric Research Program seeks to improve forecasting methods through meteorological research.

Computerized numerical forecasting techniques are being increasingly used. If the storm is found to be intensifying, a prediction of its path and speed is made and then revised as new information comes in. As the storm approaches within about 300 kilometers (190 miles) of a coast, radar instruments can monitor its speed and movement, and tide gauges fixed along the coast record changes in water level useful in predicting the arrival and height of a storm surge.

Forecasting errors increase with the duration of the forecast and distance from the coast. The average error for a 24-hour prediction of hurricane movement is 240 kilometers (150 miles). As the storm approaches the coastline and is monitored more closely, the average landfall error for a 24-hour prediction is in the order of 180 kilometers (110 miles) (Simpson, 1971a). The likelihood of reducing this drastically in the near future is not great. Simpson points out that the responsibility of the forecaster is very heavy, for a prediction error of 10° in direction of motion would shift the scene of greatest danger some 100–150 kilometers (60–90 miles) along the coast.

Warning time

At least 36 hours ahead of a tropical cyclone reaching a coast a forecaster usually tries to identify the coastal sector that should keep watch for the storm, the point of expected maximum storm surge, areas subject to excessive rain and flash floods, and indications of tornadoes. In the Caribbean Sea, the Gulf of Mexico, the north Atlantic and the Pacific Ocean, the U.S. National Weather Service under the National Oceanic and Atmospheric Administration (NOAA) distributes tropical cyclone advice to the public. Data on tropical disturbances observed by satellite for the Pacific, Atlantic, and Indian Ocean areas are available to other national agencies (U.S. Department of Commerce, 1971).

In the United States there are 24-, 12-, and 6-hour public forecasts of the position and dangerous characteristics of the storm, and up to hourly bulletins issued when needed. In Australia, warnings are issued every 6 hours when the storm is more than 100 miles from the coast, and every 3 hours when it is closer to land.

The range of adjustments and their adoption

As man has moved into hazardous areas of the earth, he has developed many adjustments to the threats surrounding him. These may be active adjustments, in

which he seeks to alter the nature of the hazard, or to protect himself and his possessions against it. They may also be passive in that he may adjust his own thinking so that he lives with the threat of hazardous events, and bears the loss from them when it comes. A range of possible adjustments to the hazard of tropical cyclones is shown in Table 30–3.

Modification of storm intensity

This work was pioneered in the United States by R. H. Simpson and J. S. Malkus (1964) in the seeding of Hurricane Esther in 1961. Decreases of 15–30 percent in maximum wind speeds were observed when Hurricane Debbie was seeded with silver iodide in 1969 (Gentry,

Table 30–3. Adjustments to the hazard of tropical cyclones

	Before the storm	During the storm	Immediately after the storm
Modifications of the extreme event	Seeding of cyclone clouds to lessen the intensity of the storm		
Modification of the damage potential	Protective shore works; levees		
	Afforestation and dune control		
	Warning system: of the approaching event; of hazardous areas	Evacuation Seeking shelter Praying	
	Evacuation		
	Construction of raised areas as refuges; designation of shelters		
	Local regulations governing debris and loose materials		
	Zoning codes		
	Building codes		
	Construction modification		
	Flood and wind-proofing		
Distribution of losses	Buying insurance		Claims on insurance
	Stockpiling of emergency supplies of water, food, and building materials		Emergency relief and reconstruction
			Bearing the loss

1970). Since the wind energy is related to the square of its velocity, even a small reduction in the intensity of the storm may significantly reduce the amount of damages from it. However, further experiment is needed to identify the relationship of probable cause and effect. There will also need to be an assessment in greater detail of the effects on the path of the storm, on precipitation, and on the general circulation of the atmosphere, before modification can be undertaken with confidence in the results.

Protective shore works

These are mainly of two types: those that attempt to bar the passage of water by artificial means such as dikes, bulkheads, seawalls, and revetments; and those that seek to strengthen the natural defenses of beach and dune such as groins and jetties, or the transfer of sand and revegetation to build or stabilize beaches and dunes. In the United States, Connecticut and New Jersey have nearly 50 percent of their coastlines protected by man-made structures. These latter structures affect movements of sand along a coast and so while providing some protection against waves, may have undesirable effects of erosion on other beaches, and may interfere with amenities such as a view of the sea (Burton, Kates, and Snead, 1969). Forests and grassy dune vegetation are a desirable feature of land use planning in storm-prone areas and mangrove swamps provide natural protection from waves. Plantings stabilize the soil against breaching by minor storms, but they offer little protection against very severe tropical cyclones, as was demonstrated in the Bay of Bengal in 1970 where forest plantings were swept away by the force of the sea surge.

The building of dikes and cross-dams, as in the Ganges delta, may open low-lying areas to human settlement. Such works may protect the occupants of the coastal region from a moderate storm, but by their encouragement of settlement may lead to heavy loss of life and damages if a severe storm overtops or breaches the protective works, as in Bangladesh in 1970.

Warning systems

Warning systems should be viewed as a combination of technical and social arrangements which allow individuals and groups affected by a storm to respond in ways most beneficial to them. On the technical side there needs to be a complex evaluation of meteorological data plus knowledge of coastal factors and engineering works which may modify the effects of the storm. On the social side the conditions for effective operation include organization for action to take place when the warning is received, issuance of the warning, directions for action transmitted as part of the warning, and evacuation where required.

Assumptions often made about warning systems are that the message is not changed in transmission, that the

recipients will understand it as intended by those who issued it, that the recipients will know what to do in their own best interest when they get the message, and that all people will receive the message. Rarely are all these conditions fulfilled; in Bangladesh, for example, the radio may be off during critical hours of the night, or women accustomed to purdah may resist evacuation to community shelters (Islam, 1970). In the United States warning systems are highly developed, with 1.9 million people put under the average warning. Despite repeated warnings, some coastal citizens in the Mississippi area hit by Hurricane Camille in 1969 did not evacuate, seriously underestimating the potential destructiveness of the storm (Wilkinson and Ross, 1970).

With present forecasting methods for tropical cyclones, an overwarning affecting an area of roughly several hundred kilometers must be expected. This overwarned area is necessarily greater where there is less forecasting equipment available, as has been the case in the Pacific region.

The capacity for nations or communities to develop warning systems is affected by population size, financial or other resources, linkages with media, use of radio and TV, government-citizen relationships, frequency of extreme events, willingness of officials to risk false alarms, relationship among governmental units, and organizational capacity.

Evacuation

More than one-half the loss of life from tropical storms is due to drownings, either from the rise in sea water inundating the land or from floods from excessive rainfall. Evacuation has proved a successful means of reducing the loss of life in the United States, as when 300,000 people were evacuated in advance of Hurricane Betsy in 1965 with only 75 lives lost, despite damages of over $1.4 billion. However, it requires considerable prior planning and coordination of warning systems and evacuation plans. It has not so far proved practicable in areas like Bangladesh where communications are poor and local attitudes not favorable (Islam, 1970).

Zoning and building codes

In many countries the use of zoning codes to restrict building in hazardous areas is not extensive, although building codes controlling the type of construction have been used in southern Florida in the United States with some success. Burton, Kates, and Snead (1969) suggest three zones of land use for the heavily settled northeast coast of the United States. These would range from (1) no structures permitted except boat docks through a zone, (2) where the type and construction of buildings is controlled, to (3) a "warning zone" where people are warned of the existence of risk. It is difficult to estimate what the effect of such codes would be, but they might be expected to influence the rate of increase in losses, and indicate a concept of acceptable risk for individuals and communities.

In Japan where much densely settled land is subject to flooding from typhoons and tsunamis, there is considerable use of government control over land use and building methods. In the city of Nagoya, for example, five land zones are designated which reflect the hazardousness of the area. In the most dangerous zone no houses or hotels are allowed. For each of the others, building regulations specify the height of the ground floor above normal tide level and the types of materials to be used (Oya, 1970).

In Queensland, Australia, work done by a university sponsored investigation of problems of building failures resulted in a toughening of the city of Townsville building codes (Oliver, 1973).

Flood- and wind-proofing

This is a combination of structural modification or protection and precise plans for the elevation or removal of materials in a building. For example the ground floor may be left open as a plaza or parking area, and floodwaters can run through it. Machinery such as pumps can be installed so that it can be quickly elevated. Shutters or boards for windows, and methods of tying down movable objects, help protect against wind action. It is estimated that much roof damage from winds could be averted by an expenditure of $25 invested in enough anchors and/or bolts to tie down the roof structure (Gentry, 1966).

Construction of raised areas as refuges

These may be made of earth or in some cases of timber to form raised platforms, and are useful in flat tidal areas such as those on the shores of the Bay of Bengal. Some stronger buildings or areas such as the roofs of schools may be designed as refuges. There needs to be considerable community planning of location, routes, and management for these to be effective. They may have the disadvantage in tropical countries that poisonous snakes or animals may seek refuge as well as people.

Insurance

In general, marine insurance has been the most available form of insurance for storm damage on a global scale. Ships are usually covered. Some special facilities such as oil rigs, where there is the possibility of a single total loss involving one platform of $20 million, may have difficulty obtaining total coverage (Stenz, 1966.) Crop insurance is available in some parts of the world such as Ceylon, where crop and animal husbandry insurance up to 50 percent of the established average yield is an element in an integrated program for increasing paddy production. A higher coverage is allowed if certain practices are followed. This covers a variety of hazards,

including excessive moisture, and is paid for by all farmers, at a uniform rate per acre (Ceylon, 1971).

In the USSR, the government pays compensation for destruction caused by wind if the wind velocity in a given region is over 55 kilometers per hour (34 mph) (Gerasimov and Zvonkova, 1972).

In the United States, wind damage has been insurable, but until recently damage from saltwater coastal flooding or inland freshwater flooding was not. This has been changed by the passage of a federal law establishing flood insurance, and there is an increasing emphasis on the possibilities of using insurance to make explicit a level of acceptable risk both for individuals and for communities.

Emergency relief

This is usually provided after any disaster by the local and national governments and by private agencies. Where there have been extreme damages it may be at least nominally international in character. There is some risk that emergency funds and supplies may be diverted to other uses, as was the case in Bangladesh in 1970 (New York *Times* Je 8, 1:3 1972).

The amounts of money spent may be substantial: after 1969's Hurricane Camille, the states affected received $173 million in federal and Red Cross assistance. This sum, plus insurance payments, meant that some $500 million was spent in a relatively short time in the stricken area (U.S. Senate, 1970, pp. 20, 1145, 2289). About half the Red Cross money in Virginia went for building repair and reconstruction, and it may be assumed that most of the insurance money was used for this purpose.

These emergency funds may have a considerable effect on policy decisions, for once the rebuilding efforts are underway changes in zoning and building regulations may become nearly impossible, and hazardous areas are quickly resettled. The experience in the postdisaster reconstruction in Anchorage, Alaska, after the earthquake of 1964 shows how quickly a hazard area is likely to be rebuilt. As the man who was planning director in Anchorage at the time points out, the response immediately after the disaster was "Planner, let's do it right." In a little over a month, this shifted to "Planner, get out of the way" (Schoop, 1969, p. 231).

Bearing the loss

This is a common personal decision of adjustment to the hazard of tropical cyclone. The losses are simply carried by the individual families or firms except as they are distributed through insurance and public relief. A study done after Hurricane Camille in the U.S. indicates that the net loss to families in Rockbridge County, Virginia, after relief and insurance payments, was well over a year's income for those in the lowest income group, ranging up to 75 percent of a year's income for those in the $10,000-and-over annual income group (U.S. Senate, 1970).

Prayer is regarded by some people as a way of warding off the danger of storms and dealing with man's feelings of frailty before the forces of nature. This has been institutionalized in the Virgin Islands where there are two holidays: Supplication Day in July, before the hurricane season, and local Thanksgiving Day in October, after it is over.

Cost of damages and adjustments

There is not, at present, data available to make an estimate of costs on a global basis. Nakano (1972) estimates the average annual property damages from typhoons in Japan at $70 million. Sugg (1972) has estimated the total cost of hurricanes to the U.S. at an annual average of over $500 million per year for the last 12 years, including in these costs damages, cost of aircraft reconnaissance, communications, protective procedures by private business and industrial interests, and evacuation costs. This is 0.05 percent of the gross domestic product (1970) for the U.S., and 0.02 percent for Japan, but since there is no uniform basis for estimating damages these figures mean very little. A comparison of two severe storms, one in Bangladesh and one in the U.S. (Table 30–4) indicates that many more lives were lost in the less developed country, while property damages had a higher monetary value in the developed one.

A comprehensive approach to direct costs of tropical cyclones would have to include, in addition to those already mentioned, the costs of protective works on the shoreline, of forecasting and warning systems, and of overhead costs of insurance, relief, and disaster assistance.

It is difficult to estimate the reduction in losses caused by any single adjustment. There is no doubt that the warning system and community organization accompanying it in the U.S. has greatly reduced the loss of life. Measures taken in the Miami area by private firms and individuals upon receipt of a warning cost them about $2 million, but Sugg (1967) suggests that in the most severe storm the best kind of warning can only reduce property damages by about 5 percent. However, these measures may be very effective for moderate storms or on the edges of areas affected by major ones.

In a developed country such as the United States, private protection costs are estimated at about $5 per capita and evacuation costs at about $50–$60 per family (Sugg, 1967). Constructing a building to withstand most wind damages adds 6 percent to construction costs in southern Florida, but construction to avoid damages from flooding would cost much more (Gentry, 1966).

Perception of hazard and adjustments

It is clear that tropical cyclones pose a major threat to lives and property in many parts of the world. It is not clear just how this threat is perceived. It is estimated that only 20 percent of the population of Miami, a high-risk urban area, has invested in protective measures and manages to put them into use during warnings (Sugg, 1967). However, loss of life is small in this area.

Table 30–4. Comparison of two severe tropical cyclones

	Estimated population affected (millions)	Lives lost	Type of damage	Amount of damage
November 12, 1970 storm, Bangladesh	4.7	300,000	Crop loss	$63 million
			Loss of cattle	280,000
			Houses damaged 400,000	400,000
			Schools damaged	3,500
			Fishing boats destroyed	99,000
Hurricane "Camille," August 17–20, 1969 U.S.	19	256	Damage, including in the coastal area:	$1,421 million
			Houses destroyed or damaged	16,805
			Shipping vessels destroyed or damaged	126
			Cattle drowned	5,000
			Trucking terminals destroyed	5
			Plus highways, oil rigs, etc.	

Sources: Bangladesh data: Frank and Husain (1971), p. 438. U.S. data: U.S. Department of Commerce (1969), p. 3; Sugg (1971), p. 52.

In Bangladesh during the disastrous storm of 1970, it is estimated that over 90 percent of the people in the area knew about the storm, but only 1 percent sought refuge (Frank and Husain, 1970). Some reasons for this have been discussed earlier.

Personal experience seems a strong factor in the perception of storm hazard. Riley (1971) suggests that there is a need to distinguish between core experience (being in the middle of a storm) and peripheral experience (being only lightly touched by one). The first experience may be cautionary; the second may lead to disregarding of warning and underestimation of danger.

Overwarning, lack of experience with such storms, and lack of knowledge of what to do may account for much of the action or lack of it that people take when threatened by a storm. The relationship between the clarity and authority of the warnings about area and severity of the storm, and the alternatives open to the people and perceived by them needs to be investigated further.

Range of public responsibility

A nation has a wide range of governmental action open to it with regard to the hazard of tropical cyclones, depending upon its resources and priorities. It can join in international efforts for forecasting services and warning systems. It can seek to modify the weather—a course which if undertaken by one nation may have international implications. It can improve warning systems, protective works for its citizens, or offer them evaluations of risk and technical advice for methods of protecting themselves and their property. In no country have all these and other measures been put by the national governments into a framework which results in a major reduction of loss of life and property damages, although some have reduced the loss of life. The most comprehensive framework for reducing loss of life and property damages from tropical cyclones has been proposed by a WMO (1972) Expert Committee in a "Plan of Action" for member countries, which if carried out would integrate meteorological, engineering, and social aspects of the problem.

Some measures such as relief payments, assistance with reconstruction, and subsidized insurance may actually increase losses, as they encourage settlement in hazardous areas, or encourage rebuilding on sites that probably will be affected by subsequent storms. Unless the system of measures is considered as a whole, it is likely that losses will continue to mount.

References

Burton, Ian, Kates, Robert W., and Snead, Rodman E. (1969) *The Human Ecology of Coastal Flood Hazard in Megalopolis.* Chicago: University of Chicago, Department of Geography, Research Paper No. 115.

Ceylon, Government of. (1971) Report of Dr. P. K. Ray. "Re-view of paddy crop insurance in Ceylon." Ceylon: Department of Government Printing, Sessional Paper No. 1.

Dacy, Douglas C., and Kunreuther, Howard. (1969) *The Economics of Natural Disasters.* New York: Free Press.

Dikkers, R. D., Marshall, R. D., and Thom, H. C. S. (1971) *Hurricane Camille, August, 1969.* Washington, D.C.: National Bureau of Standards, Technical Note 569.

Dunn, G. E., and Miller, I. B. (1964) *Atlantic Hurricanes.* Baton Rouge: Louisiana State University Press.

Frank, Neil L., and Husain, S. A. (1971) "The deadliest tropical cyclone in history." *Bulletin American Meteorological Society* 52 (6):438–44.

Friedman, D. G. (1970) "Insurance and the natural hazard." Paper prepared for the 9th ASTIN Colloquium, International Congress of Actuaries, Rangers, Denmark.

——. (1971) "The storm surge hazard along the Gulf and South Atlantic coastlines." Working paper, Travelers Insurance Co.

Gentry, R. Cecil. (1966) "Nature and scope of hurricane damage." In *Hurricane Symposium.* Houston, Tex.: American Society for Oceanography, Publication No. 1, pp. 229–54.

——. (1970) "Hurricane Debbie modification experiments, August, 1969." *Science* 168:473–75.

Gerasimov, I. P., and Zvonkova, T. V. (1972) "Natural hazards in the territory of the U.S.S.R." Paper presented at the Commission on Man and Environment meeting, Congress of the International Geographical Union, Calgary, Alta.

Harris, D. Lee. (1956) "Some problems involved in the study of storm surges." *Proceedings of the Tropical Cyclone Symposium* at Brisbane. Melbourne: Bureau of Meteorology.

Hartman, L. M., Holland, David, and Giddings, Marvin. (1969) "Effects of hurricane storms on agriculture." *Water Resources Research* 5 (3):555–62.

Islam, M. Aminul. (1970) "Human adjustment to cyclone hazards: a case study of Char Jabbar." University of Dacca, East Pakistan, Department of Geography. Toronto: University of Toronto, Department of Geography, Natural Hazards Research Working Paper No. 18.

Moore, Paul L. (1966) "Forecasting and warning systems." In *Hurricane Symposium.* Houston, Tex.: American Society for Oceanography, Publication No. 1, pp. 102–13.

Nakano, T. (1970) "Lands below sea level due to land subsidence in the urban areas of Japan." *Japanese Cities.* Association of Japanese Geographers, Special Publication No. 2, pp. 237–43.

——, et al. (1972) 'Natural hazards and field interview research." Paper presented at the Commission on Man and Environment meeting, Congress of the International Geographical Union, Calgary, Alta.

Oliver, J. (1973) Jarma Cook University of North Queensland, Australia, private communication.

Oya, Masahiko. (1970) "Land use control and settlement plans in the flooded area of the city of Nagoya and its vicinity, Japan." *Geoforum* 4:27–35.

Riehl, Herbert. (1954) *Tropical Meteorology.* New York: McGraw-Hill.

Riley, J. A. (1971) *Disaster—Storm Ahead.* Austin: University of Texas, Hogg Foundation for Mental Health.

Schoop, E. Jack. (1969) "Development pressures after the earthquake." In *Geological Hazards and Public Problems.* Office of Emergency Preparedness Conference, Region 7, Santa Rosa, Calif.

Simpson, Robert H. (1971a) *The Decision Process in Hurricane Forecasting.* Fort Worth, Tex.: NOAA Technical Memorandum NWS SR-53.

——. (1971b) "Hurricane, yes or no." *NOAA* 1 (3):12–21.

——, and Malkus, Joanne S. (1964) "Experiments in hurricane modification." *Scientific American* 211:27–37.

Stenz, Lee M. (1966) "Insurance." In *Hurricane Symposium.*

Houston, Tex.: American Society for Oceanography, Publication No. 1, pp. 313–19.

Sugg, Arnold L. (1967) "Economic aspects of hurricanes." *Monthly Weather Review* 95 (3):143–46.

———. (1968) "Beneficial aspects of the tropical cyclone." *Journal of Applied Meteorology* 7 (1):39–45.

———. (1971) *Memorable Hurricanes of the United States Since 1873.* NOAA Technical Memorandum NWS SR-56.

———. (1972) Personal communication.

Tannerhill, Ivan Ray. (1956) *Hurricanes, Their Nature and History.* Princeton, N.J.: Princeton University Press.

UNESCO. (1970) *Annual Summary of Information on Natural Disasters.* Paris.

U.S. Department of Agriculture. (1970) *After Camille: Conservation Recovery Follows Disaster in Virginia.* Richmond, Va.: Soil Conservation Service.

U.S. Department of Commerce. (1969) *Hurricane Camille.* Environmental Science Services Administration.

———. (1971) Federal Coordinator for Meteorological Services

and Supporting Research. *National Hurricane Operations Plan.* Washington, D.C.: National Oceanic and Atmospheric Administration.

U.S. Office of Emergency Preparedness. (1972) *Disaster Preparedness*, part 3. Executive Office of the President.

U.S. Senate. (1970) *Federal Response to Hurricane Camille.* Hearings before the Special Subcommittee on Disaster Relief, 91st Cong., 2d Sess., S. 3619 and S. 3745.

Wilkinson, Kenneth P., and Ross, Peggy J. (1970) *Citizens' Responses to Warnings of Hurricane Camille.* State College: Mississippi State University, Social Science Research Center, Report No. 35.

World Meteorological Organization. (1966) *International Meteorological Vocabulary.* WMO/OMM/BMO/No. 182. TP 91. Geneva.

———. (1972) *Tropical Cyclone Project: Plan of Action.* Geneva. Prepared by the Committee Panel of Experts on Tropical Cyclones.

31. Global summary of human response to natural hazards: floods

JACQUELYN L. BEYER
University of Colorado

Definition

All streams are subject to flooding in the hydrological sense of inundation of riparian areas by stream flow which exceeds bank full capacity. In arid regions the channel itself, not usually filled with water, is "flooded" at times of high runoff. The point at which the channel discharges an overbank surplus is the flood stage. This may not, however, coincide with the amount of water outside the normal channel which will cause damage to human works. It is also possible to calculate the stage of high water which is the threshold for damage to property or dislocation of human activities. Frequently the use of the term "flood stage" is based on such a perception of the event, and is therefore a definition subject to change as conditions of floodplain occupance change. This paper is not concerned with coastal flooding.

Spatial extent

Floods are the most universally experienced natural hazard, tend to be larger in spatial impact, and involve greater loss of life than do other hazards. Floods can occur on both perennial and ephemeral stream beds or in an area where no defined channel exists, such as in an arid region subject to cloudburst type storms. The problem is compounded for human adjustment by the fact that few other hazards present the ambivalent Januslike aspect of good and evil. Humans are attracted to settlement in flood hazard areas by the very characteristics—water supply and floodplain terrain—which contribute to the damage potential.

For this reason it is not surprising to find that historic attempts have been made to resolve the conflict between the need for riparian occupance and the inevitable damage, as Wittfogel describes it: "Thus in virtually all major hydraulic civilizations, preparatory (feeding) works for the purpose of irrigation are supplemented by and interlocked with protective works for the purpose of flood control" (Wittfogel, 1957, p. 24). Less elaborately organized preindustrial societies have also worked out ecological adjustments to flooding. Familiar examples of peasant adaptation to periodic flooding include the traditional agricultural organization along the lower Nile, now altered by the construction of the high Aswan dam, and the village rice culture of the lower Mekong, which will eventually be affected by flood-control components of the Mekong basin development. Another such example is the people of Barotseland in northwest Zambia where migration to higher ground is the organized response to the annual seasonal inundation of the reaches of the upper Zambezi which mark the coreland of Barotse occupance. Changes in socioeconomic patterns as such societies industrialize will undoubtedly accelerate the damage from floods. Familiar adjustments such as migration will fall outside the range of choice. Alternative workable adjustments may be inhibited by lack of knowledge, technology, and/or capital.

For industrial societies the twentieth-century concept of multiple-purpose river basin planning, now widely diffused (United Nations, 1969a), involves the consideration of flood damage reduction along with planning for beneficial use of water.

In summary, the potential for flooding is global in nature and can occur, with the proper combination of factors, whenever there is precipitation. This precipitation may range from uniform and general to sporadic and highly localized. Adjustments to hazards must be made in the context of both universality and randomness. In addition there needs to be sensitivity to beneficial uses for floodplains and water courses.

Damage potential

Types of floods are so varied in origin, duration, strength, timing, volume, depth, and seasonality that it is difficult to identify damage potential except in the most general terms. The amount of damage and the damage potential in any flood-hazard area is very closely related to the nature of occupance and to the stage of economic development as well as to the physical parameters. There also seems to be an inverse relationship between property damage as measured in monetary terms and loss of life. Societies which have much to lose in terms of structures, utilities, transportation facilities, etc., also have the technological sophistication to ensure better monitoring, warnings, evacuations, and rehabilitation—all of which contribute to lowering the human costs. Conversely, preindustrial societies, especially with dense rural populations, do not suffer large property losses but are less well equipped to provide preventative or rescue measures for people.

Clearly the main damage agent is the water itself, overflowing normal channels and inundating land, utilities, buildings, communications, transportation facilities, equipment, crops, and goods which were never meant to operate in or withstand the effects of water. In addition, high velocity of running water operates as a damage agent either directly or indirectly. In the latter case, debris carried by the water or dislodged materials batter structures, people, and goods. Debris and silt carried by the water and left behind as the water recedes operate as further damage agents.

Damage is considered to be either direct or indirect (primary or secondary). Such a classification is useful in any assessment which attempts to clarify the benefits of a damage reduction program. Loss of human life is the

most dramatic and certainly the easiest to identify as a direct result of flood events. Loss of livestock may be especially costly in rural zones.

In agricultural areas, damage involves inundation of land accompanied by erosion and/or loss of crops. It is in such cases that the season of flooding is especially significant. Water damages farm equipment, stored materials (seed, fertilizer, feed), disrupts irrigation systems and other water supply, and disrupts communication.

Urban facilities are all subject to water and force damage—buildings of all kinds, public facilities, utilities, transportation, waterway facilities, and open space. Machinery, manufactures, goods in retail establishments, household furniture can all be damaged by water, debris, and silt.

Indirect damages are generally associated with health and general welfare although such amenities as scenic values, recreational services, and wilderness preservation may also be taken into account. Normal public-health services are subject to greater pressures in the face of disruption of transportation and utilities, especially water supply. Contamination and pollution are more probable, epizootics emerge, stagnant water is left as a flood legacy, and general morbidity increases. Flooding affects the normal sources of food and shelter and hence adversely affects health conditions. Opposed to these considerations is the possibility that emergency relief operations might provide better health care and food than is normally available to some communities counteracting, to some extent, the effects of both direct and indirect damages (White, 1945).

Benefits

Rivers in flood clearly are hazardous for many kinds of human use, but a complication in planning for amelioration of the hazard is presented by the benefits of naturally flowing rivers, including overbank flooding. "Control" of flooding by protective works, especially, may negate benefits from soil nutrient renewal and fisheries. Restrictions on use of floodways will provide for damage reduction and also enhance community values through preservation of open space. In some parts of the world, exemplified by India and Bangladesh, the rhythms of agricultural production are dependent upon water brought by major storms and renewal of fertility through siltation. In such cases serious weather modification efforts should proceed only after careful assessment of the total benefit-cost pattern.

Damage assessment

Assessment includes the costs of repair, including temporary repairs; replacement and cleanup; and loss of improvements and inventories. Emergency costs are also involved. Loss of business and employment during flood disruption is also a direct cost associated with the physical damage. Similarly the transfer of public economic development funds to flood emergency and control programs represents a deferred opportunity which may be especially significant as a social cost in developing nations.

There are no comprehensive calculations of global damage from floods. In many cases, local flooding in areas remote from communications may not even be recorded. There are immense difficulties for any inventory of flood consequences including those of cost, technical expertise, comparability of data collected, allocation of losses to proper causes, and time. A pilot survey of global natural disasters by Sheehan and Hewitt (1969) provides some information for the 20 years 1947–67 for most of the world—the USSR is the major country excluded. During this period Asia (excluding the USSR) led in loss of life, with 154,000 deaths. Europe (excluding the USSR) followed with 10,540 deaths. Africa, South America, and the Caribbean area each recorded 2,000–3,000 deaths. During the same period 680 lives were lost to floods in North America, and 60 in Australia, totaling for all these regions 173,170 deaths.

This total can be compared with the 269,635 deaths attributed to all 18 other hazards in this study and if it is considered that many of the other categories—e.g., tornadoes, typhoons, hurricanes, and tidal waves—also involve flooding, the death loss is the most impressive comparison. Table 31–1 will give some idea of the magnitude of the hazard in terms of area affected, people involved, and property damage.

Damage factors

The factors which should be taken into account in assessing the damage potential for flooding in any basin include the following.

Frequency

Flood flows can occur on the average as often as once every 2 years in temperate climates and as infrequently as once in 1,000 years elsewhere. The recurrence of a particular flood flow can be predicted with reasonable accuracy over a long time span if sufficient stream-flow data are available over a long period. This emphasizes the certainty of the event, not its timing, which is not predictable since flood flows are assumed to be random events.

Frequency is a physical parameter clearly related both to perception and adjustments. The greater the frequency the more accurate is the perception of the hazard by floodplain occupants and the greater is their willingness to consider a wider range of adjustments, including alternative sites for their activities. This is demonstrated by variations in community decision making which can be correlated with frequency of flooding (Kates, 1962).

Magnitude (depth)

Magnitude of a flood may be expressed in physical, or probabilistic terms. The physical measures are rate of flow measured in cusecs, m^3/sec (cubic meters per second), or river stage in meters (or feet) above some datum (reference) point. Both of these require carefully established measuring installations to obtain reliable data. Flow can be graphically plotted versus stage to give a stage-discharge relationship which can then be useful in predicting damages. This relationship can be quite complex, and considerable care should be exercised in its use.

The probabilistic measure is a statistical method of ordering various magnitudes of flow and stating the probability that a given flow will be exceeded. Under this procedure, a 5-year flood is a flow or stage which will be equaled or exceeded 20 times, on the average, in a 100-year time span. The statistical method has validity only in areas where good flow records over a long period of time (at least 30 years) are available, although some simulation can be done. The ability to determine probability of flood magnitudes is not the same as the ability to state when such floods will occur. Some of the parameters upon which magnitude depends and which are useful in classifying the flood characteristics of a given area include the type, intensity, duration, areal extent, and distribution of precipitation; basin size and shape; floodplain topography; surface conditions of soil; and land use. There is an effort in Japan, for example, to elaborate a system of flood-hazard classification through the use of landform analysis. It is suggested that such analysis will provide for predictability of flood current, ranges of submersion, depth of stagnant water, and length of period of stagnation (Oya, 1969).

Depth can be important in terms of both the kinds of damage and possible adjustments, e.g., floodproofing of structures.

Rate of rise

This is the time from flood stage (or zero damage stage) to flood peak and is a measure of the intensity of the flood. As Sheaffer (1961) notes, this time between flood stage and flood peak represents an adjustment time during which persons affected by flooding can engage in activities to lessen the damage. Generally people will not respond to the danger of flooding until at least flood stage, so this time period is critical. It is clear that there is a relationship between the nature of the drainage system and the rate of rise—upstream areas will have a more rapid rise and shorter duration of flooding than will downstream areas.

Seasonality

This is one of the more significant factors for agricultural damage and probably the main basis for the adjust-

Table 31—1. Significant historical flood events

Date	Place	Deaths	Property damage
June 1972[a]	Eastern U.S.	100+	$2 billion
June 1972[a]	Rapid City, S.D.	215 (est.)	$100 million
May 11—23, 1970[b]	Oradea, Rumania	200	225 towns destroyed
January 25—29, 1969	Southern California	95	
July 4, 1969	Southern Michigan and Northern Ohio	33	
August 23, 1969	Virginia	100	
May 29—31, 1968	Northern New Jersey	8	$140 million
August 8—14, 1968	Gujarat, India	1,000	
January—March 1967	Rio de Janeiro and São Paulo states	600+	
November 26, 1967	Lisbon	457	
January 11—13, 1966	Rio de Janeiro	300	
November 3—4, 1966	Arno Valley, Italy	113	Art treasures in Florence and elsewhere destroyed
June 18—19, 1965	Southwest U.S.	27	
June 8—9, 1964	Northern Montana	36	
December 1964	Western U.S.	45	
October 9, 1963	Belluno, Italy	2,000+	Vaiont Dam overtopped
November 14—15, 1963	Haiti	500	
September 27, 1962	Barcelona	470+	$80 million
December 31, 1962	Northern Europe	309+	
May 1961	Midwest U.S.	25	
December 2, 1959	Frejus, France	412	Malpasset Dam collapsed
October 4, 1955	Pakistan and India	1,700	5.6 million crop acres at loss of $63 million
August 1, 1954	Kazvin District, Iran	2,000+	
January 31—February 1, 1953	Northern Europe	2,000+	Coastal areas devastated
July 2—19, 1951	Kansas and Missouri	41	200,000 homeless, $1 billion
August 28, 1951	Manchuria	5,000+	
August 14, 1950	Anhwei Province, China	500	10 million homeless; 5 million acres inundated
July—August 1939	Tientsin, China	1,000	Millions homeless
March 13, 1928	Santa Paula, Calif.	450	St. Francis Dam collapsed
March 25—27, 1913	Ohio and Indiana	700	
1911	Yangtze River, China	100,000	
1903	Heppner, Ore.	250+	Town destroyed
May 31, 1889	Johnstown, Pa.		
1887		2,000+	
	Honan, China	900,000+	Yellow River overflowed; communities destroyed
1642	China	300,000	

[a]Press reports.

[b]Adapted from Table of Disasters/Catastrophes, *New York Times Encyclopedic Almanac (1970), p. 1228*; (1972), pp. 322—33.

ments made by preindustrial riverine societies. Clearly the hazard increases where the growing season is limited and coincides with the season of flooding. Winter floods might also account for increased loss and disruption in urban areas where heating and sanitary facilities are needed to guard against increase in disease and discomfort.

Duration

The time of inundation for flood flows can vary from a few minutes to more than a month. The duration is highly correlated to the rate of rise and fall of flood crests except where drainage of land area is impeded by obstructions. Flood duration is dependent upon such parameters as source of runoff; runoff characteristics including slope and surface conditions; nature of obstructions impeding recession of waters; and man-created controls such as reservoirs, levees, and channelization.

Nature of floodplain occupance

This includes the density of settlement; types of facilities; extent of fixed facilities, buildings, and equipment;

and value of facilities. Obviously every increase in such occupance will increase the potential damage and call for some kind of adjustment, whether protective works, warning systems, and public relief capabilities, or a willingness to accept the losses.

Efficacy of forecasting and warning systems

The ability to forecast the occurrence of overbank flooding is limited to a time span in which the hydrologic conditions necessary for flooding to occur have begun to develop. The formulation of a forecast for flood conditions requires information on current hydrologic conditions such as precipitation, river stage, water equivalent of snowpack, temperature, and soil conditions over the entire drainage basin as well as weather reports and forecasts.

In small headwater regions a forecast of crest height and time of occurrence is all the information required to initiate effective adjustments since the relatively rapid rate of rise and fall makes the period of time above flood stage relatively short. In lower reaches of large river systems where rates of rise and fall are slower it is important to forecast the time when various critical stages of flow will be reached over the rise and fall. Reliability of forecasts for large downstream river systems is generally higher than for headwater systems.

Warning time for peak or overbank conditions can range from a few minutes in cloudburst conditions to a few hours in small headwater drainages to several days in the lower reaches of large river systems. As with forecasting, the time and reliability of the warning increase with distance downstream where adequate knowledge of upstream conditions exists.

Clearly the amount of information required, the data collection network necessary for collecting the information, the technical expertise required for interpretation, and the communication system needed to present the information in time to potential victims are such as to preclude many poor and developing nations from having an adequate service. The World Meteorological Organization of the United Nations, through its World Weather Watch and Global Data Processing System, hopes to coordinate efforts to improve forecasting. A recent report (Miljukov, 1969), notes that quantitative precipitation forecasts for 24–49 hours in advance are provided in parts of Australia, Byelorussian SSR, Cambodia, Canada, Czechoslovakia, France, Federal Republic of Germany, Hong Kong, India, Iraq, Japan, Mauretania, Norway, Pakistan, Philippines, Rhodesia, Romania, Sweden, Ukrainian SSR, the USSR, and the U.S. Precipitation forecasts for hydrological purposes are provided in Australia, Canada, Czechoslovakia, France, Federal Republic of Germany, India, Iraq, Japan, the USSR, and the U.S. The report also notes: "Precipitation forecasts are not accurate and reliable enough, in the present state of meteorological science, for use in the preparation of quantitative forecasts of river discharges" (Miljukov,

1969, p. 10). Most developing nations will have to rely on much less data than are ideally needed for forecasting and warning, which in turn will lessen the effectiveness of this factor with respect to flood losses.

Efficacy of emergency services

The helplessness of many small and poor societies is exemplified by the situation in Bangladesh during the tropical cyclone of 1970. Where resources are not available for planning, for the physical effort of relief and evacuation, and for coordination with other activities, little will be done outside of contributions of international aid agencies. Even in more developed areas, local conditions of transportation and public attitudes will lessen the usefulness of emergency aid, e.g., the disaster at Buffalo Creek, West Virginia, in 1972. The more elaborate and dependable such services are, however, the more there is a tendency to rely on such aid as a major adjustment and to reject consideration of less costly and more effective adjustments. There clearly must be provisions for first-level emergency aid where settlement already exists and where alternative moderations of the hazard are difficult to implement or costly, but these should normally not supplant other measures to reduce losses.

The range of adjustments and their adoption

The accompanying table (Table 31–2) suggests that the cumulative experience of centuries provides for any society or group wishing to alleviate the social and economic costs of flood losses a choice of methods, to be used singly or in strategic combinations. Much of the wisdom with respect to the need for such strategies, adapted to local conditions of basin hydrography, settlement characteristics, and economic capabilities, has been gained in industrial nations after painful and costly trial and error. It has been suggested (Goddard, 1969) that developing nations need not repeat the errors of the past, that they have models for actions and policies which would provide a much more coherent and suitable response to the flood hazard. It is clear, even with such models, that adaptation to local circumstances in any society will not be a simple matter. It is increasingly evident that various combinations of individual psychology, institutional inertia, costs, governmental policies and philosophies, and historical precedents help to condition the choice of adjustments.

Modify the flood

This category includes engineering works affecting the channel which represent the most widely accepted feasible adjustment with the possible exception of bearing the loss. Such protective works are justified where benefits exceed the costs of implementation and especially where high damage potential exists for relatively inten-

Table 31–2. Adjustments to the flood hazard

Modify the flood	Modify the damage susceptibility	Modify the loss burden	Do nothing
Flood protection (channel phase)	Land-use regulation and changes	Flood insurance	Bear the loss
Dikes	Statutes	Tax write-offs	
Floodwalls	Zoning ordinances	Disaster relief	
Channel improvement	Building codes	volunteer	
Reservoirs	Urban renewal	private	
River diversions	Subdivision regulations	activities	
Watershed treatment (land phase)	Government purchase of lands and property	government aid	
Modification of cropping practices	Subsidized relocation	Emergency measures	
Terracing	Floodproofing	Removal of persons and property	
Gully control	Permanent closure of low-level windows and other openings	Flood fighting	
Bank stabilization	Waterproofing interiors	Rescheduling of operations	
Forest-fire control	Mounting store counters on wheels		
Revegetation	Installation of removable covers		
Weather modification	Closing of sewer valves		
	Covering machinery with plastic		
	Structural change		
	Use of impervious material for basements and walls		
	Seepage control		
	Sewer adjustment		
	Anchoring machinery		
	Underpinning buildings		
	Land elevation and fill		

Source: Adapted from Sewell (1964), pp. 40–48; and Sheaffer, Davis, and Richmond (1970).

sive settlement in urban and industrial situations. In such cases the high value of fixed facilities will justify levees, dams, and channelization even when 100 percent protection cannot be guaranteed. While benefits accrue to both private and public sectors, the costs are necessarily largely public. Partly for this reason this adjustment strongly tends to encourage persistent settlement and even attracts, through a false sense of security, further floodplain encroachment. On the other hand, such engineering works are important components of multiple purpose projects which are directed toward comprehensive land and water planning and they can be complemented, for floodplain management, with other measures. A major problem is to encourage engineers and officials to think in terms of nonstructural alternatives or supplements to protective works.

Watershed treatment practices have more subtle implications with respect to flood control and are frequently more significant for their contribution to improved *in situ* land management. All such measures have their limitations with respect to major flood events. There may be some contribution to lowering the depth in small floods and to lengthening the flood-to-peak interval, but essentially the appeal of such practices is lower costs. About 90 percent of the costs for such practices are public while benefits accrue largely— about 85 percent—to private land users. Land treatment measures often complement and make more effective

protective measures but will also tend to encourage continued settlement of flood-prone areas.

Weather modification is a fairly recent technique with respect to flood control and too little is known about its effectiveness. One major problem, even given scientific certainty about effectiveness, will be the necessity to allay public fears that tampering with weather processes will increase rather than lessen floods. Recent news stories of the use of weather modification techniques in the Indochina war will not make this task easier (Shapley, 1972). The immediate postflood news reports from the 1972 Rapid City, South Dakota, flood suggest that this has already become an issue. Costs for weather modification are entirely public while benefits are about equally divided between private users and the public.

Modify the damage susceptibility

Given that there may be a need to encroach on floodplains or to accept present settlement patterns, certain measures are possible which are either less costly than protective works or bearing the loss, or which will lessen the actual damages even more. There is also a greater shift of cost bearing to private interests, especially in the case of floodproofing, with resultant increased awareness of the need for flood adjustments.

Land-use regulation, including changes in occupance, is especially suitable where there is competition for floodplain land for uses other than agricultural or recreational. The legislative and police powers of the state can be used to control and guide development of floodplains. According to Goddard, in the United States "about 35 states have adopted regulations and 500 additional places in 41 states have them in adoption process" (Goddard, 1971). Encouraging this is the 1969 Federal Flood Insurance Act which provides for governmental flood insurance subsidy to individuals in communities which agree to adopt floodplain regulation guidelines. These measures tend to encourage more efficient and less costly use of floodplains and there is a greater shared responsibility between floodplain users and authorities. Strong leadership and a commitment to long-range planning and rational allocation of land uses are also prerequisites to widespread adoption of such measures.

Floodproofing and structural changes (including land elevation) provide for even larger shifts of costs as individual users may bear all the costs and share benefits with the public on an equal basis. Such adjustments are most appropriate where flooding is not intense either in velocity or depth and where some warning time is possible. Floodproofing especially requires a network of forecasting and warning facilities along with a flood-hazard information program which will encourage preflood adjustments. Structural modifications are possible for existing structures as well as for new structures although this will increase costs. In many cases it would be too expensive to modify old buildings. Some types of buildings are better suited to modification than others but clearly damage reduction is related to size of structures and costs of modification. These adjustments tend to encourage persistent occupance and lose effectiveness where flood frequency is low. At the same time they place more responsibility on the user and thus heighten sensitivity to and knowledge about the flood hazard.

Modify the loss burden

There is much more emphasis in this category on humanitarian responses rather than calculated economic rationale, based on the inevitability of flooding and the unlikely possibility of preventing all damage by eliminating floodplain occupance. Losses will thus occur even in the face of widespread use of appropriate adjustments. When people suffer trauma and loss there can be little question of a social obligation to provide assistance. The dilemma for rational flood damage reduction, however, is that relief measures and emergency assistance unless properly designed tend to encourage persistent occupance and reluctance to accept more rational adjustments.

Insurance and tax write-offs will not decrease flood losses but there will be a spreading of loss over time and a shift of some costs to the general public. As the flood insurance program has been worked out in the United States, the insurance subsidy by the government to private carriers must be coupled with community planning for land-use regulation and other adjustments to lessen potential damage. In Hungary, where levees and flood fighting are the principal adjustments, agricultural insurance was extended in 1968 to cover flood damages (Bogardi, 1972). Whether this works as an incentive for private adjustments is not clear. There is obviously a sensitive line between encouraging further encroachment or private irresponsibility and alleviating the damages to those who must occupy floodplains. Purchases of insurance, according to recent reports after the June 1972 floods in the eastern U.S., have not been commensurate with the danger nor with the benefits to eligible individuals. Problems resulting from a hazard insurance program which was not thoroughly planned have been noted (O'Riordan, chap. 26).

Disaster relief is a necessary adjustment in order to lessen the immediate impact of a flood event and to ease the implementation of rehabilitation efforts. Whether government or private, the major disadvantage is that such measures, necessary though they may seem when disaster strikes, strongly encourage the belief that nothing else need be done.

The effectiveness of emergency measures depends largely upon the nature of the flood hazard (ideal combination of high flood frequency, low velocity and depth, long flood-to-peak interval, and short duration) and the quality of forecasting. The immediate governmental obligation is generally seen to be the removal of persons and property from flood threatened areas.

Do nothing

Bearing the loss is still the major adjustment for large numbers of floodplain occupants in developing countries (Ramachandran and Thakur, chap. 5), and is frequently modified in developed nations only by the widespread expectation of relief and emergency measures. In all cases, however, it is clear that an increasing effort to clarify public interest in floodplain situations will restrict the choice of doing nothing and management strategies will become more common (Sheaffer, Davis, and Richmond, 1970).

Reduction of loss

One element in the acceptance by any group of decision makers of a particular mix of components in a flood damage reduction program is the assessment of the comparative return from each possible choice of adjustment or combination of adjustments. If damage assessment after the fact of flooding is extremely difficult, it is even more difficult to predict what the damage will be under a set of assumptions about responses of various kinds. White and Burton have suggested methods whereby maximum damage reduction and minimum cost can be calculated for particular situations (White, 1964; Burton, 1969). Such methods may hopefully provide an additional planning tool in those situations where encroachment onto the floodplain is neither as intensive as in some industrial countries nor necessary. Some such tool is essential also to ensure the most efficient allocation of scarce resources, whether of materials, manpower, or money.

The relative contributions of each possible adjustment to reduction of potential damage can only be crudely measured at present. Such measurement is further complicated by the fact that only infrequently is a single adjustment adopted. Clearly any protective works which provide for 100 percent security under any feasible flood condition will provide 100 percent loss reduction although costs of providing such protection are likely to be unacceptable. Such security is highly improbable, both because of costs and imperfect knowledge of potential floods. The damage reduction to structures may range from 40 to 100 percent, dependent upon the size of the flood experienced and the nature of structures (White, 1964).

Watershed treatment data are inconclusive with respect to damage reduction and there are no data available for weather modification. Land-use regulation and change can provide for up to 90 percent damage reduction dependent upon the effectiveness of the regulations and the speed of application.

Data from one United States town (White, 1964) suggest that even minimal floodproofing of present structures under conditions of frequent but shallow flooding can be very effective, reducing damages by 60–85 percent. Great depths and/or high velocities would call for consideration of floodproofing as part of building design.

Emergency action increases in effectiveness where there is a long flood-to-peak interval, high flood frequency, low depths, short duration, and low velocity. Where such conditions prevail, and assuming adequate warning facilities plus personnel and equipment, emergency action can reduce damages by 15 to 25 percent. A lower range of 5 to 10 percent is more probable.

"Adequate" warning would seem to be a minimal requirement for communities subject to flood hazard, but it is not simple nor inexpensive to provide a good system. Meteorological services and communications are part of the costs. Even given an excellent network of knowledge about the physical event, it may be difficult to convey that information to persons who will have to make adjustment decisions. Factors involved in a less than optimal warning system include:

1. Reluctance of officials to give false alarms.
2. Lack of complete coverage of median used to transmit warnings (radios, telephones, etc.)—communities and individuals may not be able to afford facilities.
3. Reluctance of people to see themselves affected by distant events (storms, runoff).

4. Individual interpretation of warning messages, especially where several messages may be contradictory or the messages may be incomplete.
5. Failure to provide exact information about what recipients of warnings are to do.
6. Impossibility of warning in time for much else than rapid evacuation.
7. Dramatic warning signals triggering an influx of the curious which negates warning advantages.

Flood insurance and tax subsidies spread the burden through time and shift much of the loss to the general public but do not reduce damage.

Another indication of the relative efficacy of various adjustments in reducing damage is the importance placed on them in national and regional plans. A recent report on Hungary (Bogardi, 1972), for example, suggests that reduction of damages is to be achieved largely through levee construction and maintenance, flood fighting, and, to some extent, insurance. Recommendations for Malaysia (Flood Control, 1968) are for a flood-control program involving better data collection and improved organization for relief and evacuation, combined with structural controls, land-use regulations, and flood-resistant crops. It is estimated that these measures would reduce anticipated damage from presently known levels of flooding by 50 percent. This report does not consider dams in catchment areas justifiable for flood control alone. Engineering works are still considered primary tools for India although some attention is being given to catchment area management and weather modification. The Japanese have extended their management approach to include regulatory measures (Oya, 1969). A comprehensive summary of national efforts to cope with floods as one of many natural hazards would probably justify a comment in a report from the United Nations (1969b): "Although there is still a considerable gap in many countries between the needs for governmental action and the actual institutional framework, new administrative patterns have evolved in others which responded to the need for a more coordinated and system oriented approach to resource administration."

Perception of hazard and adjustments

The global nature of the flood hazard is suggested not only by maps of large floods and by tables of deaths, but also by reference to international interest in the problem. The special agencies of the United Nations are involved in a wide spectrum of activities, including hydrological and meteorological data collection, flood forecasting methods, world catalogue of large floods, problems of health due to floods, and relief and aid to victims. Agencies involved include the Economic Commission for Africa, Economic Commission for Asia and the Far East, World Meteorological Organization, World Health Organization, and UNESCO. In many cases small nations will have to rely on technical help and assistance through United Nations channels. There is a discernible diffusion of efforts to plan and implement comprehensive programs including Canada, Japan, United Kingdom, and the United States. Because river basin management is so popular as an economic development tool, this opens the door for widespread consideration of comprehensive flood control as a component of such programs. From the global and national institutional viewpoints there is probably adequate sensitivity to the nature of the problem, if not to the possible range of adjustments. At the individual level, it is more difficult to judge whether the knowledge gained in recent years about perception of the hazard in the United States (Kates, 1962; Burton and Kates, 1964; James, Laurent, and Hill, 1971) is applicable to individual perception in developing or industrializing nations—or even industrial nations with different social and political conditions. A summary of some of the findings of these hazard perception studies, especially of floodplain occupants in Georgia, may be listed in the form of planning guidelines (adapted from James, Laurent, and Hill, 1971):

1. It cannot be assumed that accurate knowledge of the flood hazard will inhibit all persons from moving onto the floodplain.

2. The flood hazard itself will process people over time in terms of perception of the hazard and willingness to make adjustments. Management programs can short-circuit the unhappy experiences of those who remain unaware of the hazard and reluctant to adopt adjustments by preventing their settlement (e.g., through insurance programs).

3. Prospective floodplain occupants who are initially unaware cannot be swayed by large amounts of technical information; they also tend to be people who avoid contact with public officials and are not observant with respect to natural features.

4. In contrast, people who are knowledgeable about the flood hazard and settle anyway on floodplains will be responsive to more sophisticated information than is usually presented.

5. Delineation of flood-hazard areas on a map is ineffective as a form of communication.

6. Officials who disapprove of settlement on floodplains or who think in technical terms about risk will not be effective with those who are unaware of the risk.

7. Those who know about the flood hazard will be sensitive to depths, if not to frequency, and will therefore be open to flood proofing and possibly insurance as adjustments.

8. Time reduces awareness of the hazard, especially for those moving into a hazard area where indications of past flood events are not evident.

9. The wave of concern for environmental issues has brought with it evidence that those who are unaware of the flood hazard, but who have a concern for environmental damage, may respond more to appeals that land-use regulations are ecologically sound than to information about potential property damage.

10. Flood damage sufferers who contact, or who are contacted by, officials are a biased sample in terms of response to flood hazard. Frequently this bias is associated with speculation as a motive for owning floodplain property.

11. Upstream development frequently becomes the scapegoat for downstream floodplain users threatened by floods.

12. Floodplain users who are alienated from government or authority because of other contacts are poor candidates for participation in floodplain management programs.

13. Extended delays in programs to reduce flood losses will increase alienation and make user participation more unlikely.

14. Encouragement of particular users should be part of policy, e.g., it should be made easy for those who are unaware of the hazard and/or reject adjustments to leave and be replaced by those who know something about the hazard and will be willing to adopt reasonable adjustments, including insurance and flood proofing or structural change. Where even these adjustments are too costly in the light of potential damage the policy should be to consider purchase and reversion to open space and recreational use.

It is hard to believe that persons would vary much with respect to a number of factors involved in determining the degree of knowledge about flood events, anticipation of future events, and willingness to consider various possible adjustments. Confirmation of this belief awaits further investigations of human response in diverse societies.

References

Baroyan, O. V. (1969) "Problems of health due to floods." Tbilisi, USSR: United Nations Inter-regional Seminar on Flood Damage Prevention Measures and Management.

Bogardi, I. (1972) "Floodplain control under conditions particular to Hungary." International Commission on Irrigation and Drainage, 8th Congress.

Burton, Ian. (1969) "Methods of measuring urban and rural flood losses." Tbilisi, USSR: United Nations Inter-regional Seminar on Flood Damage Prevention Measures and Management.

_____, and Kates, Robert W. (1964) "The perception of natural hazards in resource management." *Natural Resources Journal* 3 (2):412–41.

Flood Control, Report of the Technical Subcommittee for. (1968) Government of Malaysia: Director of Drainage and Irrigation.

Goddard, James E. (1969) "Comprehensive flood damage prevention management." Tbilisi, USSR: United Nations Inter-regional Seminar on Flood Damage Prevention Measures and Management.

_____. (1971) "Flood-plain management must be ecologically and economically sound." *Civil Engineering—ASCE* 000:81–85.

James, L. Douglas, Laurent, Eugene A., and Hill, Duane W. (1971) *The Flood Plain as a Residential Choice: Resident*

Attitudes and Perceptions and Their Implications to Flood Plain Management. Atlanta: Georgia Institute of Technology, Environmental Resources Center.

Kates, Robert W. (1962) *Hazard and Choice Perception in Flood Plain Management.* Chicago: University of Chicago, Department of Geography, Research Paper No. 78.

Miljukov, P. I. (1969) "Review of research and development of flood forecasting methods." Tbilisi, USSR: United Nations Inter-regional Seminar on Flood Damage Prevention Measures and Management.

Oya, Masahiko (1969) "Flood plain adjustments, restricted agricultural uses, zoning, and building codes as damage prevention measures." Tbilisi, USSR: United Nations Inter-regional Seminar on Flood Damage Prevention Measures and Management.

Sewell, W. D. F. (1964) *Water Management and Floods in the Fraser River Basin.* Chicago: University of Chicago, Department of Geography, Research Paper No. 100.

Shapley, Deborah. (1972) "News and comment." *Science* 176: 1216–20.

Sheaffer, John R. (1961) "Flood-to-peak interval." In Gilbert F. White, ed., *Papers on Flood Problems.* Chicago: University of Chicago, Department of Geography, Research Paper No. 70.

——, Davis, George W., and Richmond, Alan P. (1970) *Commu-*

nity Goals—Management Opportunities: An Approach to Flood Plain Management. Chicago: University of Chicago, Center for Urban Studies. Report by Institute for Water Resources, Department of the Army, Corps of Engineers.

Sheehan, Lesley, and Hewitt, Kenneth. (1969) "A pilot survey of global natural disasters of the past twenty years." Toronto: University of Toronto, Natural Hazards Research Working Paper No. 11.

United Nations. (1969a) *Integrated River Basin Development.* New York: rev. reprinting.

——. (1969b) "Some institutional aspects of adjustments to floods." Tbilisi, USSR: Resources and Transport Division, Department of Economic and Social Affairs, United Nations Inter-regional Seminar on Flood Damage Prevention Measures and Management.

White, Gilbert F. (1945) *Human Adjustment to Floods: A Geographical Approach to the Flood Problem in the United States.* Chicago: University of Chicago, Department of Geography, Research Paper No. 29.

——. (1964) *Choice of Adjustment to Floods.* Chicago: University of Chicago, Department of Geography, Research Paper No. 93.

Wittfogel, Karl A. (1957) *Oriented Despotism: A Comparative Study of Total Power.* New Haven, Conn.: Yale University Press.

32. Global summary of human response to natural hazards: earthquakes

THOMAS C. NICHOLS, JR.
University of Colorado

Definition

An earthquake is the sudden release of accumulated strain energy that takes the form of shock waves and elastic vibrations (seismic waves) that are transmitted through the earth in all directions (Fig. 32–1). The resulting oscillatory and sometimes violent movement of the earth's surface is often associated with widespread permanent deformations, which may include (1) tilting of segments of the earth's crust, (2) offset along faults, (3) compaction of loose or unconsolidated soil materials, (4) landslides and mudflows, (5) liquefaction of soils, (6) avalanches, and (7) new fracture developments in rock masses. The manner in which earthquake strain energy is released is not very well understood. From the new global tectonics, Isacks, Oliver, and Sykes (1972) conclude that earthquake activity is generally explained

as the result of interactions and other processes near the edges of large mobile lithospheric plates spreading from ocean ridges. Presently it is thought that very shallow earthquakes (less than 14.5 kilometers or 9 miles) may be caused (1) by slippage along large fault segments that contain accumulated strain energy (Cook and Anderson, 1972; Nur, 1972; Scholz, Wyss, and Smith, 1969; Wu, 1972), (2) by injection of pore fluids or change of pore pressure that mechanically and chemically disrupts a locally delicate stress equilibrium condition (Byerlee, Wilson, and Peselnick, 1972; Raleigh et al., 1970), (3) by detonation of nuclear devices, and (4) as a result of volcanism (Endo, 1972). Very little is known about energy release mechanism of intermediate (60–300 kilometers or 40–190 miles) and deeper (below 300 km or 190 miles) earthquakes. It has been suggested that (1) phase changes in molten magmas (Griggs and Handin, 1963), (2) creep

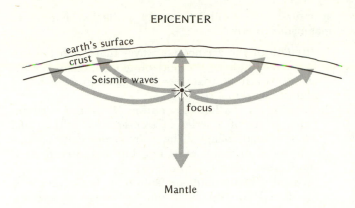

EPICENTER

earth's surface

crust

Seismic waves

focus

Mantle

Fig. 32-1. Emission of seismic waves from earthquake source (focus)

instabilities in plastically deforming magmas (Orowan, 1963), and (3) velocity discontinuities along subduction zones (Verhoogen et al., 1970) are possible mechanisms releasing deep earthquake energy.

Damage potential

The earthquake hazard is complex in that it may have numerous direct and induced effects on the earth's surface. Among the direct surface effects are ground motion from seismic waves or tectonic (fault motion) movements of the surface. The induced effects are subsidence, compaction, landslides, soil failure, local sea waves, tsunamis (seiches), fire, and snow avalanches. This multiple hazard extorts a high toll in loss of life and damage to human habitation. Saarinen (1970) ranks earthquakes third in total loss of life attributed to natural hazards from 1947 to 1967. During that 20-year period, 56,100 people were killed by earthquakes throughout the world. These deaths are 12.7 percent of the total lives lost to natural hazards during the past two decades. Only floods and cyclones exceed the toll of deaths caused by earthquakes during that period.

No estimate is available for the monetary cost of annual damage from all earthquakes. Measured in terms of property lost or structures needing restoration and cleanup, the San Fernando earthquake of February 9, 1971, which had a magnitude of 6.6 on the Richter scale, cost over half a million dollars (Grantz, 1971). Yet this was not even a major earthquake. The 6.7 San Fernando quake had about one-eightieth of the energy of the 8.3-magnitude San Francisco quake of 1906. There are 60 to 70 earthquakes every year with magnitudes comparable to the San Fernando earthquake. According to Mukerjee (1971 p. 10), "Between 1970 and 2000 we may expect a major earthquake in the San Francisco Bay area of magnitude around 8 Richter and duration of more than a minute. The expected damage to assets would run to 25 billions of dollars in 1970 prices and loss in life may be in the hundreds or thou-

sands depending on the time of day the quake strikes and the adjustments that are made prior to the disaster and afterwards." The worldwide trend toward urbanization on or near areas of high seismic risk would indicate that the potential for earthquake damage and deaths is increasing rapidly and without adequate human adjustments to the hazard.

Extent of damages

The major impact of earthquakes upon civilization is the extent of physical harm, anxiety, and personal loss suffered by the populace, which in turn is related to structural damage and subsequent potential fire and flood in a stricken area. Structural damage in any area, however, is related not only to magnitude, frequency, and wave types of an earthquake but also to structural homogeneity, integrity, and foundation conditions of the buildings. Damage generally occurs because horizontal forces are exerted against buildings of structural design intended to resist only vertical forces (Iacopi, 1964) or because of the uneven resistance of structural elements to destructive forces. A grim example of damage caused by horizontal shaking and bad ground conditions occurred in the Caracas earthquake of 1967. The earthquake's Richter magnitude was only 6.5, but damage in Caracas was heavy, even though confined to several areas selective to certain types of structures and foundation conditions. Four high-rise buildings collapsed in a pancake fashion, killing 200 people. Other high-rise buildings around the four collapsed buildings were so badly damaged that many had to be evacuated (Steinbrugge and Cluff, 1968). The severe structural damage apparently was related to the modification of seismic waves caused by the thick alluvial deposits (40-120 meters or 130-400 feet) throughout the Caracas valley. The fundamental wave period of ground movements within the alluvial deposits nearly coincided to the fundamental wave period of high-rise buildings between 10 and 20 floors, thus creating harmonic motion of the buildings that caused severe damage. Similar buildings on firm ground did not experience harmonic oscillation, as the fundamental wave period of ground motion was much less than the fundamental building periods. The difference in behavior of ground motion in firm rock versus that in alluvial deposits was not attributed to distance from the focus but was assumed to be caused by changes of the seismic wave characteristics from firm to soft ground (UNESCO, 1967a). The total structural damage in the Caracas quake was estimated to be more than $100 million, and 250 people lost their lives.

Similarly, in recent Turkish, Japanese, and Philippine earthquakes much of the severe damage occurred in structures located on alluvial deposits.

In the Mudurnu Valley earthquake in Turkey, July 22, 1967, the horizontal displacements, wherever found, were always larger in alluvium than in bedrock. The controlling factors of damage were foundation stability and type of construction rather than proximity to the fault break. The earthquake destroyed 5,000 houses, killed 86 people, and injured 332 (UNESCO, 1967b).

The Philippines earthquake, August 2, 1968, of magnitude 7.3 Richter, caused large damage in the Manila area. The damage, however, was concentrated in a relatively small part of greater Manila that lies near the mouth of the Pasig River and probably includes the deepest and most recent alluvial deposits in the city. The damage in this area was estimated to be near $4 million, with 268 deaths and 261 injuries (UNESCO, 1968.

The Nügata earthquake in Japan of June 16, 1964, was characterized by damage to reinforced concrete buildings resulting from liquefaction of sandy layers of soil. The damaged buildings were concentrated in an area of recently abandoned water courses, where fine sandy soils saturated with ground water are thickly deposited. Twenty-six people were killed, 447 injured, and over 15,000 houses were damaged by motion, fire, or flood (Nakano, 1972).

Slosson (1972) says that the severity of failures in the Los Angeles earthquake of February 1971 was generally associated with the type of foundation materials. The earthquake had a Richter magnitude of 6.6, nearly destroying 730 homes and damaging 20,000 homes. Those homes suffering damage, in order of the most to the least, were built on (1) older (pre-1963) poorly engineered fill deposits, (2) loose alluvial material, (3) semiconsolidated alluvial materials, and (4) properly engineered fills and bedrock.

From these few examples it is evident that geologic foundation criteria are very critical factors in earthquake damage. These factors are only beginning to be understood.

It has been apparent in practically all earthquakes that the type of building construction is a pronounced factor contributing to earthquake damage. For instance, in the previously mentioned Mudurnu Valley earthquake of 1967 there was a striking contrast of nearly total destruction to unreinforced brick-walled structures as compared to only slight damage to nearby timber frame houses.

In the San Fernando earthquake of 1971, Richter magnitude 6.6, nearly all of the deaths attributed to structural failure occurred in old buildings constructed before the earthquake-design provisions were made mandatory. In general, buildings constructed after these provisions were made mandatory performed much better than the older buildings. Total estimated damage of this earthquake was more than one-half billion dollars and a loss of 64 lives (Grantz, 1971).

Besides the critical factors of foundation conditions and structural design detailed above, other factors also contribute to earthquake damage.

According to Iacopi (1964), at least five factors, structural and natural, strongly influence damage to manmade structures.

1. Strength of the earthquake waves that reach the surface. Magnitude of the horizontal component is especially critical in that few structures provide safety against horizontal shaking.
2. Duration of earthquake motion. It is the accumulating effect of a series of tremors that is the usual cause of wall collapse. The main tremor may substantially weaken many buildings, but less intense aftershocks may cause collapse.
3. Proximity to the fault or fault zone. Structures resting directly on the fault zone are obviously precariously located if there is movement along the fault. Otherwise damage is not necessarily a function of the distance from the fault.
4. Geologic foundation. To many engineers and insurance experts this is the most important factor in earthquake damage. Studies have shown damage to buildings on soft ground to be much more extensive than to buildings on hard ground. Buildings partly on soft ground and partly on hard ground are even more precariously located with respect to earthquake response.
5. Building design. The necessary design strength of structures in resistance to both horizontal and vertical seismic motions can be accomplished by adequate bracing and structural continuity.

It can be seen by the building classification rating devised by the Pacific Fire Rating Bureau that structural design is paramount in the assessment of insurance rates.

Areal extent

Larger earthquakes are felt over considerable areas, in some cases over 4 million square kilometers (1.5 million square miles). However, the degree of shaking varies considerably over the felt area. While the degree of shaking tends to fall off with distance from the center of the disturbance, it varies greatly, depending on the solidness of the ground on which structures stand. The intensity of shaking and consequently the destructive effects of an earthquake are much greater in areas distant from the epicenter composed of alluvium, beach sands, artificial fill, or other unconsolidated foundation than they are on firm rock closer to the epicenter.

A major earthquake in Kansu Province, China, in 1920 near the Tibetan border struck with destructive force over an area of 40,000 square kilometers (15,000 square miles). It was reported to have been felt throughout 4 million square kilometers (1.5 million square miles).

The New Madrid, Missouri, earthquake of 1811 shook more than two-thirds of the United States over an area of 2.6 million square kilometers (1 million square miles) and was felt as far away as the East Coast and Canada. It stopped clocks in Boston and set bells ringing in Virginia. It was responsible for profound changes in the level of land for thousands of square kilometers, raising and lowering areas as much as 6 meters (20 feet), draining swamps, altering the course of the Mississippi River, and creating new lakes, such as Lake St. Frances, west of the Mississippi, and Reelfoot Lake to the east, in Tennessee (Tufty, 1969).

The Alaska earthquake of 1964 caused significant damage to the ground and structures throughout an area of 130,000 square kilometers (50,000 square miles). Marked changes of water levels were noted in recording wells as far away as Georgia and Florida in the United States. Few earthquakes have had such marked effect on the crust of the earth and its mantle of soil (Hansen et al., 1966).

Regional and spatial distribution

In Fig. 32–2 it can be seen that the majority of earthquakes occur in the seismically active circum-Pacific belt, along the Sunda arc, and as a diffuse band through the younger (Tertiary) mountain ranges of Asia and Europe (Verhoogen et al., 1970), also referred to as the Mediterranean and trans-Himalayan zone (Gilluly, Waters, and Woodford, 1968). Most earthquakes appear to follow global tectonic boundaries.

Although most earthquakes occur within the well-defined earthquake zones, there have been notable exceptions in the past, i.e., the New Madrid (1811) and Charleston (1886) earthquakes in the United States, the Agadir earthquake (1960) in Morocco, the Koyna earthquake (1967) in India, and others.

Fig. 32–2. Major seismic belts of the world. Source: U.S. Geological Survey

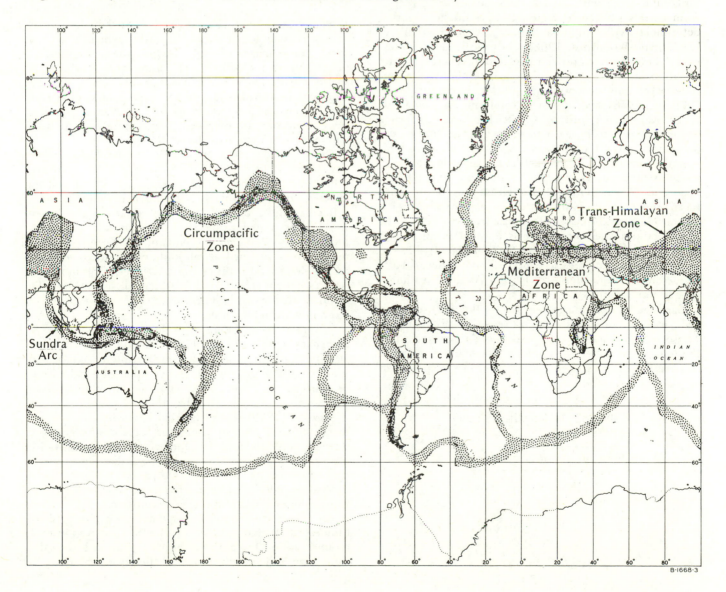

B-1668-3

Magnitude

Seismologists use several magnitude scales, depending on the type of earthquake and type of data available. The Japanese use a scale of seven magnitudes developed by K. Wadati. It was from this scale that Richter adapted his logarithmic scale of magnitudes, which measures disturbances from less than 1 to over 8, implying a range of over 10 million to 1 (Richter, 1969). Other scales have been developed by seismologists for local earthquakes in many parts of the world, notably in New Zealand, Italy, and Japan (Richter, 1958).

The "magnitude" scale, expressing the size of earthquakes, which was developed by the American seismologist C. F. Richter, is based on the amplitude of the largest horizontal trace written by a standard seismograph located 100 kilometers (60 miles) from the epicenter (point on the earth's surface directly above the focus of an earthquake). The variation of this largest horizontal trace with distance and focal depth (depth from earth's surface to region of earthquake origin) is determined with empirical tables and graphs (Gilluly, Waters, and Woodford, 1968). The magnitudes so determined have been empirically related to energy by the equation $\log E = 11.4 + 1.5\,M$, where M is the magnitude based on the amplitude of the horizontal trace (Richter, 1958) and E is the total energy. Using this relationship, each unit of the Richter scale indicates energy released to be 31.6 times as great as that of the preceding unit. Other empirically derived relations indicate the energy release of each magnitude unit to be as much as 60 times as great as the preceding unit. Thus, a magnitude 2 earthquake will release 30 to 60 times more energy than a magnitude 1 earthquake, and a magnitude 8 earthquake will release from 8×10^5 to 12×10^6 times the energy released by a magnitude 4 earthquake. It is estimated by Richter (1958) that the largest earthquakes release energy on the order of 10^{25} ergs, roughly equivalent to 12,000 of the Hiroshima-type atomic bombs.

Earthquakes of magnitude 1 on the Richter scale ordinarily are detected only by sensitive seismographs. Those of magnitude 2 are perceptible to persons near the epicenter under favorable circumstances. Damage (intensity VI or VII; see Table 32-1) rarely appears in earthquakes of less than magnitude 4.5. For convenience, seismologists speak of earthquakes of magnitude 7 and over on the Richter scale as major earthquakes, while those of magnitude 8 and over are evidently great earthquakes (Richter, 1969).

The San Francisco earthquake of 1906 had an estimated Richter magnitude of 8.25, and the Tokyo earthquake of 1923 had a magnitude of 8.1. The Colombia earthquake of 1906 and the Assam, India, earthquake of 1950 both had magnitudes of 8.6, the largest yet determined by the Richter method. The Alaska earthquake of 1964 had a magnitude estimated to be from 8.4 to 8.6 (Gilluly, Waters, and Woodford, 1968). The largest known earthquakes from 1897 to 1965 can be seen in Fig. 32-3. It is interesting to note that these earthquakes, all of Richter magnitude greater than 8.0, are predominantly shallow.

Intensity

Intensity is a qualitative measure of earthquake effects at a particular location, and is based primarily on the extent of damage, loss of life, and the perception of people to earthquake effects. Intensity is arbitrarily scaled upward from an initial condition of barely noticeable shaking through various levels of human perception and response, damage, and ground motion to the ultimate condition of complete panic, total destruction, and massive ground disturbance.

DeRossi and Forel were the first to develop an intensity scale (values from I to X) in the 1880s (Adams, 1964). In 1902 the Italian seismologist Mercalli set up another scale, from I to XII, which provided a more refined analysis of major damage than did the Rossi-Forel scale. Finally, in 1931 the Mercalli scale was modified by Wood and Neumann, taking into account modern features such as tall buildings, modern vehicles, and underground utilities (Iacopi, 1964). The modified scale is known as the Modified Mercalli scale (MM). To avoid confusion with magnitudes, intensity ratings are given in Roman numerals.

A more recent intensity scale was compiled by S. Medvedev and others in 1963 (Adams, 1964). The scale classifies intensities using categorized modifying definitions to describe structural types, quantity and extent of damage, and nature of perception. The classification is probably more objective than the MM scale.

Barosh (1969) suggests that microregionalization maps, showing relative intensities for different geologic settings, could be used with intensity-epicentral-distance curves to predict intensities in particular geologic settings. He suggests that this technique can be used also to estimate potential seismic effects on manmade structures from underground nuclear explosions.

An engineering intensity scale has been proposed by Blume (1970) that is based on the damped response of ground motion for both earthquakes and nuclear explosions. Some of the advantages of this scale, according to

Table 32-1. Magnitude-intensity relations

Magnitude	Maximum intensity
2	I-II
3	II
4	V
5	VI-VII
6	VII-VIII
7	IX-X
8	XI

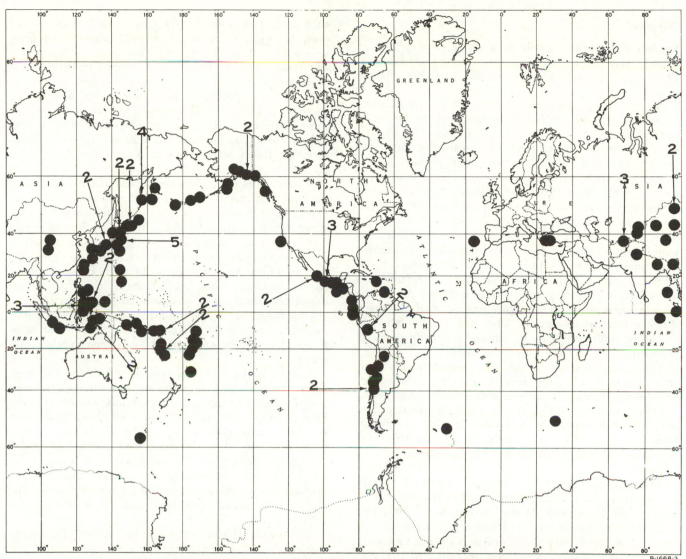

GREAT EARTHQUAKES 1897-1965 RICHTER MAGNITUDE 8.0 and ABOVE
ESSA-C&GS NATIONAL EARTHQUAKE INFORMATION CENTER

Fig. 32–3. Great earthquakes 1897–1965, Richter Magnitude 8.0 and above. Numbers indicate frequency of occurrence. Source: U.S. Geological Survey

Blume, are that it is an index of the effect of ground motion on all types of structures, it avoids subjective evaluations, it is simple and easily applied, and it will remain valid no matter how future buildings change.

Frequency

More than one million earthquakes shake the earth each year, on an average of about two each minute. Some scientists say there may be about 5 million a year, counting all the microearthquakes and tremors that are picked up only on highly sensitive seismographs (Tufty, 1969).

According to Richter (1958), a million earthquakes a

year is a conservative estimate for the whole world, because it does not include all earthquakes of the very smallest magnitudes.

From 1900 to 1964 there was a yearly average, the world over, of about 20 major earthquakes, one or two of which were usually great. Since 1964, the year of the great Alaskan earthquake, the annual total of major earthquakes has been distinctly less, with none exceeding magnitude 8 until mid-January 1971 (Richter, 1969).

A recurrence curve recently plotted for the San Andreas Fault (Wallace, 1970) shows an 8 magnitude earthquake occurring every 102 years, and smaller shocks at shorter intervals.

Duration

Earthquakes usually occur as a series of shocks that include foreshocks, the main shock, and aftershocks. There can be endless variety in the way these shocks occur.

All the shocks are essentially similar to one another, except in magnitude. The main shock is the largest in magnitude.

There can be a long series of foreshocks that occur before the main shock, or there may be only a few. Sometimes these foreshocks are so delicate that people do not feel them. Birds and animals, however, sometimes seem sensitive to even the slightest tremor and show signs of nervousness or alarm (Tufty, 1969).

Other foreshocks are so strong that they are almost major earthquakes in their own right. The Matsushiro earthquake swarm in Japan started in August, 1965. Activity was so high at one time that 600 earthquakes per day were felt. Very small shocks had been occurring, without exception, in the epicentral area of the main shock a few months earlier. The time interval between the ultramicroearthquake swarm and the main shock of magnitude 5 served as a forecasting tool.

The main shock rarely lasts for a minute in any local area—generally for only several seconds, although to people experiencing it the time seems much longer.

Strong shaking from a major shock frequently lasts only 30–60 seconds. The major shock of the 1906 San Francisco earthquake lasted only 40 seconds. However, the major shock of the Alaska earthquake lasted 3–4 minutes.

After the main shock has passed, aftershocks or smaller shocks may continue intermittently for days, weeks, months, or years. The Alaska earthquake of 1964 had some 12,000 aftershocks of Richter magnitude greater than 3.5 in the 69-day period following the main shock, and several thousand more were recorded in the next year and a half. In the first 24 hours there were 28 aftershocks, 10 of which exceeded magnitude 6, and 55 aftershocks within the first 48 hours of a magnitude greater than 4 (Hansen et al., 1966).

Forecast capability

The capability for areal definition of hazard for earthquakes is well developed, i.e., most earthquakes occur in distinct regional areas of the world, and within these broad regions (Fig. 32–2) seismic regionalization maps exist.

In 1969 a seismic risk map of the United States was developed by a group of research geophysicists headed by S. T. Algermissen of the Coast and Geodetic Survey of the U.S. Department of Commerce. The map (Fig. 27–1 above), a revision of an earlier 1951 version, shows areas of the United States most vulnerable to earthquakes and the types of damage that can be expected in each area. The zones on the map, based on studies of past earthquakes, do not, however, indicate frequency with which damaging earthquakes occur. Algermissen and Perkins (1972), however, do discuss distribution in time and space of occurrence of earthquakes.

A Japanese program on earthquake prediction did succeed in 1966 in a long-range forecast of the Matsushiro earthquakes (Hagiwara and Rikitake, 1967). Specialists from the Japanese Earthquake Research Institute, the Japan Meteorological Agency, the Geographical Survey Institute, and other governmental institutions met approximately once a month and discussed in detail the observed seismic data from Matsushiro, which was experiencing hundreds of microearthquakes every day. Whenever they thought that the probability of earthquakes of some magnitude was great, warnings along with the analyzed information were issued to the public by the Japan Meteorological Agency. These warnings did not attempt exact predictions of time, place, and magnitude, mentioning only the dangerous period (usually a range of a few months), a rough idea about location, and possible maximum magnitude. The situation was much the same as for long-range weather forecasting.

The basis for warnings was provided, for the most part, by repetition of leveling surveys, microearthquake and ultramicroearthquake observation, and observation by water-tube tiltmeters. It would have been difficult to supply information to the public if the observation network (array) had not been well established before the violent quakes hit. To some extent, it was possible to predict the Matsushiro earthquakes even though nothing was known about the underground process. The violent activities of April and August 1966 were successfully foretold.

The people of Matsushiro had become conditioned to earthquakes. Reaction to the warnings was varied. Local governments worked hard to prevent possible earthquake damage by repairing school buildings, strengthening fire brigades, and so-forth. This was also the case for the national and private railways. Operators of hotels and inns foresaw business problems due to lack of tourists.

In the United States, scientists have not yet developed a warning system. Recently (1973), however, on the basis of swarms of microearthquakes, the U.S. Geological Survey ventured a prediction of a 4.5 magnitude earthquake 20 miles southeast of Hollister, California. Some National Center for Earthquake Research scientists believe that long-range prediction of earthquakes in California will be possible in the near future, provided an adequate monitoring system is achieved (Pakiser et al., 1969). Richter (1969) believes that earthquakes may not be predicted in the same way as is the weather for another century.

An ad hoc panel sponsored by the Office of Science and Technology has proposed a 10-year program of earthquake research which includes research in prediction of geophysical phenomena.

Numerous monitoring techniques are available to measure ground tilt, strain, seismic activity, fluctuations in the magnetic field, and rock stress in drill holes. Other phenomena that could be measured to predict earthquakes are changes in water levels in wells and the occurrence of lightning. In some parts of the world, earthquakes are often accompanied by various forms of lightning. It has been recently suggested (Finkelstein and Powell, 1970) that this correlation may be due to the ability of quartz in the earth's crust to develop electrical charges when subjected to elastic deformation, as in a quake. Observations of such electrical precursors to earthquakes, it is proposed, may be useful in prediction. German seismologists are monitoring subterranean gases, such as methane, which are released in abnormal amounts from the fault preceding a quake. In March 1969 a German geologist successfully predicted an aftershock to an earthquake in southern Germany by observing an abnormal concentration of methane gas in a research shaft. A rare radioactive gas, radon, has been used in a similar manner by Soviet geophysicists (Purrett, 1971).

Documentation of man-made earthquakes has greatly increased the understanding of earthquake mechanisms and may lead to eventual control of earthquakes. In 1945, Carder documented approximately 600 local tremors during the 10 years following the formation of Lake Mead in Arizona and Nevada. Since then it has been found that the filling of other reservoirs has been accompanied by tremors.

The recent earthquakes near Denver and Rangely, Colorado, are thought to be man-made, resulting from pore-fluid pressure increases during underground fluid injections at these locations (Healy et. al., 1968; Raleigh et al., 1970).

Other "man-made" earthquakes have been detected as a result of nuclear testing. Small numbers of moderate-sized earthquakes have been detected after nuclear explosions at the Nevada test site (Hamilton and Healy, 1970; Boucher, 1971). It is thought that nuclear explosions cause the release of natural tectonic strain energy.

In addition to providing new insight into the mechanisms of earthquakes, these discoveries provide a potential means for modifying or controlling them. In some cases fluid injections may gradually release built-up strain in a series of minor harmless quakes (Byerlee, Wilson, and Peselnick, 1972). A U.S. Geological Survey project now underway at Rangely, Colorado, relates changes of fluid pressures to earthquake occurrences (Raleigh and Healy, 1972).

In order to estimate how near the breaking point the San Andreas Fault is, it was proposed that fluids be injected into an area under very controlled conditions and that the pressure be increased just to the point where tiny tremors occur. Researchers could then extrapolate to find the amount of stress needed to trigger a major quake.

Available adjustments

Possible adjustments to earthquake hazards include (1) warning systems; (2) earthquake prevention; (3) structural protection, including resistance to fire as well as shock; (4) insurance; (5) land-use change; (6) fire-prevention measures; (7) relief; (8) rehabilitation; and (9) loss bearing on inaction.

Warning systems and earthquake prevention technology are not yet available adjustments, because there is very little understanding of earthquake mechanisms. As previously discussed, the Japanese have used, with some success, a warning system based on frequency of micro-earthquakes. However, the success of a prediction system depends on the amount of warning time prior to the earthquake event, and the Japanese system as yet cannot define very accurately the time, place, or expectancy of earthquakes. Both the United States and Japan have proposed 10-year research programs for the prediction of earthquakes (Mukerjee, 1971).

Poorly understood earthquake mechanisms presently do not permit any planned modifications of earthquake hazards. However, future research and some of the present work being done, injecting fluids into rocks at depth, may lead to successful modification patterns.

The best adjustment to earthquakes presently available to man is learning where to live in order to avoid the hazard and how to live with the hazard when it is unavoidable. Land use that avoids settlement of high-risk areas by dense populations and complex structures is one available adjustment that is too seldom used. Seismic regions often combine the very best advantages of trade, communications, and strategic location with the very worst earthquake risks to life and property. Seismic risk maps are too seldom the endorsed basis of zoning maps drawn by city planners.

One recent example of this failure to use land so as to avoid earthquake risk is Japan's choice of paradoxical sites for a new steel mill and six new earthquake antidisaster centers. The new Kawasaki steel mill is being built on landfill in Chiba Prefecture, where part of Tokyo Bay has been reclaimed. This landfill, only a few feet above sea level, is subject to hazards of earthquake, tsunami, flood, and typhoon. At the same time, six new disaster centers are being constructed to aid the 600,000 residents of the low-lying Koto area of Tokyo suffering from floods and fires caused by earthquakes. Instead of avoiding seismic sea-level sites, industry in Japan is seeking them out and creating them where none formerly existed.

In the United States, the land-use problem is similar. By the year 2000 it is estimated that 15 percent of the United States population will be living near earthquake zones in California alone. Even with research emphasizing earthquake prediction factors, many engineers and scientists are concerned about the present safe development of these areas of high risk (Pakiser et al., 1969).

Earthquake-resistant structural design is another available adjustment in high-risk areas. Japan has a national network of 250 strong motion seismographs, installed specifically to record the reactions of buildings to all tremors. Responses of buildings are now predicted by feeding seismographic data into electronic computers. This information has led to changes in construction techniques and to the revision of the building code in Tokyo to permit the construction of structures over 36 stories high.

Los Angeles has undergone similar changes in building code and construction techniques. The February 9, 1971, earthquake revealed that not all of the adjustments made to earthquake-proof buildings were adequate. The high-rise buildings in Los Angeles (20–50 stories) performed well during the earthquake, with no significant structural damage. Many, however, did suffer nonstructural damage, such as cracked plaster, windows, and ceilings, and damaged electrical and mechanical equipment. Other new structures designed to resist earthquakes were damaged in the region of strong shaking. Notable examples were the six-story Olive View Hospital and the two-story Olive View Psychiatric Day-Care Center. Both of these buildings suffered major structural damage (Earthquake Engineering Research Institute Committee, 1971).

Sharing the cost of quakes through insurance is another available adjustment. In Los Angeles, earthquake insurance is available on a special risk basis, but it is so expensive that few buy it. The current rates of most policies are in the range of $1.50 to $2.50 per $1,000 and have a 5 percent deductible provision (Earthquake Engineering Research Institute Committee, 1971). Thus, on coverage of a $30,000 home, the first $1,500 of damage would be excluded.

In New Zealand, the Earthquake and War Damage legislation is a national natural hazard insurance scheme. The act provides a national fund through taxation that compensates property owners for property damage incurred because of natural hazards. By having such a nationalized fund, building codes and land-use measures (zoning) can be more easily established and enforced (O'Riordan, 1970).

Aid to disaster victims often comes in the form of relief and rehabilitation funds from federal governments and large international organizations. For example, in the United States, federal disaster assistance has increased greatly in the past decade, and the Red Cross contributes much aid in emergency assistance and rehabilitation. There are nearly 50 federal organizations involved in domestic relief and rehabilitation (Mukerjee, 1971).

The logical adjustment to earthquakes is research to understand them better. As cities grow and populations expand, more lives are threatened each day. The loca-

tion of nuclear power plants, oil refineries, fuel tanks, reservoirs, dams, and pipelines in seismic areas presents a growing potential disaster to more people every day. If research is to keep pace it must be adequately funded on both a national and worldwide scale.

The USSR has a well-developed state seismological service registering movements of the earth's crust, and research is primarily directed toward methods of forecasting (Gerasimov and Zvonkova, chap. 29). The Institute of Geology of China has an earthquake program involving 10,000 trained personnel who spent more than 60 percent of their effort in 1972 on earthquake prediction (Chao, 1973).

The proposed national program for the United States would last 10 years and cost $137 million. The Worldwide Network of Standard Seismographs is one example of multiple use by a federal agency—the National Oceanic and Atmospheric Administration—of seismological research. Originally designed to detect underground nuclear explosions, its standardized seismographs and international communications network now serve as the largest earthquake reporting service in the world.

Perception of hazard and the adoption of adjustments

Numerous examples can be given of people who have failed to perceive the hazard of earthquakes or have chosen repeatedly to take the risk and bear the losses. One of the classic examples is the persistence with which the population of Turkey continues to remain and rebuild along the Anatolian Fault, which has suffered 11 damaging earthquakes in the past 30 years.

The United Nations is sponsoring a $4.2 million program that may produce quakeproof housing in Turkey and four other southeast European countries. Special building codes are planned, and data from newly established seismic observatories will be used to produce maps delineating zones where earthquakes are likely to occur. Hopefully, they will be used by designers, engineers, and architects to plan towns, housing, and public works that are located on the safest ground available and built in the most earthquake-resistant fashion.

A positive precedent has been set by the Alaskan city of Valdez in selecting a new and safer townsite after the 1964 earthquake destroyed its port facilities and much of the commercial and residential district. The problems associated with evacuating a former townsite are complicated by economic, social, and political issues that had to be resolved. The residents were reluctant, the town council was cautious, and the move was slow and piecemeal. The sentiment seemed to be "You go first." The fact that the relocation was completed by October 1967 and that the old town was abandoned gives some hope that people in other high-risk areas will perceive the earthquake hazard and take similar action when it becomes apparent that other alternatives are simply a repetition of past mistakes.

The mental distress caused by aftershocks may be greater than that caused by the main earthquake. More than one month after the February 9, 1971, San Fernando earthquake, a strong new aftershock caused heavy damage in Granada Hills. Some residents said that the aftershock was worse than the original quake. The aftershock was rated 4.0 on the Richter scale by the California Institute of Technology, but the University of California at Berkeley said that it was 5.0. According to psychiatrists and psychologists treating San Fernando earthquake cases, some 300 aftershocks are worse on some individuals' mental health than one large shock. Individuals suffering these aftershocks developed a sort of "preparatory vulnerability" that caused them to freeze with fear instead of seeking a safe place during aftershocks. This effect seemed greater in children and newcomers to the area. Aftershocks drove some residents to leave the state permanently, even though they were able to tolerate the main shock without making such a decision.

Since earthquake prediction seems to be a goal of the Japanese scientific community, and since both Russia and the United States include earthquake prediction in their research program proposals, it seems necessary to research at the same time how various populations will respond to warnings. The adoption of a broad research program on earthquakes seems to be the best adjustment that any nation can make to the hazard. The funding of these research projects and the implementing of the results into positive action in the form of land-use laws, building codes, insurance policies, and public information services seem to be the most pressing need. The cost of the San Fernando earthquake would have funded earthquake research for the next 10 years in the United States.

References

Adams, W. M. (1964) *Earthquakes—an Introduction to Observational Seismology*. Boston: D. C. Heath.

Algermissen, S. T. (1969) *Seismic Risk Map of Conterminous U.S.* Environmental Science Services Administration.

——, and Perkins, D. M. (1972) *A Technique for Seismic Zoning: General Consideration and Parameters*. Proceedings of International Conference on Microzonation for Safer Construction Research and Application, Seattle, pp. 865–78.

Barosh, P. J. (1969) "Use of seismic intensity data to predict the effects of earthquakes and underground nuclear explosions in various geologic settings." *U.S. Geological Survey Bulletin* 1279.

Blume, J. A. (1970) "An engineering intensity scale for earthquakes and other ground motion." *Seismological Society of America Bulletin* 60:217.

Boucher, G. (1971) "Evidence for secondary seismic events within three seconds following the underground nuclear explosion Faultless." *Geological Society of America Abstracts with Programs* 3:86.

Byerlee, J. D., Wilson, M. G., and Peselnick, L. (1972) "Elastic shock activity and fluid injection." *Geological Society of America Abstracts with Programs* 4:945.

Carder, D. S. (1945) "Seismic investigations in the Boulder Dam area, 1940–41, and influence of reservoir loading and local activity." *Seismological Society of America Bulletin* 35 (4).

Chao, E. C. T. (1973) "Contacts with earth scientists in the People's Republic of China." *Science* 179:961–63.

Cook, K. L., and Anderson, J. M. (1972) "Correlation of earthquakes with changes in strain and tilt near the Wasatch fault, Utah." *Geological Society of America Abstracts with Programs* 4:139.

Earthquake Engineering Research Institute Committee. (1971) "Damage to the Olive View Hospital buildings." In *The San Fernando, California, Earthquake of February 9, 1971. U.S. Geological Survey Professional Paper* 733:88–90.

Endo, E. T. (1972) "Focal mechanisms for the May 15–18, 1970, shallow Kilauea earthquake swarm." *Geological Society of America Abstracts with Programs* 4:155.

Finkelstein, D., and Powell, J. (1970) "Earthquake lightning." *Nature* 228:759–60.

Gilluly, James, Waters, A. C., and Woodford, A. O. (1968) *Principles of Geology*. San Francisco: W. H. Freeman.

Grantz, A. (1971) "The San Fernando, California, earthquake of February 9, 1971—Introduction." *U.S. Geological Survey Professional Paper* 733:1–4.

Griggs, D., and Handin, J. (1963) "Observations on fracture and a hypothesis of earthquakes." *Geological Society of America Memoir* No. 79.

Hagiwara, T., and Rikitake, T. (1967) "Japanese program on earthquake prediction." *Science* 157:761–68.

Hamilton, R. M., and Healy, J. H. (1970) "Aftershocks of underground nuclear explosions." *Geological Society of America Bulletin* 2:556.

Hansen, W. R., et al. (1966) "The Alaska earthquake, March 27, 1964." *U.S. Geological Survey Professional Paper* 541.

Healy, J. H., et al. (1968) "The Denver earthquakes." *Science* 161:1301–10.

Iacopi, R. (1964) *Earthquake Country*. California: Lane Book Company.

Isacks, B., Oliver, J., and Sykes, L. R. (1972) "Seismology and the new global tectonics." In *Plate Tectonics*. American Geophysical Union.

Mukerjee, T. (1971) *Economic Analysis of Natural Hazards: A Preliminary Study of Adjustments to Earthquakes and Their Costs*. Toronto: Natural Hazards Research Working Paper No. 17.

Nakano, T. (1972) *Natural Hazards and Field Interview Research—Reports From Japan*. Commission on Man and Environment, 22d International Geographical Congress, Calgary, Alta., Paper No. 48.

Nur, A. (1972) "Fault creep and tectonic stress." *Geological Society of America Abstracts with Programs* 4:211.

O'Riordan, T. (1970) *The New Zealand Earthquake and War Damage Commission—Study of a National Natural Hazard Insurance Scheme*. Boulder: University of Colorado, Institute of Behavioral Science, Natural Hazard Research, Working Paper No. 20.

Orowan, E. (1963) "Mechanism of seismic faulting." *Geological Society of America Memoir* 79:323–45.

Pakiser, L. C., et al. (1969) "Earthquake prediction and control." *Science* 166:1467–74.

Purrett, Louise. (1971) "The possibilities of earthquake prediction." *Science News* 99:131–33.

Raleigh, C. B., et al. (1970) "Earthquakes and waterflooding in the Rangely oil field." *Geological Society of America Abstracts with Programs* 2:660–61.

——, and J. H. Healy. (1972) "Faulting and crustal stress at Rangely, Colorado." *American Geophysical Union Geophysical Monograph* No. 16.

Richter, C. F. (1958) *Elementary Seismology*. San Francisco: W. H. Freeman.

——. (1969) "Earthquakes." *Natural History* 78:37–45.

Saarinen, T. F. (1970) *Environmental Perception*. Washington, D. C.: NCSS.

Scholz, C., Wyss, M., and Smith, S. W. (1969) "Seismic and aseismic slip on the San Andreas fault." *Journal of Geophysical Research* 74:2049–69.

Slossen, J. E. (1972) "Damage to residential housing by the earthquake of February 9, 1971." *Geological Society of America Abstracts with Programs* 4:238.

Steinbrugge, K. V., and Cluff, L. S. (1968) "The Caracas, Venezuela, earthquake of July 29, 1967." *Mineral Information Service* 21:3–13.

Tufty, B. (1969) *1001 Questions Answered About Natural Land Disasters*. New York: Dodd, Mead.

UNESCO. (1967a) "The Caracas earthquake of July 29, 1967."

——. (1967b) "The Mudurnu Valley earthquake, July 22, 1967."

——. (1968) "Philippines earthquake, August 2, 1968."

Verhoogen, J., et al. (1970) *The Earth*. New York: Holt, Rinehart & Winston.

Wallace, R. E. (1970) "Earthquake recurrence, San Andreas fault." *Geological Society of America Bulletin* 81:2875–89.

Wu, F. T. (1972) "Rupture velocities and seismic source mechanisms." *Geological Society of America Abstracts with Programs* 4:266.

Index

Abbott, A. T., 153
Adams, W. M., 278
Algermissen, S. T., 160, 280
Alpert, M., 195, 199
American Association of Geographers, Commission on Geography and Afro-America, 119
Amiran, D., 15
Anderson, J. M., 274
Archer, P. E., 14, 230
Arey, D., 27
Arrow, K., 207–8
Australia, New South Wales, Queensland: drought-affected properties of, 128; resources of, 132; soils of, 129; weather prediction, 296; see Drought
Avalanche areas: *Norway,* adjustments include buildings sheltered by natural obstacles, transverse abutments over roads, tunnels under hazardous areas, 176, 178–79; government policy of building regulations, emergency funds, mapping avalanche areas, road planning, 176–78; property damage, estimate of, 176; *Peru,* hazard, perception of, 11 (Table 1–1); *USSR,* glacier hazard in the Caucasus, Pamirs, Tian Shan Mts., 246; heavy snow hazard in the Aiknaiventchorr, Caucasus, Kamchatka, Khibin, and Ural areas, 246; survey, method of conducting, 246

Bakan, P., 192
Baker, E. J., 12
Bangladesh Water and Power Development Authority, 19, 22
Barlow, J., 151
Barnes, C., 90
Barosh, P. J., 278
Baroyan, O. V., 273
Barton, A. H., 187
Bartz, F., 161
Bates, F. L., 30
Baumann, D. D., 12 (Table 1–1), 14, 27–28, 30, 36, 187
Becker, G. M., 199, 208, 210, 213
Berry, L., 11 (Table 1–1), 98
Bi-State Metropolitan Planning Commission Iowa–Illinois, 54
Blume, J. A., 278
Blumer, H., 75
Bogardi, I., 271, 272
Bogdan, R., 75
Boone, L. M., 230
Borton, T. E., 200
Boucher, G., 281
Bowden, M. J., 12 (Table 1–1), 161–62
Boyd, D. W., 193, 202
Bradley, D., 197, 211
Brazee, R. J., 162
Brill, R. C., 83
Brinkmann, W. A. R., 80 (Table 10–1), 86
British North American Act, Canadian Constitution, 219
Bronson, W., 162
Brooke, C., 98
Brooks, C. E. P., 83
Brooks, R. H., 11 (Table 1–1), 142
Bruyn, S. T., 75

Burton, I., 4, 14, 15, 26, 60, 63, 66, 68 (Table 8–4), 91, 127, 145, 149, 187, 190, 192, 219, 220, 261, 272–73
Byerlee, J. D., 274

California State Legislature, seismic safety, 160
Campbell, D., 128
Campbell, D. T., 31
Canada, 269, 273; see Insurance
Canadian Department of Environment, National Committee Water Resources Research, insurance study, 219
Carder, D. S., 283
Changnon, S. A., Jr., 226
Chao, E. C. T., 282
Chapanis, A., 192
Chapman, J. P., 193
Chapman, L. J., 193
Cheatle, R. J., 12 (Table 1–1)
Chi-square test, 33, 158
Clarke, W. B., 162
Cloud, W. K., 162
Cluff, L. S., 275
Coastal erosion, California: *Bolinas,* environment of, 70; erosion of Franciscan Formation, Merced Formation, Monterey Formation, and Point Reyes Peninsula, 71; individual and community reliance, perception of, 11 (Table 1–1), 13, 70, 73–75; property damages, protective construction, solution by, 73 (Table 9–1); River and Harbor Act, assistance by, 78; San Andreas Fault, effect on, 71; Standard Oil Company, oil spill, 77; U.S. Army Corps of Engineers, erosion study, 73, 78; U.S. Coast and Geodetic Survey, responsibility of, 72
Cohen, J., 162
Collins, B. E., 31
Cook, E., 30
Cook, K. L., 274
Coombs, C. H., 188
Cory, H., 98
Cotter, G. M. M., 98
Cyert, R. M., 189, 200, 207

Dacy, D. C., 25, 201, 213, 225, 264
Dams: Aswan, Egypt, 265; Clywedog, United Kingdom, 50–51; multipurpose, benefit-cost analysis, 220
Davis, F. K., 82
Davis, G. W., 270–71
Dawes, R. M., 188
Day, R., 206–7
Decision processes: data, collection of, 192; data processing of bias, 196, crop allocation, 198, and health effects, 197; decision making, seeking alternatives, 188–89; drought, 194, 200; earthquake building code, 201; earthquake probabilities, 200; flood control, 190, 195–96, 200; floodplain development, 197; information and response, limitations of, 197–98; landslide probabilities, 200; National Environment Protection Act, resource allocation, 200; representative selection, theory of, 188 (Table 24–1), tropical cyclone, modification of, 193, 202–3
DeGroot, M. H., 199

Dempsey, J. T., 171
Deutcher, I., 31
Dikkers, R. D., 257
Dillon, J. L., 188
Drought, definition of, 119, 128
Drought hazard: *Australia, South,* government policy of farm density, 136, loans, 128, 136, marginal lands, 132–33, water, and wheat, 134; land use, the Mallee area, 130 (Table 16–1); questioning of inhabitants, 11 (Table 1–1), 134–35; *Brazil, Northeast,* hazard, perception of, 11 (Table 1–1); *Kenya, East,* economy based on barter, dry farming, lack of transportation, 88; environment, 87–88 (Table 11–1); Kamba adjustments of land-reform technology and new crops, 91, 114; Kamba government policy of famine relief method, and resource diversion, 87–90, 96; perception of animal losses, crop losses, deaths, and nutrition, 11 (Table 1–1), 92–93 (Tables 11–2, 11–3); philosophy of self-reliance, 94 (Table 11–4); *Mexico, Oaxaca,* adjustments of irrigation, 125–26, proportional planting, 125; climate of, 128; economy of maize and bean production by small farms, 120, 125; Mitla piedmont, variability of rainfall, 119, 120–22 (Table 15–1), 125; questioning, method of, 13–14 (Table 1–2), 121–22; water projects, valley development by, 119, 121; *Nigeria, Northwest,* adjustments of fishing and making crafts to supplement subsistence farming, 116, 118 (Table 14–1) hazard, perception of, 11 (Table 1–1), 117–19; questionnaire, use of, 115, 117; *Tanzania, Northeast,* adjustments of communal farm, irrigation, surpluses, 105, 111 (Tables 13–3, 13–4, 13–5), 114; economy of, 105–6, 109–10, 113 (Table 13–6); environment of, 105–6; perception of drought variability and production losses, 11 (Table 1–1), 105, 107 (Table 13–1), 108–10, 111 (Table 13–2), 113–14 (Table 13–7); questionnaire, use of, 106; *Tanzania, Sukumaland,* adjustments of drought-resistant crops and insurance, 102 (Table 12–3), 104 (Table 12–4); economy of, 101 (Table 12–2), 102 (Table 12–3); rain stations, history of, 99 (Table 12–1); questioning, method of, 98; risk, perception of, 11 (Table 1–1), 104
Dunn, G. E., 258
Dupree, H., 11 (Table 1–1), 119
Dyer, J., 208

Earney, F. C. F., 12 (Table 1–1), 174
Earthquake hazard: Building codes, construction technique, and estimates of potential damage in California, San Fernando Valley, U.S. Department of Housing and Urban Development, 206, and San Francisco, 162–63, 164 (Table 20–2), 165 (Table 20–3), 275, and Japan, 14 (Table 1–2), 232 (Table 28–1), 233 (Table 28–2), 234 (Table 28–3), 235, 238, 276, 278, 280, and in the Philippines, 276; earthquake detection, Japan and U.S., 280, and USSR, 244; earthquake intensity, scale of, 278 (Table 32–1); hazard, perception of, 11 (Table 1–1), 166; hazard, predic-